Automata and Computability
A Programmer's Perspective

Automata and Computability
A Programmer's Perspective

Ganesh Lalitha Gopalakrishnan

CRC Press
Taylor & Francis Group
Boca Raton London New York

CRC Press is an imprint of the
Taylor & Francis Group, an **informa** business

CRC Press
Taylor & Francis Group
6000 Broken Sound Parkway NW, Suite 300
Boca Raton, FL 33487-2742

© 2019 by Taylor & Francis Group, LLC
CRC Press is an imprint of Taylor & Francis Group, an Informa business

No claim to original U.S. Government works

Printed on acid-free paper
Version Date: 20190206

International Standard Book Number-13: 978-1-138-55242-5 (Hardback)

Visit the Taylor & Francis Web site at
http://www.taylorandfrancis.com

and the CRC Press Web site at
http://www.crcpress.com

Printed and bound by CPI Group (UK) Ltd, Croydon, CR0 4YY

Contents

Foreword   xiii

Preface   xv

Acknowledgments   xix

1   What Machines Think   **3**
  1.1   Problems Without Algorithms   4
  1.2   How to *Define* a Computer?   5
  1.3   Practical Application: Syntax Definition/Checking   7
  1.4   Simplified Turing Machines as Parsers   11
  1.5   Automata and Computability for Lifelong Learning   13

2   Defining Languages: Patterns in Sets of Strings   **15**
  2.1   Symbol, Alphabet, String, Language   15
    2.1.1   Symbol   15
    2.1.2   Alphabet   15
    2.1.3   String or Word   16
    2.1.4   Various Notions of Zero and One   17
  2.2   Language   18
    2.2.1   Language Concatenation   20
    2.2.2   The Zero and One for Language Concatenation . . .   21
    2.2.3   Zero and One of Language Concatenation in Python   21
    2.2.4   Exponentiation of a Language   22
    2.2.5   Python Encoding of Language Exponentiation   23
    2.2.6   Union and Intersection of Languages   24
  2.3   Useful Results, Slippery Roads   25

3   Kleene Star: Basic Method of Defining Repetitious Patterns   **27**
  3.1   Three Ways to Describe Star   27
  3.2   Additional Definitions and Properties of Star   28
  3.3   Language Complementation   31
  3.4   Other Language Operations   32
    3.4.1   Symmetric Difference, Subtraction   32
    3.4.2   Reverse of a Language   32

3.5   String/Language Homomorphisms 33

     3.5.1   Taking Star Repeatedly 35

3.6   Enumerating Strings in a Language 35

4   Basics of DFA   **41**

4.1   DFA Everywhere 41

4.2   Elements of a DFA 42

4.3   Formal Structure of DFA 43

4.4   The Language of a DFA 44

     4.4.1   DFA as String Classifers 44

     4.4.2   Basics of Designing a DFA 45

4.5   Formal Definition of DFA Language Acceptance 45

4.6   "Lasso" Shape of DFA and the Pumping Lemma 47

     4.6.1   General Statement of the Pumping Lemma 48

4.7   Proving a Language to Be Non-Regular 50

     4.7.1   Why All Splits of x, y, z? 51

4.8   Grossly Abusing the Pumping Lemma 52

     4.8.1   Inability to Prove with this Pumping Lemma 53

4.9   Regularity-Preserving Transformations Aid Proofs 54

5   Designing DFA   **57**

5.1   Understanding the Language to Be Realized 57

     5.1.1   The Language of Equal Changes 57

     5.1.2   Best Practices to Correct DFA Design and Verification 58

5.2   Examples of Designing DFA 59

     5.2.1   The Language of Blocks of 3 59

     5.2.2   DFA for "Ends with 0101" 60

     5.2.3   DFA for "MSB/LSB-first Binary Number is Divisible

            by 3" 60

     5.2.4   DFA for "Third-last bit is a 1" 62

5.3   Automd: A Markdown Language for All Machines 64

     5.3.1   Markdown for DFA 64

6   Operations on DFA   **67**

6.1   Complementation of DFA 67

6.2   Union and Intersection of DFA 67

6.3   Language Equivalence and Isomorphism 71

6.4   DFA Minimization and Myhill-Nerode Theorem 72

     6.4.1   Fully Worked-out Example of DFA Minimization . . 72

     6.4.2   Salient Code Excerpts 75

6.5   Examples of Language Design and Manipulation 76

     6.5.1   Use of Union, Minimization, and Language

            Equivalence 77

     6.5.2   Use of DeMorgan's Law 78

7 Nondeterministic Finite Automata   **81**

  7.1 Overview of NFA 81

  7.2 Formal Description of NFA 83

  7.3 Language of an NFA: Example Driven 84

      7.3.1 Simulations of the NFA of Figure 7.5 84

      7.3.2 Simulations of the NFA of Figure 7.3 85

  7.4 Language of an NFA: via Eclosure 85

      7.4.1 Defining Eclosure 86

      7.4.2 Definition of δ and $\hat{\delta}$ 86

  7.5 NFA to DFA Conversion through Subset Construction . . . 87

  7.6 Brzozowski's DFA Minimization Algorithm 90

      7.6.1 Reversal of a DFA Yields an NFA 90

  7.7 A Complete Illustration of Brzozowski's Minimization . . . 92

8 Regular Expressions and NFA   **95**

  8.1 Regular Expressions 95

  8.2 RE to NFA Conversion: Examples, Algorithm Sketch 96

  8.3 A More Extensive Example 100

  8.4 Regular Expressions: Ubiquitous, yet Error-Prone 101

  8.5 Anatomy of the RE to NFA Converter 103

  8.6 Example: Designing an Error-Correcting DFA 106

      8.6.1 Error-correcting RE for "within Hamming Distance

             of 2" 106

      8.6.2 NFA-based Design of "within Hamming Distance of 2"108

  8.7 Minimal DFA and Isomorphism 108

  8.8 DFA Ultimate Periodicity to Solve the Postage Stamp

     Problem 111

      8.8.1 Ultimately periodic sets and lengths of members of

             a regular language 111

      8.8.2 Stamp Problem and Ultimate Periodicity via Jove . . 112

      8.8.3 Applying to numbers that are not relatively prime . 113

      8.8.4 Solving for three stamps 113

      8.8.5 Lengths of strings accepted by DFA 113

9 NFA to RE Conversion   **115**

  9.1 NFA to RE Conversion Algorithm 115

  9.2 Illustration on Pedagogical Example 116

  9.3 Illustration on Non-trivial Example 118

  9.4 Checking the Conversion 121

  9.5 DFA, NFA, and RE Are Equally Powerful 122

  9.6 Implementation of NFA to RE 123

  9.7 Closure Results Pertaining to Regular Languages 123

10 Derivative-Based Regular Expression Matching   **127**

  10.1 Introduction to RE Derivatives 127

10.2 Definitions 129
    10.2.1 Derivative Rules 129
    10.2.2 Nullability Rules 132
10.3 Implementation of Derivative-Based String Matching 134
    10.3.1 Derivatives: Closing Thoughts 134

**11 Context-Free Languages
and Grammars**   **137**
11.1 Context-Free Language Examples 137
11.2 Context-Free Grammars and Parse Trees 138
    11.2.1 Elements of Context-Free Grammars 139
    11.2.2 Parse Trees, Language of a CFG 139
11.3 Avoiding Mistakes in Designing CFGs 140
    11.3.1 Completeness and Consistency 141
11.4 The Design of CFG, and the Hill/Valley Plot
for Arguing Consistency and Completeness 142
11.5 Ambiguous Grammars, and Disambiguation 144
    11.5.1 Disambiguation 145
    11.5.2 Disambiguation Is Crucial! 147
    11.5.3 Impossibility Results 147
11.6 Inherently Ambiguous Languages 147
11.7 Expressing DFA via CFGs 149
    11.7.1 *Purely* Right Linear Grammars 150
    11.7.2 Closure, *Purely* Left Linear, and Mixed Linearity . . 151
11.8 Historical Importance of the Theory of Parsing 152
    11.8.1 Combating Inherent Ambiguity 153
11.9 A Pumping Lemma for CFLs 154
    11.9.1 Application of the CFL Pumping Lemma 157
11.10 The Complement of a Non-CFL Can Be a CFL 157
    11.10.1 Growing "Inside-Out" 158

12 Pushdown Automata   **161**
12.1 Pushdown Automaton Basics 161
12.2 Formal Description of PDA 164
    12.2.1 Acceptance, Deterministic PDA 165
12.3 Exploring the PDA for L_{Dyck} Using Jove 165
12.4 PDA Behavior Through Examples 167
    12.4.1 Rerunning pda6 with Larger Stack Allowed 171
12.5 Toward More Practical PDA 173
12.6 CFG to PDA Conversion 175
    12.6.1 Disambiguation 179
12.7 Practical Knowledge Imparted by Jove: Three Parsers . . . 181

13 Turing Machines   **183**
13.1 Brief History of Turing Machines 183

13.2 Universal Computing Devices 184

13.3 Formal Definition of Turing Machines 186

13.4 Examples of Simple TMs 189

 13.4.1 A Simple DTM that Flips Bits 189

 13.4.2 TMs that check if a string contains 101 189

13.5 A DTM for $w\#w$ and an NDTM for ww 193

 13.5.1 A DTM Recognizing $w\#w$ 193

 13.5.2 A Nondeterministic TM Recognizing ww 193

 13.5.3 Nondeterminism does not increase a TM's

  Expressive Power 194

13.6 Example: A TM that Works on the Collatz Problem 200

 13.6.1 Markdown for the Collatz Problem TM with

  Comments 200

13.7 The Chomsky Hierarchy 203

 13.7.1 Recursively Enumerable and Recursive Languages . 204

13.8 An Alternate Notation for Instantaneous Descriptions . . . 205

14 Interplay between Formal Languages   **209**

14.1 Why Study Impossibility Results? 209

 14.1.1 Definitions: Procedure and Algorithm 209

 14.1.2 Formal Languages Associated with Turing Machines 210

 14.1.3 Allaying Confusion: Language vs. Language Family 210

14.2 One Example of a Recursive and an RE Language 211

 14.2.1 A Recursive Language 211

 14.2.2 Prerequisites to Defining the Notion of RE 212

 14.2.3 Combining Semi-deciders for L and $\overline{L}$ 213

 14.2.4 The "no wimp" Clause 213

 14.2.5 A Recursively Enumerable Language 213

14.3 Other Examples of Recursive and RE Languages 215

 14.3.1 RE Languages that are not Recursive 216

 14.3.2 Why is it called Recursively **Enumerable**? 217

 14.3.3 Alternate proof of A_{TM} being RE 218

 14.3.4 Some More RE Languages 218

14.4 Summary of Decidability/Semi-Decidability Results 219

 14.4.1 Existence of Non-RE languages 220

14.5 RE and Recursive Sets: More High-Level Proof Sketches . . 221

 14.5.1 Language of DFA D where $L(D) = \Sigma^*$ is Recursive . . 221

 14.5.2 Language of CFG G where $L(G) = \emptyset$ is Recursive . . 222

 14.5.3 Language of LBA that halt on input w is Recursive . 223

 14.5.4 Language of Turing machines whose first output is '3' 224

15 Post Correspondence, and

Other Undecidability Proofs   **227**

15.1 Post Correspondence: "Drosophila" for Decidability 227

15.2 Proof Sketch of the Undecidability of PCP 229

    15.2.1 Tile Construction Basics 230

    15.2.2 Proving Grammar Ambiguity by Reduction from PCP 231

    15.2.3 PCP in Jove 231

15.3 Undecidability of the Acceptance Problem 233

15.4 Halting ($Halt_{TM}$) is Undecidable 234

15.5 Mapping Reductions 236

    15.5.1 Undecidable problems are "A_{TM} in disguise" 239

16 NP-Completeness   241

16.1 What Does NP-Complete Mean? 241

    16.1.1 Grouping Problems: Solving One Implies Solving All 242

    16.1.2 Some Historical Notes 243

16.2 NPC Notions Defined Based on NDTMs 244

    16.2.1 P-time 244

    16.2.2 NP-time 246

    16.2.3 NP Verifier 246

    16.2.4 Examples of P-time and NP-time Deciders 247

    16.2.5 Decider versus Verifier Views 247

16.3 Introducing SAT Problems 248

    16.3.1 A Warmup: 2-SAT 249

    16.3.2 2-SAT: Examples and Algorithm 251

16.4 3-SAT and Its NP-Completeness 253

16.5 3-SAT Is NP-Complete 254

    16.5.1 3-SAT is in NP 255

    16.5.2 Every Language in NP Reduces to 3-SAT 255

    16.5.3 How P=NP is Obtained if 3-SAT $\in$ P? 256

16.6 Show that Clique Is NPC: Reduction from 3-SAT 257

    16.6.1 Clique is in NP 257

    16.6.2 Some Language in NPC Reduces to Clique 257

16.7 Complexity Classes, Closing Caveats 258

    16.7.1 NP-Hard Problems can be Undecidable (Pitfall in
            Proofs) 258

    16.7.2 The CoNP and CoNPC Complexity Classes 260

16.8 SAT in Practice 261

17 Binary Decision Diagrams as Minimal DFA   267

17.1 Boolean Functions in Computing Theory and Practice . . . 267

17.2 Boolean Functions as Minimal DFA of Their On-Sets 268

17.3 The Importance of Variable Ordering 271

    17.3.1 Finding a better input variable order 272

    17.3.2 Functions with linearly sized BDDs 273

17.4 From Minimal DFA to BDD: Intuitive Presentation 275

17.5 On BDD Sizes 277

18 Computability Using Lambdas   **279**

18.1 The History of Lambda Calculus 279

18.2 Lambdas from a Programmer's Perspective 280

18.3 Syntax and Semantics of the Lambda Calculus 282

18.4 Illustration: Church Numerals in Python 284

18.5 Illustration: Booleans, Pairs, Other Functions 284

18.6 Handling Recursion 287

18.7 Obtaining Fixpoints from Fixpoint Equations 288

    18.7.1 Y: A Fixpoint Finder 288

    18.7.2 The Y Combinator 288

    18.7.3 Expression Recursion using Y 289

    18.7.4 Reason for an alternate fixpoint finder Ye 290

18.8 Illustrating the Use of Fixpoint Combinators 292

18.9 Combinators 292

Appendices

Appendix A A Recap of Discrete Math   **297**

A.1 Sets 297

    A.1.1 Set Builder 297

    A.1.2 Powerset 298

    A.1.3 Complement 299

    A.1.4 Equivalence Classes, Partitioning 299

A.2 Mathematical Logic 299

A.3 Proof Methods: Using Contrapositive, by Contradiction . . . 300

A.4 Cartesian Product, Binary Relations, Functions 301

A.5 Functions: Signature, Onto, Into, Total 301

A.6 Trees 302

Appendix B Catalog of Jove Functions   **303**

B.1 Jove's Top-Level Functions 303

    B.1.1 Chapters 2 and 3 304

    B.1.2 Chapters 4 through 6 305

    B.1.3 Chapters 7 through 9 305

    B.1.4 Chapter 10 306

    B.1.5 Chapter 11 307

    B.1.6 Chapter 12 307

    B.1.7 Chapter 13 308

    B.1.8 Chapter 15 308

    B.1.9 Chapters 16 and 17 308

    B.1.10 Chapter 18 309

B.2 Jove's Use of Python, Including Lambda Basics 310

Appendix C   There Are More Languages than RE Languages 313

C.1 Gödel Hash 313

C.2 Cantor-Schröder-Bernstein (CSB) Theorem 314

C.3 Cantor's Diagonalization Proof about Languages 315

C.4 $|Real|$ Is Higher than $|Nat|$ 316

Selected References   **319**

Index   **323**

FOREWORD

Compared to other disciplines, theoretical computer science is a relatively young field. Nevertheless, more than seventy years have already passed since Alan Turing introduced formal definitions of a computer and an algorithm in 1936; for his fundamental contributions Turing is generally considered to be the father of not only theoretical computer science, but also of artificial intelligence. The late 1950s and the 1960s saw a great increase in the activities related to theoretical computer science, and this activity continues to the present day. Many books have been written on this subject: texts suitable for a first course as well as advanced monographs. The reader may well ask: Is another introductory text really needed? The answer is a clear yes, because of the special features of this book.

As the book's title states, this text is written from the programmer's perspective. Along with the printed book, Gopalakrishnan provides a suite of Jupyter notebooks for dealing with automata and languages, packaged under the name Jove. This, and the frequent relations the author establishes between theoretical concepts and their applications should make the theory attractive to software engineering students, as well as to the more mathematically oriented computer science students.

Many texts on theoretical computer science begin with a preliminary chapter on mathematical notions such as sets, relations and functions. In contrast to this, the present book begins with a brief history of computer science, and makes the case that automata and computability are at its very core. This provides a much stronger motivation for the study of these theoretical topics than do sets, relations and functions. Incidentally, the discrete mathematics background is provided in the book, but as an appendix.

The book is very readable. New concepts are illustrated by numerous examples and further clarified by sets of exercises, both theoretical and programming-oriented. The students have an opportunity to verify their understanding by using Jove programs. Numerous footnotes, often humorous, provide references to further reading in publications and websites.

The coverage of topics in theoretical computer science is quite comprehensive. After the short history of computing, there is a gentle introduction to formal languages and operations on them. Deterministic and non-deterministic finite automata, regular expressions, context-free languages, pushdown automata, and Turing machines then follow. The book concludes with the much more challenging concepts of undecidability, NP-completeness and lambda calculus. Altogether, this is a very valuable up-to-date addition to the literature.

Janusz A. Brzozowski
University of Waterloo
Waterloo, Ontario, Canada

PREFACE

This is a book on **the central theoretical pillars of computer science**! We begin with the theory of syntax processing based on machines and grammars and finish with a study of the absolute limits of computing, asking questions such as *"what can and cannot be algorithmically solved using computers?"* We assume that the reader already has basic working knowledge of sets, relations, functions, and predicate logic (Appendix A provides a refresher).

Half our educational material comes in the form of Jupyter notebooks (software) that accompany this book (being called *Jove*).[1] Jupyter is a compendium of runnable code in Python[2] documentation, and may also include web references and Youtube videos.[3] Jupyter notebooks are the modern equivalent of *Engineers' Notebooks,* and find use in real-world projects of immense importance. For instance, some of the scientists who successfully detected gravitational waves recently employed Jupyter notebooks in their research to perform some of their calculations.[4] Some of my Jove notebooks employ *Jupyter widgets* that provide sliders and pull-down menu selections to incrementally expand language constructions, nondeterministic machine executions, etc. "By Jove," please take advantage of these interactive facilities!

I recommend that you first luxuriate by simply running the already provided notebooks (especially those that begin with Youtube videos) and execute their commands one by one. After that, try to peek into Appendix B that summarizes all of Jove's functions, and imitate some of my examples. Then, *read the Python code that powers these notebooks.* At that point, you will have a firm grasp of the covered topics.

In Appendix B, we also provide an overview of how to use lambda functions in Python. Jove's Python functions are written largely using a subset of Python based on functional programming. In our codes, we employ recursion and also higher order functions such as map, reduce, fold, and filter. Given that all modern programming languages (even C++) nowadays include some elements of functional programming, these are valuable concepts to carry with you beyond this course.

In the study of computability, it is common practice to define machines, grammars and languages solely in mathematics.[5] Unfortunately, such definitions possess subtle nuances that can often stump students.[6] Augmenting the definitions with actual runnable code can greatly facilitate one's ability to grasp such slippery topics.

Providing alternate definitions is also *vastly safer practice* in real life. Mathematical definitions are like *code that has never been run even once,*[7] and unfortunately can be incomplete or contain bugs.[8] Having alternate definitions increases the chances of spotting mistakes by noticing disagreements.

---

[1] Jove is a synonym of Jupiter, the Roman god of the sky. According to Wikipedia, the Roman practice of swearing by Jove to witness an oath in law courts is the origin of the expression "by Jove!"

[2] Jupyter also supports other languages. We use Python exclusively, but employing only a very small subset that is quite easy to learn.

[3] Many of our notebooks start with videos that describe the rest of the notebook.

[4] A 2017 Physics Nobel prize winning discovery.

[5] In fact, imparting the requisite mathematical maturity to understand such presentations is one of the goals of this course.

[6] Not to mention automata-theory book authors such as myself! A good example is the notion of acceptance of a string by a nondeterministic push-down automaton (PDA, Section 12.2.1). By interacting with a PDA using Jove, one can minutely study the execution history and understand every associated detail.

[7] Knuth once famously wrote, **"Beware of bugs in the above code; I have only proved it correct, not tried it."**

[8] "Running" mathematical definitions means examining them within a theorem-prover—something that very few have the time or skill-set to properly carry out.

Book Highlights: This book serves "not just the standard fare of automata theory" but also topics that help with your overall CS education, and also some that are more modern, and are transitioning to practice.

Part I of this book begins with a historical perspective in Chapter 1, "What Machines Think." Discussing the history of computer science is supremely important because ours is an area that, despite providing the society countless bounties, still struggles to safeguard the everyday citizen from the perils of software bugs. In his Turing award lecture, Dijkstra's closing paragraph goes as follows:[9] *We shall do a much better programming job, provided that we approach the task with a full appreciation of its tremendous difficulty, provided that we stick to modest and elegant programming languages, provided that we respect the intrinsic limitations of the human mind and approach the task as Very Humble Programmers.*

In fact, it is very easy to feel humble, as Dijkstra espouses. Just 60 years ago, people did not know how to write even simple lexers and parsers. They used crude methods that were so terrible to look at and analyze, and were untrustworthy. The CS community conquered this complexity through formal methods. Formal methods are central to all advances in computer science. They include just about any method that systematically guides us toward the construction of *precision software components*.[10] Viewed this way, the very first formal methods in computer science indeed were the theory of regular expressions, context-free grammars and recursively enumerable sets!

[10] By way of analogy, think of *precision machining*. If one can build ball-bearings with a tolerance of 0.1mm, they are perhaps good for putting into bullock-cart wheels, but not into automobile wheels. If one can achieve micron-level precision, one can put them into automobiles or even jet-engine turbines. Initially, precision machining increases product costs and does not immediately benefit the society. But eventually, through mass production, we curb the costs.

Some may think that syntax processing is "well-understood old stuff"— but this is simply not true! Today, we are in a fast-changing world where syntax processing is ever more important. Today's markdown languages, JSON files, web forms, and countless other tools all employ elements of lexing and parsing. To design them robustly and reliably, one needs to truly understand the subtle nuances of language processing. Bugs introduced while reading-in and processing someone's command inputs or web forms are difficult to track down, and may cause severe loss of money and may even threaten life (for example if one evaluates a formula gathered through a web form pertaining to medical drug proportions).[11] The *formal methods* introduced in this course are thus very relevant to modern software design.

[11] Regular-expression processing bugs do occur in the field.

We cover syntax-processing beginning in Chapter 2, "Defining Languages: Patterns in Sets of Strings" and Chapter 3, "Kleene Star: Basic Method of Defining Repetitious Patterns." Part II then takes over, discussing the crucially important topic of DFA over three chapters: "Basics of DFA" (Chapter 4), "Designing DFA" (Chapter 5), and "Operations on DFA" (Chapter 6). Nondeterministic automata and regular expressions are then discussed over three chapters: "Nondeterministic Finite Automata" (Chapter 7), "Regular Expressions and NFA" (Chapter 8), and

"NFA to RE Conversion" (Chapter 9).

Chapter 10 discusses "Derivative-Based Regular Expression Matching." This approach allows us to handle the complementation of regular expressions without paying the exponential cost of determinization. This is an elegant and important algorithm lost in the mist of time, but publications on this topic have appeared in top conferences even as recently as 2016 [2]. Our presentation of derivative-based parsing rules also resembles how one writes *operational semantic definitions*. This is another example of how formal methods help quality conscious companies develop specifications of complex pieces of software.

We provide plenty of material that increases the students' understanding of how lexers and parsers are used together in full applications. This is achieved as follows. First, in Chapter 11 ("Context-Free Languages and Grammars") and Chapter 12 ("Pushdown Automata"), we cover context-free grammars and parsing. Then we provide the definition of a simple *lexer* (based on regular expressions) and *parser* (based on context-free grammars) using the widely used tool, "PLY."[12] Our first lexer/parser pair fits under *half a page of Python code* (Figure 10.4)! We then follow it up with another lexer and parser that turn regular expressions into NFA using PLY (again, in about a page, Figure 8.1).

We finally provide a more realistic lexer and parser for a markdown language that we design (see Section B.1.5) for writing clear descriptions of NFA, DFA, PDA, and Turing machines accompanied by comments.[13]

Chapter 13 discusses Turing machines. We make the study of Turing machines (TM) enjoyable by equipping them with "fuel tanks" that users can load with some number of "gallons" of fuel. This gives us a chance to concretely talk about time-complexity (the amount of fuel consumed before termination). This sets the stage for Part III that begins with Interplay between Formal Languages (Chapter 14) and proceeds to give you a domino game in Chapter 15 ("Post Correspondence, and Other Undecidability Proofs") that you can play within Jove using the PCP solver written by Ling Zhao. This deceptively simple game actually encompasses the essence of Turing undecidability—a fact that is at the same time satisfying and bewildering! Chapter 16 introduces NP-Completeness—a foundational aspect of anyone's CS education.

Chapter 17 ("Binary Decision Diagrams as Minimal DFA") gives you the ability to play with the PBDD tool written by Tyler Sorensen, launching it from within Jove. Chapter 18 ("Computability Using Lambdas") gives you the ability to play with lambda calculus within Jove.

Exercises are included within chapters as well as towards their end.[14] Proofs of the stated theorems are either sketched or cited. A solution manual for instructors will be made available as supplementary material. Solutions to selected problems will be included as supplementary

---

[12] PLY stands for "Python Lex and Yacc." Lex and Yacc were celebrated lexers and parsers created in Bell Labs in the 1970s. Written by Dr. David Beazley, PLY is available from `http://www.dabeaz.com/ply/`.

[13] We call this markdown "automd" standing for "Automaton Markdown."

[14] We number the exercises and theorems the same as the section (or subsection) containing them. This idea (proposed by my late colleague Kris Sikorski) makes locating them easier, we hope.

material on the book's website `https://bit.ly/Automata_Jove`. Jove notebooks will be kept updated on the github page of Jove, namely `https://github.com/ganeshutah/Jove`.

We provide many examples of formal methods and functional programming. For example, we encourage you to construct DFA using two distinct approaches and check whether their minimized forms are *isomorphic* (Chapter 5). Equivalence checking is a powerful formal methods idea, and is more comprehensive than "plain old testing." During equivalence checking, *all* behaviors are covered – not just those for which a user might care to write tests for.

We present formal methods for Boolean reasoning using Binary Decision Diagrams (BDD, Chapter 17). Knuth calls BDDs *one of the only really fundamental data structures that came out in the last twenty-five years*. We show how BDDs and minimal DFA are, essentially, one and the same. BDDs are routinely used by the hardware industry to detect and eliminate serious bugs from microprocessors.

Most books on automata theory deal with Turing machines but do not discuss lambda calculus. However, *both* these perspectives are crucially important in computer science. Besides, ideas from functional programming are powering many industries, increasing programmer productivity and helping reduce bugs. Functional programming can be quickly and elegantly introduced by demonstrating how to encode computability concepts using lambdas. We provide a glimpse of the power of the lambda notation by showing how familiar recursive functions such as factorial and fibonacci can be written using lambdas and a fixpoint finder, namely the "Y combinator." Essentially, we provide a Jove notebook that tells you how to "compute everything" using lambdas.

This is for those who love to code: Please make an attempt to not just use Jove, but actually *extend it!* Please stick to the extremely simple subset of Python that I've chosen, so that this material remains accessible to the widest possible audience.

You may find my use of lambdas and higher order functions to be a departure from simplicity—but in fact, *they are there to make the code resemble mathematics a bit more*, and also *they help with lambda calculus*. Hence please don't convert these codes to imperative for-loops (loops may be more readily understood, but such a conversion can be a step back from the intended pedagogy). To remain self-contained, I provide tutorials on lambdas and higher order functions in Section B.2.

Last but not the least, this is largely a book on programming and interacting with abstract machines. Section B.1 documents *almost all the Jove functions*, as if Automata Theory is just "an advanced math library." That is the Programmer's Perspective that the book's title boasts.

Acknowledgments

I wish to acknowledge all those who selflessly give to others. Stories such as that of scientist and professor Gordon Hamilton `https://www.nytimes.com/2016/10/25/science/gordon-hamilton.html` make one instantly feel humble, giving one a different perspective on life. Here are some excerpts from this article:

> Many climate scientists spend most of their days behind a desk. ... Yet for a whole cadre of climate scientists, the work entails real physical risks. Thousands of specialists – glaciologists, geologists, geodetic engineers, wildlife biologists and many others – must travel to remote regions to better understand the effects of warming on the natural world. Gordon Stuart Hamilton, 50, a glaciologist at the University of Maine, was killed when on a scientific expedition to Antarctica. He was surveying a trail to find the crevasses that can make working on glacial ice so dangerous, and his snowmobile plunged into one of them. He died doing a job whose urgency and importance, whose implications for the fate of all humanity, he understood as well as anyone. Yet he had carried out his work with a sense of wonder.

Coming to computer science, clearly, one feels an immense sense of gratitude for scientists who pursued the true meaning of computability despite their personal poverty and lack of recognition. You will be reading about many of these scientists in this book. The name of Moses Schönfinkel comes to mind: please read about him. `https://en.wikipedia.org/wiki/Moses_Schonfinkel` Here are some excerpts from this article:

> In 1929, Schönfinkel had one other paper published, on special cases of the decision problem (Entscheidungsproblem), that was prepared by Paul Bernays. After he left Göttingen, Schönfinkel returned to Moscow. By 1927 he was reported to be mentally ill and in a sanatorium. His later life was spent in poverty, and he died in Moscow some time in 1942. His papers were burned by his neighbors for heating.

Alan Turing's, Marie Curie's and Stephen Hawking's personal lives also evoke a similar sense of awe and respect. Our scientific fields are replete with stories of such struggles. *Please be sure to look up the personal lives of scientists whose names you hear about! It often heightens the power of their inventions, revealing the dedication of the woman or man behind these inventions we may otherwise take for granted.*

Specific Acknowledgements: My special thanks to all those who gave me input on this book, read it minutely and found typos, contributed various items to this book, and the authors of existing books in this area.

My most heartfelt thanks to a few select individuals who I admire deeply. The first is Prof. John Brzozowski for his very careful proof-reading, his writing of a foreword for me, and last but not least, his **immensely impactful** contributions to computer science! Second in my list is Professor Rick Neff of BYU-Idaho. Rick is the most astute reader I have encountered in my life! As I've once heard, "the ability to spot ugliness is the beginning of wisdom!" Rick literally X-rays into typos and marks them beautifully in red ink.[15] Third in my list is Professor Geoffrey Draper of BYU-Hawaii for drawing all the cartoons in this book *as well as the back-cover cartoon*! Geoff impressed me as a student in my CS 6100 class during Fall 2005 when, while taking my class, he also drew delightful cartoons that were the highlight of my 2006 Springer book "Computation Engineering." Geoff's cartoon of Alan Turing stumbling upon his invention has gone far and wide around the world!

Next, I would like to specifically thank these individuals: Shriram Krishnamoorthy for his astute comments; David Dill, Ching-Tsun Chou and Vivek Sarkar for contributing to the back-cover quotes of this book; Randi Cohen for proposing that I write this book; both Randi and Karen Simon who gave considerable feedback; Callum Frazer for his help with this project; Ian Briggs for testing Jove, creating many of the interesting machines that can be executed under Jove, and writing the Lambda Calculus code used in Chapter 18; Heath French for testing Jove; Steve Siegel for his encouraging feedback; Michael Bentley for setting up Jupyter hub for my class, and proof-reading; Paul Carlson and Arnab Das for their extensive feedback; and all my TAs and students who suffered through the early versions of this book that got used in the CS 3100 classes I've taught. I thank Mark Van der Merwe for integrating Ling Zhao's PCP solver to support the experiments in Chapter 15, Mate Soos for the web-based SAT solver mentioned in Chapter 16, and Tyler Sorensen for contributing the BDD package used in Chapter 17. I thank Dr. David Beazley for creating the excellent PLY lexer and parser package, and Prof. Fernando Perez, creator of IPython and co-founder of Project Jupyter for creating Jupyter notes and also encouraging me to post my work on the Jupyter Gallery of Notebooks. I also thank the students who took my class *Models of Computation* (CS 3100) at Utah during Fall 2017 and 2018 for their feedback. I apologize if I left out anyone inadvertently, and wish to be gently reminded.

Last but not least, I wish to thank my amazing family (whose initials are captured by the regular expression $(kg)^* \cap ([a-z][a-z])^3$) for their love and support. I thank my late parents (whose names I have taken on as my middle and last names) who saw to it that I lacked nothing at all, and my amazing brothers and sister-in-laws who took care of my parents. I also dedicate this work to the many inspiring books that brought me into science, and to the noble souls who gifted them to me.

[15] While I've never been to the Red Sea, thanks to Rick, I did traverse it! Any remaining typos are those that I carelessly missed despite Rick's comments.

Part I: Foundations

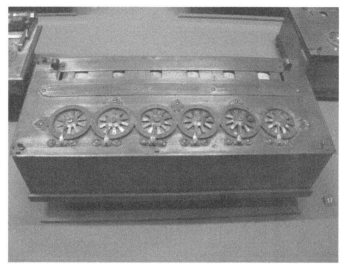

Pascal's calculator
(Photographed by this author, Aug 2012
Musée des Arts et Métiers in Paris France)

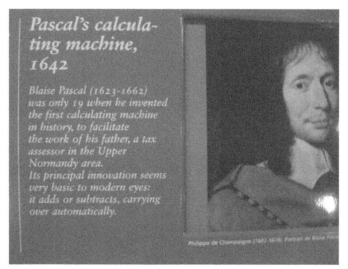

Biography of Pascal
(Musée des Arts et Métiers)

1
What Machines Think

Computers are everywhere and their power seldom ceases to amaze even those deeply embedded within the field! Computers seem to do (or be involved with) everything: from processing payroll to making telephone calls and even brushing our teeth (hiding within electric toothbrushes). They have also beaten humans at (the quiz show called) *Jeopardy!* One may therefore ask in all honesty: "can computers do *everything?!*"

There are obviously things computers can't do.[1] But what all can they do, and what does "can do" mean in the context of a *machine*? Remember that these questions were raised during the early part of the 20th century when actual computers weren't around; therefore, it isn't clear what answers flew around. It is quite possible that people relied on their immediate experiences to provide answers—much the same way as people imagined airplanes to be before the Wright brothers flew one.[2]

Coming back to the subject of computing, early computing successes were marvels of engineering. Blaise Pascal created an impressive machine around 1642 to assist his father with his work (Page 1). According to Wikipedia, Pascal conceived the idea while trying to help his father in his job as a tax collector. Pascal's calculators could add and subtract two numbers directly and multiply and divide by repetition. About 200 years later, Babbage designed the Analytical Engine (trial design of 1837 shown in Figure 1.1). He even hired the first ever programmer—-Ada Lovelace—who was introduced to Babbage when she was just 17 years of age, and got "hooked onto computing!"[3]

At the turn of the 20th century, people started pondering the question of the fundamental limits of computing. They began formulating this question in terms of the types of problems in mathematics and logic that can be "solved" using a machine. It wasn't clear why there should be a fundamental limit to the power of computers. After all, authors such as Jules Verne were at this time writing science fiction pieces pertaining to going to the moon in spaceships (even before planes arrived).

[1] Like screw up something as badly as humans alone can do!

[2] Some early models of airplanes actually flapped their wings (imitating birds) and invariably crashed (unlike most birds).

[3] Ms. Lovelace described the first algorithm to compute Bernoulli numbers using the Analytical Engine of Babbage.

Figure 1.1: Babbage's Analytical Engine. By Science Museum London / Science and Society Picture Library. Reproduced under Creative Commons Attribution-Share Alike 2.0 Generic license. http://creativecommons.org/licenses/by-sa/2.0.

Figure 1.2: The website http://aturingmachine.com and the Youtube Video https://youtu.be/E3keLeMwfHY of a Mechanical Turing Machine by Mike Davey. Please search for *An Interview with Mike Davey about his Homemade Turing machine* also.

[4] The distinction between a procedure and an algorithm is taken up in Chapters 13 through 15. Basically, a procedure is anything that is mechanizable, and it is also an algorithm if it halts on all of its expected inputs. We already gave you an example of these ideas in the context of Fermat's problem.

[5] For a detailed discussion (that tends to be rather technical), kindly see https://plato.stanford.edu/entries/goedel-incompleteness/. This, however, does not mean that humans cannot guide the creation of useful proofs. See an early paper that proves facts about a tool that synthesizes logic circuits [1].

1.1 Problems Without Algorithms

One can always state **problems** without knowing how to solve them. For instance, Fermat posed the problem of finding a, b, c that are natural numbers greater than 1 such that $a^n + b^n = c^n$ for $n > 2$. This problem remained open for over 300 years till 1995 when it was conclusively settled by Andrew Wiles in the negative: no such a, b, c can exist [46]. Until that year, one could have defined a **procedure** that could have searched for every possible a, b, c in a systematic way for a given n (say 3). For instance, the procedure could list all a, b, c that add up to 6, then list all a, b, c that add up to 7, etc., till such an a, b, c, were found that satisfied this equation. However, since 1995, thanks to the proof by Andrew Wiles of a theorem called *Fermat's last theorem*, we know that this procedure would go on forever, not finding any a, b, c. In fact, we now have an **algorithm**: just print "impossible to find such a, b, c" and halt.

In the same vein, people in the 20th century were in hot pursuit of algorithms for pretty much every problem that can be posed in mathematics! Most notable in this quest was a German mathematician named David Hilbert who challenged the mathematics and logic community with 23 problems pertaining to logic, mathematics and computability. One of Hilbert's challenges was to prove the following:[4]

> There should be an algorithm—a systematic and mechanical procedure that also terminates on any input—to decide the truth of **any** logical statement in mathematics.

This was such a bold quest: Hilbert wanted *any* mathematical question to be algorithmically solved (always halt with "here is the solution" or "you can't have a solution.") A long line of famous mathematicians and logicians that includes Gödel, Church, and Turing showed that this goal was *impossible* to realize! They showed that many mathematical systems are *undecidable*: there isn't an algorithm to decide the truth or falsity of statements made in them! They also showed that many logical systems are *incomplete*: there are logical systems that are so powerful that one cannot *prove* known truths in them.[5]

> For concreteness, Hilbert's tenth problem was to devise an algorithm for finding solutions (over integers) for *Diophantine equations*—equations of the form
> $$3x^2 - 2xy - y^2z - 7 = 0.$$
> Another Diophantine equation is
> $$x^2 + y^2 + 1 = 0.$$
> It turns out that the former has the solution $x = 1, y = 2, z = -2$

while the latter has no solution.

A series of results developed in the 1940s through 1970 by Julia Robinson.[6] The joint work by Yuri Matiyasevich, Julia Robinson, Martin Davis, and Hilary Putnam helped settle Hilbert's 10th problem in the negative through the so-called MRDP theorem (see a nice historical account at [34]). Their work showed that alas, we cannot write a single computer program that always halts and either prints out the solution for such an equation (if one exists) or prints out that such a solution does not exist. **A third possibility** is to be admitted *in case the equation has no solution:* **any** *such program must necessarily loop and never halt!*

In this book, we shall be building some of the foundations toward appreciating all this momentous work. Some of the later chapters will also detail the topic of the fundamental limits of computing.

The vast majority of this book, however, discusses the *serendipitous outcomes* of these early pursuits seeking the fundamental limits of computing. In the end, these pursuits did help settle many open theoretical questions, but along the way, they spun off many fundamental ideas that form the bedrock of modern Computer Science. Fundamental developments in this young field are led by a brand new community of researchers who go by the name *computer scientist* (and not "mathematician" or "logician").[7]

1.2 How to Define a Computer?

Even to get started on Hilbert's program, one had to clearly define "a computer." Remarkably, within a few decades of the early 20th century, mathematicians managed to arrive at **the** *definition* of a computer (*i. e.*, an *ultimate computing device*). This work was spearheaded by many famous scientists including Alan Turing of England as well as Alonzo Church, Emil Post, Stephen Kleene and John von Neumann of the United States.[8]

Alan Turing's approach was to assume basic representations for numbers (in terms of 0's and 1's) and describe how one performs operations on numbers. His recipe for expressing an algorithm went something like this:

- At each step of the algorithm, if a character c (say 0) is read under the Turing machine head (or "cursor"), change the character to a different one (say 1). Then move the head one step left, one step right, or stay at the same spot. Then advance to the next step of the algorithm.
- The algorithm is deemed to have halted when it reaches one of the previously selected "halting steps." In that case, the compu-

[6] We are using a system called Jupyter to illustrate ideas underlying this book. "Jupyter" stands for Julia and Python, where the language Julia is named after the same Julia Robinson who also helped settle Hilbert's 10th problem. She is considered one of the preeminent computational theory researchers.

[7] It is not a big stretch to say that your iPhone exists due to the aforesaid scientists, plus of course countless excellent engineers who followed!

[8] Turing as well as von Neumann were also involved in the construction of some of the early computers.

[9] It does not matter how long it takes to compute something; all that matters is "can it be done at all, in a *finite* amount of time?" For instance, in Mike Davey's Turing machine, about 3 Turing machine operations such as moving left or erasing can be done in one second. Saving a megabyte of data takes 870 of the 1000-foot tape rolls shown in Figure 1.2.

Figure 1.3: Prof. Geoff Draper's imagination of how Alan stumbled upon his invention. This is one of many cartoons donated by Professor Draper for my 2006 book [21] (for all of Geoff's cartoons included in that book, kindly see http://draperg.cis.byuh.edu/cartoons).

[10] Alonzo Church served as a PhD advisor for Alan Turing at Princeton University in 1938.

[11] In this book, we shall adhere to Turing machines as our notion of a computer (and "computation").

[12] Metaphorically, all "roads of thought" lead to "the Rome of universal computability."

[13] One *can* bail a swimming pool with a teaspoon.

tation ends, leaving behind the contents of the tape as the final result.

It was soon shown that anything that is mechanically computable can be described in elementary terms such as this.[9] See a modern Turing machine detailed in Figure 1.2 and a cartoonist's view of the historical moment in Figure 1.3. An excerpt from Sir John Dermot Turing's book [44] recounts a conversation between Aspray and Church [4]:

Aspray: Can you tell me something about his (Turing's) personality?

Church: I did not have enough contact with him to know. He had the reputation of being a loner and rather odd.

(Aspray notes in [19] "Of course, the same often was said of Church.")

Turing was not alone in the process of trying to define the concept of computation. His contemporary, Church,[10] meanwhile proposed the Lambda Calculus. Another contemporary, Post, also proposed Semi-Thue Systems around the same time. It was also shown around this time that things written in the Lambda Calculus of Church or Semi-Thue Systems of Post could also do no more and no less than a Turing machine![11] This convergence[12] was taken as further evidence that the notion of what is mechanically computable is unique. It also happens to be one of the most fundamental of insights that humanity has achieved in terms of understanding what "computation" means.

Paraphrasing Prof. Phil Wadler (now of the University of Edinburgh),

"when researchers come up with the same powerful idea even when starting from multiple alternative perspectives, the idea itself is fundamental—"existed out there all the time", and *happens to have been discovered—not invented.*"

Turing Machines Define a Computer

From a practical perspective, it is clear that a supercomputer is more powerful than the computer in your cellphone. It may take a second to do on a supercomputer what might take years to solve on your cellphone. Yet (if you allow enough time and provide enough memory), anything that can be solved on a supercomputer can be solved on your cellphone. More specifically, the *same types of problems* can be solved on these machines.[13] The reason we do not consider absolute time in our study of computation is that it is an entirely ephemeral notion; computers keep getting faster as well as different from each other all the time. However, an entire area called *complexity theory* is in fact devoted to the study of "how much time computations take." We will be studying one of the most central of approaches for measuring complexity—namely, the notion of NP-completeness—in Chapter 16.

1.3   Practical Application: Syntax Definition / Checking

One of the main applications of the material we are going to study in this book is in *defining* program syntax and in *checking* adherence to program syntax definitions.

Beginning in the late 1950s, computer scientists began seeking practical ways of writing computer programs. They began developing rich and expressive syntactic constructs, making programs easier to create as well as *readable*. As soon as programming languages having formal syntactic rules were created, it became clear that one had to *parse* these programs (*i.e.*, check for syntactic errors).[14]   Now, parsing is the act of checking for "allowed patterns," and flagging errors when disallowed patterns are seen. For instance,

- The identifier of a program variable is comprised of letters followed by digits or special characters such as _; for example, `Head1_ptr`.
- Multiple variables can be declared in one line, separated by commas:

```
int a, b, c, d, e;
```

- A real number in standard fractional form has an optional sign followed by the whole part, a period, and the fractional part; for example, `-1.03`.
- A nested block is a collection of matching braces with things between braces. Matched braces themselves are as follows: { }, {{ }}, {{{ }}}.
- A prototype declaration and a function definition match as follows (example):

```
char func(int, float);

char func(int a, float b) { body }
```

As soon as one defines a syntax, one has to entertain the notion of syntactic errors. A compiler relegates the low-level work of checking for syntactic errors in the declaration of identifiers and numbers to what is called the *scanner*, which is nothing but a highly simplified parser. The results of scanning (when successful) are then fed to the main parser which checks for the overall program well-formedness, and generates code.

As an example, a scanner turns something like

[14] If there are no syntactic errors, the code generator would typically be called next to generate the object (machine) code.

```
int My_var1, My_Var2, My_Var3;
int My_var4 = -1.03;
int main(){{}}
```

[15] It stashes away information regarding which exact identifier is being declared, which keyword is being processed, etc., in a data structure called a *symbol table* for later reference.

into (respectively) the following simpler patterns;[15] a parser must now recognize these patterns as being legal:

```
keyword id, id, id;
keyword id = number;
keyword keyword lpar rpar lbrace lbrace rbrace rbrace
```

It is now clear that

- A scanner must reject a number of the form 30.-1 and accept a number of the form -1.03;
- A parser must reject a variable declaration statement of the form
  keyword ; id id id,,
  and must accept something of the form
  keyword id, id, id;
- There are usually no a priori limits on the lengths of identifiers or numbers, nor limits on how many declarations one can have, or even the level of nesting allowed using braces. Thus,
  -12222244343434343.4343434343434343566689991
  must be accepted as a legitimate number, and
  {{{{{{{{}}}}}}}}
  must be allowed to be a legitimate nested-brace structure.
  In fact, you can check that the following is a syntactically correct C program:

```
int main(){{{{{{{{}}}}}}}}
```

whereas if you omitted one right-brace, you will get a syntactic error, essentially saying this:

[16] This is akin to how we define infinite sets such as integers, but in practice may not go beyond a certain size. The known universe itself has only about 10^{82} atoms, according to https://www.universetoday.com/36302/atoms-in-the-universe/.

```
error: expected '}'
note: to match this '{'
int main(){{{{{{{{}}}}}}}}
           ^
1 error generated.
```

- Even if a system likes to impose limits, it is best to design scanners and parsers as if they had no such limits.[16]

Pattern Classes in Program Syntax

From the aforesaid examples, we can see four classes of patterns arising in programming languages. Parsers must reckon with these patterns and declare them to be legal or illegal, as the case may be:

Regular Patterns: We introduce the notion of regular patterns through examples:
- **Finite and fixed-size** patterns arise in programming. For instance, keyword `main` is just four letters in that order.
- These days, we see password rules: *"Must be between four and 12 characters in length, with at least one upper-case and one lower-case and one special character and one number."* Even these are finite and fixed-size patterns.
- Identifiers in programming languages consist of a letter followed by one or more letters, digits, or special characters. All identifiers are of fixed length (they can't be empty, but it might be a billion characters long). We classify them under **finite but unbounded.**
- We have other interesting recurring patterns as in `keyword id, id, id;` The pattern is
  "keyword followed by one or more ID+Comma pairs, ending with ;"
  These are also finite but unbounded. The structure of these patterns is
  - a single keyword, followed by
  - a finite (but unbounded) number of ID+Comma pairs, and
  - terminated by a single ;
- Still another regular pattern that typifies repetition is

$$01001010010100101001....$$

Notice that this is a rather boring alternation of 01 and 001.

Context-Free Patterns: We introduce context-free patterns through examples:
- We have the *properly nested pattern*, which is context-free:
  `{{{{{{{{}}}}}}}}`
  There are places where we have other variants too; for instance
  `((()))((())())`
  This is a pattern of perfectly nested parentheses, and has the following properties:
  - The same number of left parentheses as right parentheses, and this is again a *finite but unbounded* number;
  - "Proper nesting," *i.e.*, when we sweep the parentheses left-to-right, then

* at every intermediate point, the number of left parentheses '(' is greater than or equal to the number of right parentheses ')'.

* At the end of the string, we will have an equal number of parentheses of both kinds.

- Yet another context-free pattern—that actually resembles proper nesting—is:

$$01001000100001000001000001\,100000010000010000100010010$$

Notice that this is a *palindrome*.[17]

Context-Sensitive Patterns: We now introduce context-sensitive patterns through examples:

- The pattern contained in function prototypes is context sensitive:
  - A finite but unbounded listing of things in some order, appearing in the argument list (focus on `int` and `float`);

```
char func(int, float);
```

  - This is then followed by the same listing of things in the actual function declaration (again focus on `int` and `float`):

```
char func(int a, float b) { body }
```

  - Yet another context-sensitive pattern is

$$01001000100001000001000001\,0100100010000100000100000001$$

Note that this is not a palindrome, but rather, some odd-looking pattern of 0's and 1's going up to some length, which is then followed by a copy of the same pattern *without any reversal*.

Recursively Enumerable Patterns: The next natural set of patterns defined in computer science corresponds to the notion of computation itself.[18] It is too early in this book to be telling you what this class of patterns is, other than through some examples. We shall present two examples (merely for the sake of completeness) and then move on, revisiting this issue much later in this book.

Consider the familiar sorting algorithm—any sorting algorithm at all. We know that given an input array, such a sorting algorithm produces an output array where all the elements in the input array are present but (say) in ascending order. Now, think of this as a pattern (A,B) where A is the input array (written out according to some conventions in binary)

[18] The term "recursively enumerable" was coined in an era when actual computers weren't around. It means *computable*—and not so much whether a function calls itself (colloquial usage of the term "recursion.")

Also, some of you may know that there are other pattern classes such as *recursive* sets, which are a special case of recursively enumerable. The Chomsky hierarchy (discussed on Page 12) offers four grammars that match these pattern classes, and so we limit our current discussions to these four pattern classes.

and B is the output array. Well, we can consider a collection of such (A, B) pairs as defining *one specific example* of a recursively enumerable pattern class.

A more interesting example of a recursively enumerable pattern class is this: *The set of all (M, w) pairs such that M is the text of a program and w is an input on which program M can be run* **such that** *M when run on w will not infinitely loop.* These are the kinds of recursively enumerable sets that we shall actually be studying later. One interesting fact is that if we write another program, say P, that attempts to pick out *exactly* those (M, w) pairs where M won't loop on w, then we will find that even the best such "P" we write has this behavior:

- When given a specific (M_1, w_1) pair where M_1 won't go into a loop[19] on w_1, program P would actually print "yes, it won't loop" and halt.
- For a general (M_1, w_1) pair where M_1 loops on w_1, P itself will go into a loop (in the process of its checking). You might build special-case checks into P such as *Check for M_1 being a particular program such as "while(1);"*. However, in general, you can't "parse out" all the looping programs. Your best bet to program P that finds out whether M_1 infinitely loops on w_1 is to have P run M_1 on w_1! Well, you know that such a P will loop in the case that M_1 loops on w_1.

> Recursively enumerable sets contain complex patterns that are intimately tied to computations carried out by Turing machines (or devices equivalent to them). Contrast these two statements:
> - *The set of all (M, w) pairs where M is a program and w is an input on which M can be run.*
> - *The same set of (M, w) pairs* **such that** *M when run on w will not loop.*
>
> The "such that" part adds a detail directly tied to computations, and dramatically jacks up the richness of the pattern class. Without that part, we merely have the superficial structure of programs and inputs—not what they can do when "*executed.*" For this reason, recursively enumerable sets are also known as *Turing-recognizable sets.*

[19] The notion of whether M_1 loops on w_1 is well-defined, and has a Boolean answer; either it would infinitely loop or it won't. The issue we are facing is *can we write a computer program that discovers what the actual answer is?*

1.4 Simplified Turing Machines as Parsers

In the 1950s and 1960s, scientists working on practical aspects of computer science ended up developing three **simplified versions of Turing machines**:

1. Finite automata (FA) are machines defined by Rabin and Scott around 1957 [40]. These are machines that can help parse *regular languages*.
2. Pushdown automata (PDA) defined primarily by Ginsburg [20]

[20] Push each left parenthesis arriving from the input stream onto the stack. Whenever a right parenthesis arrives, pop a left parenthesis off of the stack to match it. If the stack is empty in the end, and was not found empty when a right parenthesis arrives, then we have perfectly nesting parentheses.

[21] It can be shown that if instead of a stack we allow an unbounded queue to be implemented and accessed, we will straightaway obtain the power of a Turing machine. One can call these machines "PQA" standing for push-down queue automata. Even having two stacks instead of one gives us the power of a Turing machine! So the *single stack assumption* is crucial to be able to limit the programming power to that of a push-down automaton. Also, to remain as a PDA (and not become more powerful) you cannot access the stack in an undisciplined manner (such as reaching into the stack and examining the items). Instead, you must employ the push and pop operations.

and Greibach [23] in the early 1960s; these are machines that can help parse *context-free languages*. Pushdown automata can be thought of as finite automata augmented with a single unbounded stack. As one can imagine, the stack helps match perfectly nested parentheses using an obvious algorithm.[20]

3. Linear bounded automata (LBA) defined by Kuroda [30] and others, also in the early 1960s; these are machines that can parse *context sensitive languages*.

Note that Turing machines are the ones that define the structure of recursively enumerable languages.

What Programming Restrictions are Modeled by TM Simplifications? The aforesaid Turing machine simplifications directly map to restrictions in programming. We discuss this issue, illustrating it on C.

- Suppose you are asked to program in the C programming language within function `main()` where you are only allowed to declare a finite number of *finite* variables (variables of finite type such as a bit or a char that can hold only a finite object). You are not allowed to allocate any heap storage or engage in any recursive calls of functions. While this sounds limiting, this is exactly the power of finite automata that help parse regular languages—very much relevant in practice.

- Now suppose you are given a little bit more freedom and allowed to employ a collection of functions each of which can allocate only a finite number of finite variables—but the functions can call each other recursively. Such a programming language gives you the power to describe push-down automata, with the recursion stack simulating the *single* stack of a push-down automaton.[21]

- Now if we are allowed unbounded memory that can be accessed more freely (*e.g.*, heap allocated memory or arrays with no a priori bounds), then programs in such programming languages attain the power of Turing machines.

- Linear bounded automata can be arrived at by limiting the manner in which we access this unbounded memory. Specifically, we *snip* the tape of the Turing machine beyond the portion on which the initial input is written. This "truncated-tape Turing machine" is what a linear bounded automaton is.

In summary, the three machine types defined earlier – finite automata, push-down automata and Turing machines – can be arrived at simply by limiting one's ability to program in specific ways. These limitations are all related to how memory can be allocated or accessed.

Machines, Patterns, and Chomsky's Grammars

In a completely disconnected thread of work, Chomsky [16] was working on the topic of *formal grammars*. These grammars were stratified into four types of grammars by Chomsky (called Type-0, Type-1, Type-2, and Type-3). These grammars correspond to the kinds of patterns/languages recognized by FA, PDA, LBA, and TM![22] Thus, it was firmly established that formal language machines, patterns and grammars were not isolated ideas, but facets of something more fundamental and deeper. We will be revisiting the (so-called) *Chomsky hierarchy* in Chapter 13.

[22] Recall that Turing machines were discovered in the 1930s, finite automata in the late 1950s, and the other machine types in the 1960s.

1.5 Automata and Computability for Lifelong Learning

Automata and computability are essential for your lifelong learning as a computer scientist.

For instance, it is important to know the theory of *deterministic finite-state automata* (DFA) well before you can appreciate the topic of Hidden Markov Models (HMM) in Machine Learning. Almost all speech recognition systems and word-completion systems in cellphones are HMMs, which are nothing but DFA with their state transitions governed by probability values.

You may not be a machine learning fan but a computer networking fan. Even then, "you can't escape DFA!" You might want to write a deep packet inspection facility to detect malware flying over the internet. Well, you won't be able to do this effectively unless you know the theory of DFA as well as efficient implementation methods thereof [33].

You may neither be a machine learning person nor a networking person, but merely interested in processing HTML documents efficiently. Even in this endeavor, you will need to know the topics of this book rather well, in order to handle the complex syntax-spaces that need to be correctly handled during parsing.

The Birthing Struggles of Computer Science   The creation of computers that enjoy their present-day power and prominence did not happen overnight.

- Before we learned to incorporate operator precedence rules into grammars, various ad hoc parsing methods were in use; these were quite dreadful and impossible to reliably understand. Around 1962, Knuth wrote a paper [27] pointing out how early compilers used to replace each + with a)))+(((and each * with a))*((, and added enough compensating parentheses. According to Knuth, this scheme "seemed to work quite miraculously" (clearly, this is too crude and ad-hoc, and runs into trouble often).

- In the late 1960s, a spacecraft sent to Venus was allegedly lost [24] because of a period typed in place of a comma (see §11.8).[23]

[23] We use '§' to mean "Section."

- Attempts to parallelize finite automata appear even in recent prestigious conference papers [36].

In short, there isn't a better time in computer science to be studying *Mod-*

els of Computation. We in fact plan to make your journey ultra pleasant by taking a hands-on approach through the introduction of the Jove software.

2

Defining Languages: Patterns in Sets of Strings

> **Chapter Gist:** *We begin with how strings are formed from symbols over an alphabet (§2.1), and how languages (sets of strings) can be defined and operated upon (§2.2). We assume basic knowledge about sets (see §A.1 for a refresher). We denote whole numbers by Nat (usually written $\mathbb{N} = \{0, 1, 2, \ldots\}$) and positive and negative whole numbers by Int (usually written $\mathbb{Z} = \{0, 1, -1, 2, -2, 3, -3, \ldots\}$).*

2.1 Symbol, Alphabet, String, Language

Four notions are basic to our study of computations: Symbol, Alphabet, String, and Language.

2.1.1 Symbol

Symbols are things with which humans and machines communicate with other humans or machines. They are taken to be primitives in each context. In a very general sense, examples of symbols include English characters, entire English words (when digging deeper into the characters in the word is irrelevant in a context), musical symbols, flashes of light, or even smoke signals. **We hereafter require symbols to be strings of length 1**, *i.e.*, single characters such as a, b, 0, or 1. In our Python encodings in Jove, strings of unit length will be regarded as symbols.[1]

2.1.2 Alphabet

An **alphabet** is a *finite* and *non-empty set* of *symbols*. We will use Σ to denote an alphabet. In each context, we will pick an alphabet and then stick with it (*i.e.*, we usually do not change the alphabet in the middle of

[1] The manner in which we quote a string does not matter; thus, 'a' and "a" mean the same.

a construction). Here are examples of alphabets:

- $\Sigma = \{0,1\}$ in mathematics.

  In Python, this alphabet will be written {"0","1"}

- $\Sigma = \{a\}$ in mathematics.

  In Python, this alphabet will be written {"a"}

- $\Sigma = \{a\}$ is an example of a *singleton alphabet*.

> An alphabet is like your entire keyboard, and each symbol is like one key.

2.1.3 String or Word

A **string** or **word** is a *finitely long* and *possibly empty sequence of symbols*.[2] Strings are read *left-to-right*.[3] An empty string is denoted ε in mathematics. In Python, we express an empty string as either "" (opening and closing double-quotes) or as '' (opening and closing single-quotes). By definition, the length of the empty string ε is 0.

> A string is said to be "over" an alphabet. Thus the string "ateAnApple" consists of eight distinct symbols, namely $\{a,t,e,A,n,p,l\}$. This string is over an alphabet that includes these eight characters. Any string could be considered to be *over* a larger alphabet than the set of characters present in it; thus, "dad" could be viewed to be over $\Sigma = \{a,b,c,d\}$. This is because the user may not choose to press some of the keys. But all the symbols in a string must be present in the alphabet (or, your keyboard). Note: Instead of saying "string s is over alphabet Σ," we may instead say "alphabet Σ underlies string s."

[2] In this book, we never deal with infinite strings. All the machines we build operate on finite strings. In many research areas (especially Formal Methods), automata over infinite words are important; one popular class of such a machine is that of Büchi automata.

[3] Unlike in languages such as Arabic where strings may be read right-to-left.

We **concatenate** two strings to obtain a new string. Concatenation is expressed through juxtaposition. For example, ab concatenated with cde is written $ab\,cde$ and results in string $abcde$. In Python, "ab" + "cde" denotes string concatenation, resulting in "abcde".

Exercise 2.1.3, Language Operations

1. As per our definitions, can *Nat* be an alphabet? Why or why not?
2. Consider the string "Hello there!". What are the symbols present in this string, and what is the smallest alphabet underlying this string?
3. Are all palindromes the concatenation of a string with its reverse? If not, why not? □

2.1.4   Various Notions of Zero and One

In mathematics, the notion of **Zero** and **One** (or "Unit") are fundamental. The basic idea is that when you have an operator resembling "multiplication," then:

- multiplication of x by Zero yields Zero for any x, while
- multiplication of x by One yields x.

In this book, we shall help you see what the Zero is and what the One is for various situations (*i.e.*, different algebraic systems). This will serve you as a powerful memory aid for many language-theoretic rules that are otherwise difficult to intuitively understand. In almost all our presentations, we will show you how Zero and One work for numbers, and then contrast it with how they "ought to work" for language-theoretic ideas.

Zero and One for Nat**,** Int**, and** $Real$**:**   We are all familiar with the following over Nat, Int, and $Real$: $0 \times 65 = 0$; $65 \times 0 = 0$ (multiplication is commutative); $1 \times 65 = 65$; $-65 \times 1 = -65$; $0.0 \times 0.557 = 0.0$; $0.557 \times 1.0 = 0.557$; $\pi \times 1 = \pi$; and $0.0 \times \pi = 0.0$. We state these obvious facts to motivate similar thinking for strings and languages.

- **One for string concatenation**: If string concatenation is viewed as multiplication, then ε must be the One element, because:
  - $\varepsilon ab = ab\varepsilon = ab$
  - In Python, `"" + "ab" = "ab" + "" = "ab"`.
- **Zero for concatenation**: There is no Zero for string concatenation (no "Zero string" such that when any string is concatenated with it, we get back the Zero string).

Exponentiation: Now that we are viewing concatenation as multiplication, and since it is quite natural to view repeated multiplication as exponentiation, we will introduce the idea of exponentiation into the algebra of strings, as follows:

- For a symbol a, we have $a^2 = aa$, $a^3 = aaa$, etc.
- For a string ab, we have $(ab)^2 = abab$, and $(ab)^3 = ababab$.
- As another example, $(abacaea)^2 = abacaeaabacaea$.
- In Python, we can write these as follows:
  - `"a" * 2 = "aa"`
  - `"a" * 3 = "aaa"`
  - `"ab" * 2 = "abab"`
  - `"ab" * 3 = "ababab"`
  - `"abacaea" * 2 = "abacaeaabacaea"`

Exponentiation by Zero: We know that for any real number x, we have $x^0 = 1$. This is because we want to support a recursive definition for exponentiation. For instance,

- $x^n = x \times x^{n-1}$

But, as in all good recursive definitions, we must specify the basis case. Thus, we fix

- $x^0 = 1$.

This way, we can calculate things like 4^3:

- $4^3 = 4 \times 4^2$;
  $= 4 \times 4 \times 4^1$;
  $= 4 \times 4 \times 4 \times 4^0$;
  $= 4 \times 4 \times 4 \times 1$;
  $= 64$.

In other words, picking x^0 to be 1 is crucial for a recursive definition of exponentiation to work out. *In the same vein*, exponentiation of a string 0 times must yield the unit for string concatenation, which is ε. Let us check this out:

- $s^0 = \varepsilon$ (exponentiating s 0 times yields the unit of concatenation for strings, namely ε)
- $s^n = ss^{n-1}$ ("s concatenated with an $n-1$-fold repetition of s");
- Thus, for string a,
  - $a^3 = aa^2$;
    $= aaa^1$;
    $= aaaa^0$;
    $= aaa\varepsilon$;
    $= aaa$.

Clearly, this notion works out smoothly for the string algebraic system also.

Exercise 2.1.4, Zero, One, Exp

1. Consider a string $s = abacaca$. Now consider the string exponentiation s^4. How many a's are there in s^4? How about b's, c's and d's?

2. Let *Nat* be regarded as the universal set in some domain of discourse. Now consider subsets of *Nat*, and let us view *set intersection* of such sets as "multiplication." In that case, what is the *One* element of this multiplication operator? What is the *Zero* element? Recall that for any $s \subseteq Nat$, these One and Zero elements must satisfy:

   (a) $s \cap One = s$

   (b) $s \cap Zero = Zero$ $\square$

2.2 *Language*

A **language** is a *possibly empty* and *possibly infinite set of strings* (each string is finite).

Two examples:

- $EmptyLang = \emptyset$

In Python, EmptyLang = set({}) (or even set())

- $Mylang = \{\varepsilon, aa, abc\}$

  In Python, MyLang = {"", "aa", "abc"}

- Using the notation of set builder, we can describe languages more conveniently.

- Example: Let us specify all strings of "mindless repetitions" of the 01 pattern:[4]

$$L_{01Rep} = \{(01)^i : i \geq 0\}$$

$L_{01Rep} = \{\varepsilon, 01, 0101, 010101, 01010101, \ldots\}$

- As another example:

$$L_{a\_lt\_b} = \{a^i b^j : i, j \geq 0, and\ i < j\}$$

We list some elements: $L_{a\_lt\_b} =$
$\{b, abb, aabbb, abbb, aabbbb, bbbb, bbbbb, \ldots, a^5 b^{55}, \ldots\}$

> [4] Note that (and) are not part of our alphabet; they are used to group 01 for the purpose of applying the exponent i.

Exercise 2.2, Languages

1. Why is ε not a member of $L_{a\_lt\_b}$?
2. Which inequality in the definition of $L_{a\_lt\_b}$ would you alter in order to induce ε into this language?   □

Convention: By saying $i, j \geq 0$ we will tacitly assume that i and j are natural numbers. Since for every natural number p there is a natural number $q > p$, we can easily conclude that La_lt_b is infinite.

Finite Approximations of Infinite Languages in Python: Note that most languages discussed in this book are infinite, and hence cannot be printed out in their entirety. Thus, when we study these languages in Python using Jove, we approximate these languages by printing out all the strings under a given length. This gives a rough idea of what the language is about, without seeing all of its content.[5]

Let us employ the set builder notation in Python and define a finite approximation of the La_lt_b. We will call it La_lt_b_9 to capture the fact that this language will include all strings of length 9 and under. Note that La_lt_b_9 is being defined below using Python's set comprehension operation.

> [5] Similar to how we write down $\pi = 3.1415926$ even though that is only an approximation of π.

```
Mylang = {'', 'aa', 'abc'} # a language over Sigma = {'a','b','c'}


La_lt_b_9 = { "a"*i + "b"*j
              for i in range(10) for j in range(10) if i < j }


>>> La_lt_b_9           # Approximated to <= 9 i's and j's
```

```
{'aaaaaabbbbbbbbb', 'abbbbb', 'aaabbbbbbbbb', 'aabbbbb', 'aabbb',
'aaaabbbbbbbbb', 'bbbbbbb', 'aaabbbbbbbbb', 'aaabbbbbb',
'aaabbbbb', 'bbbbbb', 'abb', 'aaaaaaaabbbbbbbbbb', 'abbbbbb',
'aaaabbbbbb', 'bbb', 'aaaabbbbbb', 'aabbbbbb', 'aaaabbbbb',
'aaaaabbbbbbbbb', 'aabbbbbbbbb', 'abbbbbbbbb', 'aaaaaabbbbbbbb',
'abbb', 'bb', 'aaaaaabbbbbb', 'abbbbbb', 'aaaaaaabbbbbbbb',
'aabbbbbb', 'aaaaabbbbbbb', 'aaaaabbbbbb', 'bbbbbbbbb', 'bbbbb',
'aaabbbb', 'b', 'abbbb', 'abbbbbbbb', 'bbbbbb', 'bbbb', 'aaabbbbbb',
'aaaabbbbbbbb', 'aaaaabbbbbb', 'aabbbbbbbb', 'aaaaaaabbbbbbbbbb',
'aabbbb'}
```

Exercise 2.2, Languages (Python)

1. Write a one-line list comprehension in Python to generate the set of all substrings of s="abc". You can assume that s="abc" is a statement issued prior to your set comprehension.

2. Write down a one-line set comprehension in Python to generate a set of strings of the following form:[6]

[6] Your printout may list the different lengths in a "strange" order, but as you recall, the order within sets does not matter.

```
{'', '(((((())))))', '(((())))', '()', '((()))', '(())'}
```

We want this set to include all strings consisting of n left parentheses followed by n right parentheses for $0 \le n \le 5$. Note that for $n = 0$, we are generating ε (or '' in Python).

[7] Notice that str[::-1] is an idiom to reverse string str. For details, please consult any Python tutorial that discusses the topic of strings adequately.

3. Is this true in Python for two strings p and q?[7] Explain! Argue (in one paragraph) why this works for any two strings p and q.

    (p+q)[::-1] == (q[::-1] + p[::-1])

4. In a sentence or two, describe the contents of L_1, L_2, and L_3 in English.

$$L_1 = \{ (^n)^n : n \ge 0 \}$$

$$L_2 = \{ w : w \text{ is a string of balanced parentheses} \}$$

$$L_3 = \{ w : w \text{ has an equal number of parentheses} \}.$$

Note that the alphabet consists of the two symbols (and). By "balanced parentheses," we mean that in any string over (and), every occurrence of a right parenthesis must be matched by a left parenthesis that appears earlier in the string. In your answer you should also mention all possible language inclusions that exist among L_1, L_2 and L_3 (i.e., if $L_x \subset L_y$ for $x, y \in \{1, 2, 3\}$, mention that in your answer). □

2.2.1 Language Concatenation

We begin with the notion of **Language Concatenation**. It is defined as follows. For two languages L_1 and L_2, their concatenation, written $L_1 L_2$, is

$$L_1L_2 = \{xy : x \in L_1 \wedge y \in L_2\}$$

Let us use Python to illustrate this idea:

```python
def cat(L1,L2):
    """Concatenation of two languages.
       If  A = set(['ab', 'bc']) is one language,
       and B = set(['11', 'ab', '22']) is another language,
       then cat(A,B) returns set(['abab', 'bc22', 'ab11', 'ab22',
                                  'bcab', 'bc11'])
    """

    return set({x+y for x in L1 for y in L2})
```

2.2.2 The Zero and One for Language Concatenation

Viewing language concatenation as multiplication, what is the Zero (L concatenated with it gives back Zero) and One (L concatenated with it gives back L)?

Zero of a Language for Language Concatenation: It so turns out that the empty set (empty language) $\emptyset$ is **the** Zero of language concatenation. This is because

- The language concatenation $\emptyset L$ as well as $L\emptyset$ yields $\emptyset$. Here is why:
  - Look at the definition of concatenation:

$$L_1L_2 = \{xy : x \in L_1 \wedge y \in L_2\}$$

  - Now, if L_1 is $\emptyset$ or L_2 is $\emptyset$, the result is $\emptyset$ (we cannot "find an $x \in \emptyset$").

One of a Language for Language Concatenation: It so turns out that the set $\{\varepsilon\}$ is **the** One of language concatenation. This is because the language concatenation $\{\varepsilon\}L$ as well as $L\{\varepsilon\}$ gives back L.

Exercise 2.2.2, Zero and One for Concat

1. What is the difference between $\{\varepsilon\}$ and $\{\emptyset\}$? Which of these is a language *over the alphabet* $\{2,3\}$?
2. Prove that $\{\varepsilon\}$ is indeed the One element for language concatenation by showing that it left-multiplies or right-multiplies any language L to give back L. □

2.2.3 Zero and One of Language Concatenation in Python

The aforesaid ideas are now more crisply introduced, also providing Python encodings.

- *Empty Language or "Zero" of Language Concatenation: ∅ or {}*
  We call it the "zero" language because it is like the 0-element for concatenation.

```python
def Phi():
    """This is the ZERO language for concatenation
       (concatenation is viewed as multiplication)
    """
    return set() # Don't write it as {} because
                 # this is ambiguous - dict or set?
```

- *Unit Language, or "One" of Language Concatenation:* The Unit language is {ε}.

```python
def Unit():
    """This is the UNIT language for concatenation,
       when concatenation is viewed as multiplication.
    """
    return {""} # Set with epsilon
```

2.2.4 Exponentiation of a Language

We now proceed as before, and define the exponentiation of a language, which simply means "repeated concatenation" of a language.

Language Exponentiation:
- We have the required basis case:
  $L^0 = \{\varepsilon\}$, the Unit language
- We also have the general recursive case:
  $L^n = LL^{n-1}$

Hugely important observation: We observe that L^0 is the unit language $\{\varepsilon\}$. Note that this Unit language is generated by the Python function Unit().

- This assertion is true of **any** language L—even for empty language ∅.
- Note that the Zero language ∅ is generated by the Python function Phi().
- Thus we have $\varnothing^0 = \{\varepsilon\}$
  The zeroth exponent of the empty language (zero language) is the unit language.[8]

[8] How do we read and unravel this? Here is how. You may protest "but, but.." 0^0 is undefined in mathematics whereas we can take the zeroth exponent of the Zero language and get back the One language??

Fear not! That is fine. We are not taking number 0 raised to number 0, but only Language Zero raised to Number Zero. This does not get into the technical reasons as to why 0^0 is undefined in mathematics.

Why is 0^0 undefined in mathematics? Consider a series that converges toward 0. If you let it converge to 0 and then exponentiate, you will be tempted to say "it must be 0." If you take the intermediate series elements and raise it to 0, you will be tempted to say "it must be 1." It can't be both! This is well-explained in Prof. Peter Alfeld's page http://www.math.utah.edu/~pa/math/0to0.html.

2.2.5 Python Encoding of Language Exponentiation

```
def exp(L,n):
    """Exponentiate a language.
       If A = set(['ab', 'bc']) is a language, then
       exp(A,2) --> set(['abab', 'bcab', 'bcbc', 'abbc'])
    """

    return Unit() if n == 0 else cat(L, exp(L, n-1))
```

Exercise 2.2.5, Languages (review)

It is crucial that you do the following exercises before you move onto the next chapter.

1. Suppose $\Sigma = \{0,1\}$—commonly called *"the alphabet"* is treated as a language. All alphabets are special cases of languages; there is nothing surprising here! The alphabet $\Sigma = \{0,1\}$ is a language of two strings, each of length 1. Now, write out the contents of the following language exponents in their entirety. To avoid confusion, we have written out the answer for one case:

   (a) $\Sigma^2 = \{00, 01, 10, 11\}$

   (b) $\Sigma^0 =$?

   (c) $\Sigma^1 =$?

   (d) $\Sigma^3 =$?

2. Suppose a language $M = \{0, 10\}$ is given. What are the following language exponents? **We work out one case in detail for you.**

   (a) $M^2 = \{00, 010, 100, 1010\}$. Here is the explanation:

   - $M^2 = MM$
   - This means we must select two *random strings* from M (repetitions allowed) and concatenate them.
   - The random selections can yield
     - 0 and 0, whose concatenation is 00;
     - 0 and 10, whose concatenation is 010;
     - 10 and 0, whose concatenation is 100;
     - 10 and 10, whose concatenation is 1010.

   (b) $M^0 =$?

   (c) $M^1 =$?

   (d) $M^3 =$?

3. On Page 20, we defined three languages L_1, L_2, and L_3. Answer these questions with respect to these languages.

   (a) List the three shortest strings in L_1^3.

   (b) List a string of length 6 in L_2 that is not in L_1^n for any n.

   (c) What is the shortest string common to L_1^0, L_2^0, and L_3^0, and why? □

2.2.6 Union and Intersection of Languages

Language union and **intersection** are nothing but set union and intersection:

$$L_1 \cup L_2 = \{x : x \in L_1 \lor x \in L_2\}$$

$$L_1 \cap L_2 = \{x : x \in L_1 \land x \in L_2\}$$

```
def lunion(L1,L2):
    """Return the language union of L1 and L2."""
    return L1 | L2


def lint(L1,L2):
    """Return the language intersection of L1 and L2."""
    return L1 & L2
```

Notes: (1) Language complementation is defined only in Chapter 3, as we need to define the notion of a universal set over an alphabet before we can perform complementation. (2) One may consult §B.1.1 to see all the Jove functions corresponding to language operations.

Exercise 2.2.6, Languages (identities)

1. On Page 20, we defined three languages L_1, L_2, and L_3. Answer these questions with respect to these languages.
   (a) Does $L_1 \cup L_2$ match any of these three languages? Which one, why?
   (b) Repeat for $L_1 \cup L_3$ and $L_1 \cap L_2$.

2. (This question can be answered with the hints given below, even though we are introducing "star" officially only in the next chapter.) Let us define a function *star* with the following definition:
   (a) $star(L, 0) = L^0$
   (b) $star(L, 1) = L^0 \cup L^1$
   (c) $star(L, 2) = L^0 \cup L^1 \cup L^2$

   Now write down the contents of $star(L, n)$ for various L and n. Again, to avoid confusion, we have written out the answer for some number of cases:
   (a) $star(\{0, 1\}, 2) = \{\varepsilon, 0, 1, 00, 01, 10, 11\}$
   (b) $star(\{0, 1\}, 0) = ?$
   (c) $star(\{0, 1\}, 1) = ?$
   (d) $star(\{0, 1\}, 3) = ?$
   (e) $star(\{0, 10\}, 2) = \{\varepsilon, 0, 10, 00, 010, 100, 1010\}$
   (f) $star(\{0, 10\}, 0) = ?$
   (g) $star(\{0, 10\}, 1) = ?$
   (h) $star(\{0, 10\}, 3) = ?$

(i) $star(\{0,1,00,\varepsilon\},2) = \{\varepsilon,0,1,00,01,000,10,11,100,001,0000\}$

(j) How many elements are there in $star(\{0,1\},n)$? Explain your answer.

(k) Suppose we define

$$star(\{0,1\}) = star(\{0,1\},\infty)$$

That is, when we drop the second argument of the over-loaded *star* function, we assume that its meaning is the same as the *star* function of two arguments where the second argument is set to ∞.

Question: How would you describe one random string in $star(\{0,1\})$?

Think of a general way of characterizing it; here is a start: An arbitrary string that is finite/infinite and each symbol in the string is a .... □

2.3 Useful Results, Slippery Roads

We now discuss many situations that trip up new students on these topics. **It is crucial that you study these situations.** To help you think through these items, we present them as exercises.

> **Exercise 2.3**, Slippery concepts

1. Show that L_E is the set of even-length strings over alphabet $\{0\}$.
   $L_E = \{0^{2i} : i \geq 0\}$

2. Show that $L_E = \{(00)^i : i \geq 0\}$ (the parentheses are used to group the two 0's and are not part of the alphabet).

3. Let $L_O = \{0^{2i+1} : i \geq 0\}$. Show that $\{0\}^* = L_O \cup L_E$.

4. Describe this language in English:
   $Eq_{01} = \{0^n 1^n : n \geq 0\}$

5. Which of the following languages is Eq_{01} equal to, and why/why not:

   (a) $L_1 = \{0^i 1^i : i \geq 0\}$

   (b) $L_2 = \{0^n : n \geq 0\} \{1^n : n \geq 0\}$

   (c) $L_3 = \{00^i 11^i : i \geq 0\}$ (the exponentiations apply to only the single 0 and the single 1 respectively)

   (d) $L_4 = \{00^i 11^i : i > 0\}$

   (e) $L_5 = \{00^i 11^i : i \geq 0\} \cup \{\varepsilon\}$

   (f) $L_6 = \{0^i 1^j : i,j \geq 0, \text{ and } (i = j)\}$

6. Consider the language
   $L_7 = \{0^i 1^j : i,j \geq 0\}$

   Is it true that
   $L_7 = \{0^i : i \geq 0\} \{1^i : i \geq 0\}$?

   Explain, providing reasons.

[9] Chapter 3 is where we "officially" define language complementation.

7. Someone proposes that the complement of L_6 (written $\overline{L_6}$) is defined as follows.[9]

$L_8 = \{0^i 1^j : i, j \geq 0, \text{ and } (i \neq j)\}$.

Assume that the alphabet is $\Sigma = \{0, 1\}$.

(a) If true, argue why.

(b) If not true, list four strings in $\overline{L_6}$ that are not in L_8.

(c) Describe all the strings in $\overline{L_6} - L_8$ (set subtraction of L_8 from L_6), dividing them up into conveniently specifiable classes (explain each class first in English, and then using set comprehensions).

(d) Are there strings in L_8 that are not in $\overline{L_6}$? Explain.   □

3

Kleene Star: Basic Method of Defining Repetitious Patterns

Chapter Gist: *We begin with star, one of the most important of language operators (§3.1). Using star, we define the universal set of finite strings (§3.2) and then language complementation with respect to the universal set (§3.3). This is followed by other operators, namely symmetric difference, reverse, and homomorphism (§3.4). We introduce two ways of enumerating strings from a language, namely lexicographic and numeric orders (§3.6).*

3.1 Three Ways to Describe Star

Given any language L, there is an operator called **star** (written L^*) that obtains a new language derived from L. Called the "star of L," this language amounts to the repeated selection of strings from L followed by concatenation.[1] Star is an extremely important language builder, and most of this chapter is devoted to its deep study. After we present star, we also introduce a few additional language operators in this chapter.

We present three ways, *all equivalent*, to define star:[2]

Definition 3.1:

Star, Definition 1: $L^* = L^0 \cup L^1 \cup L^2 \cup \ldots$

Star, Definition 2: $L^* = \bigcup_{i=0}^{\infty} L^i$

Star, Definition 3: $L^* = \{x : \exists k \in Nat, x \in L^k\}$

Definition 2 is just a compact way of denoting Definition 1. Definition 3 says

[1] Also known as the Kleene-star in honor of the mathematician Stephen Kleene.

[2] Your experience solving the exercise on Page 24 will come in handy for you now! It is the same *star* operator mentioned there.

> Choose some k from Nat. Then, one string you may choose to include in L^* is a member x of L^k. Include all such x.

The role(s) of the Star operator

In our studies, star serves two (related) purposes that are now elaborated:
Capture the notion of zero or more repetitions: From Definition 1, we can see that star helps perform concatenations of a language L with itself. That is, we can make $i \in Nat$ random selections of strings from L (repetitions allowed) and concatenate those strings.

Define the universal language: Suppose we have a keyboard with one key, say 1, and we are asked to tap it a *finite* number of times. We would generate all these strings: $\{\varepsilon, 1, 11, 111, 1111, \ldots, \}$. This set would contain string 1^k for any $k \geq 0$. We can consider this to be a *universal* language generated for the alphabet $\Sigma = \{1\}$. This language is nothing but $\{1\}^*$ according to the definitions of Star just now introduced.

If we had two keys, *i.e.* $\Sigma = \{0, 1\}$,[3] we would generate

$$Sigma^* = \{\varepsilon, 0, 1, 00, 01, 10, 11, 000, 001, 010, \ldots, 111, \ldots,\}$$

.

> Σ^* is the **universal language** over the alphabet Σ.

3.2   Additional Definitions and Properties of Star

Let us define the notion of "star upto n" (written L_n^*) as follows:

$$L_n^* = L^n \cup L_{n-1}^*$$

with

$$L_0^* = \{\varepsilon\}$$

The following Python function encodes L_n^* in Jove.[4]

```python
def star(L,n):
    """Star a language, bounding the iteration to the given n.
    If A = set(['ab', 'bc']) is a language, then
    star(A,2) --> set(['abab', 'bcbc', 'ab', 'abbc', '',
                       'bc', 'bcab']).
    """
    return Unit() if n == 0 else lunion(exp(L,n), star(L,n-1))
```

Unfortunately, $L^* = L_\infty^*$ cannot be[5] computed using Python, as it requires

---

[3] Please note that $Sigma^*$ is not the same as the language $Nplicate$ defined as follows:

$$Nplicate = \{a^i : a \in \Sigma, i \geq 0\}$$

Reason: In $Nplicate$, we pick an $a \in \Sigma$ and $i \in Nat$, and then repeat *that* a i times. Thus, $Nplicate$ has 00000, 1111, etc., but not even 01 or 10. In Σ^*, star allows you to **change the selection of** a **at every step**, and so 01, 010100, etc. are present in it.

[4] Keep in mind that Python may list strings in an order that depends on its internal implementation details.

[5] At least straightforwardly...

modeling the notion of an infinite exponent of a language.

> **Hugely important observation:**   We observe (once again) that
> L^0 is $\{\varepsilon\}$ for **any** language L. Now, since $L^* = L^0 \cup \ldots$, we have
> these facts:
>
> - $\varepsilon \in L^*$ for **any** language L.
> - In particular
>
> $$\varepsilon \in \emptyset^*$$
>
>   because $\emptyset$ is also a language (it is the Zero language).
> - But note that $\emptyset^1 = \emptyset$, and so $\emptyset^i = \emptyset$ for all $i \geq 1$.
> - Thus, we can assert
>
> $$\emptyset^* = \{\varepsilon\}$$
>
> - Thus we have
> $\emptyset^* = \{\varepsilon\}$
>   or in other words,
>   *The star of the empty language (zero language) is the unit language.*

Generating *All Strings Over an Alphabet* via Star: A moment's
reflection should convince you that using star, one can generate *all possible strings over an alphabet*. Let us see this explicitly through some
experiments:

```
>>> Sig01 = {'0','1'}  # Initialize the alphabet
>>> star(Sig01,0)
{''}
>>> star(Sig01,1)
{'', '0', '1'}
>>> star(Sig01,2)
{'', '0', '00', '01', '1', '10', '11'}
>>> star(Sig01,3)
{'', '0', '00', '000', '001', '01', '010', '011', '1', '10',
 '100', '101', '11', '110', '111'}
```

Observe that with respect to the alphabet $Sig01$ defined to be $\{\,'0','1'\,\}$,
each invocation of star with n as the second argument generates all possible strings over the alphahet $\{0,1\}$ of length up to n. The mathematical
definition of Star corresponds to setting n to infinity.

> $\{0,1\}^*$ contains all possible finite-length strings over 0 and 1.

Two more examples to drive the point home:

1. Consider a language $L = \{10, 01, 100\}$ and let's define $M = L^*$. Again, it is easy to see that M is a union of L^i for various i, and so each string in M is finite (obtained by concatenating L with itself *a finite number of times*). However, M itself is an *infinite* set because there is no upper limit to the length of strings within M. In other words, for every string $s \in M$, there is a longer string $s' \in M$.

2. Finally, consider language $N = M^* = L^{**}$. Now, N consists of strings from M^i for various i. Since every string in M is finite, strings in M^i are also finite. Thus even N has only *finitely* long strings.

Here are additional fun facts (some are in the exercise below):

> $L^* = L^{**}$ for any language L (proved in §3.5.1).

Exercise 3.2, Star properties

1. Describe the language below by listing six different strings from it. Pick as many different kinds of strings—i.e., avoid obtaining the next string simply by putting parenthesis around your previous selection:

$$L_2 = \{w : w \in \{(\,,\,)\}^*, \text{ and } w \text{ is well parenthesized}\}$$

2. Consider the language
$$L_7 = \{0^i 1^j : i, j \geq 0\}$$
Is it true that
$$L_7 = \{0\}^* \{1\}^*?$$
Explain, providing reasons.

3. There are exactly two languages (call them L_1 and L_2) over any[6] alphabet Σ such that their stars are finite. Which are these languages?

4. Let our alphabet be $\Sigma = \{0, 1\}$. Let w^R be the reverse of a string w. Consider these languages:
$$L_{P0} = \{w : w \in \Sigma^*\}$$
$$L_{P1} = \{ww^R : w \in \Sigma^*\}$$
$$L_{P2} = \{waw^R : a \in (\{\varepsilon\} \cup \Sigma), w \in \Sigma^*\}$$
$$L_{P3} = \{waw^R : a \in \Sigma, w \in \Sigma^*\}$$
$$L_{ww} = \{ww : w \in \Sigma^*\}$$

   (a) Which language (L_{P1} through L_{P3}) denotes the set of all palindromes over Σ?

   (b) Which of these languages are regular? context-free? context-sensitive? Explain at a high level, using the intuitions presented in Chapter 1 (no formal proofs are necessary).[7]

5. Let us define these languages, where (and) are meta symbols (for grouping) and not part of the alphabet:

$$L_E = \{(00)^i : i \geq 0\}$$

[6] Recall that alphabets cannot be empty.

[7] Have a dialog with your instructor or TA and obtain some hints.

and

$$L_O = \{0(00)^i : i \geq 0\}$$

Let L_{P2} be as in Question 4. Answer these questions:

(a) Is $L_E \cup L_O = \{0\}^*$? Explain.

(b) Is $L = LL$ true for any of the above languages taking the place of L? Explain.

(c) Is $L = L^*$ true for any of the above languages taking the place of L? Explain.

(d) Is $L_O L_O = L_E$? If so, explain. If not:

    i. What is $(L_O L_O) - L_E$?

    ii. What is $L_E - (L_O L_O)$?

(e) Is $L_E^* = \{0\}^*$? Explain.

(f) Is $L_O^* = \{0\}^*$? Explain.

6. Write a proof outline for why $L^* = L^{**}$. Your approach should be as follows:

- We have to argue language equality of the form $A = B$.
- Since languages are sets, this boils down to showing $A \subseteq B$ and $B \subseteq A$.
- Argue this through extensionality; *e.g.*, for $A \subseteq B$, pick an $x \in A$ and argue that $x \in B$ follows.   □

3.3 Language Complementation

Languages are sets, and therefore **language complementation** is subtraction of the language from the universal language over an alphabet. We already defined the notion of *all strings formable over a certain alphabet*, namely Σ^*, and referred to it as the universal language for a given Σ.

> Thus, the complement of a language L is
>
> $$\overline{L} = \{x : x \in (\Sigma^* - L)\}$$
>
> which is all the strings typeable on a keyboard with keys in Σ that do not fall within L.

Given that complement involves star, we must finitely approximate complement before we can express it directly in Jove:

```
def lcomplem(L, alph, n):
    """Complement L relative to alphabet alph. alph is also
       given as a set of strings. We subtract from the "star
       up to n" of the alphabet alph, the language L.
    """

    return lminus(star(alph, n), L)
```

3.4   Other Language Operations

There are many more useful language operations, and we now discuss them at a high level.

3.4.1   Symmetric Difference, Subtraction

Given the definition for complementation, we can now define the **symmetric difference** of two languages:

$$(L_1 - L_2) \cup (L_2 - L_1)$$

where $L_1 - L_2$ is defined to be $L_1 \cap \overline{L_2}$ and $L_2 - L_1$ is defined to be $L_2 \cap \overline{L_1}$. **Language subtraction** can also be defined as

$$L_1 - L_2 = \{x : x \in L_1 \wedge x \notin L_2\}$$

These definitions can be compactly written in Python as shown below:

```
...  return L1 ^ L2  # for symmetric difference


...  return L1 - L2  # for subtraction
```

Exercise 3.4.1, Language puzzles

1. Let L_{eqabc} be the subset of $\{a,b,c\}^*$ where each $s \in L_{eqabc}$ has the same number of a, b, and c. Let $L_{as} = \{a\}^*$, $L_{bs} = \{b\}^*$, and $L_{cs} = \{c\}^*$.

   (a) Describe the language $L_x = L_{eqabc} \cap (L_{as}L_{bs}L_{cs})$ in English.

   (b) Describe L_x through set comprehension.

   (c) Describe $L_y = L_{eqabc} \cap (L_{cs}L_{as}L_{cs}L_{bs}L_{cs})$ in English.   □

3.4.2   Reverse of a Language

Reversing a language reverses every string in the language. $rev(L) = \{rev(s) : s \in L\}$

```
# In Python, there isn't direct support for reversing a string.
# The backward selection method implemented by S[::-1] is what
# many recommend. This leaves the start and stride empty, and
# specifies the direction to be going backwards.
# Another method is "".join(reversed(s)) to reverse s
```

```
def revs(S):
    """Reverse a string.
       revs('ab') --> 'ba'
    """
    return S[::-1]


def revl(L):
    """Reverse a language.
       revl(set(['ab', 'bc'])) --> set(['cb', 'ba'])
    """
    return set(map(lambda x: revs(x), L))
```

3.5 String/Language Homomorphisms

Sometimes we want to change the strings in a language in simple ways. Imagine wanting to mildly obfuscate messages: instead of sending `Hello there`, you'd send `Ifmmp!uifsf` by shifting all letters by one position in their ASCII encodings. Now, given a homomorphism from Σ^* to codomain Γ^*, it can be applied to a language $L \subseteq \Sigma^*$ to produce a language $G \subseteq \Gamma^*$. Language G is defined in the obvious manner: *"apply the homomorphism to every string in the language"*. Here, h is overloaded in its use, and becomes the **language homomorphism** function $h(L) = \{h(x) : x \in L\}$.

These can be accomplished through functions `shomo` and `lhomo` (language homomorphism) below:

```
def shomo(S,f):
    """String homomorphism wrt lambda f.
       Suppose hm = lambda x: chr( (ord(x)+1) % 256 )
       Then shomo("abcd",hm) --> 'bcde'
    """
    return "".join(map(f,S))


def lhomo(L,f):
    """Language homomorphism wrt lambda f.
       Suppose rot13 = lambda x: chr( (ord(x)+13) % 256 )
       Then lhomo("Hello there", rot13) --> 'Uryy|-\x81ur\x7fr'
    """
    return set(map(lambda S: shomo(S,f), L))
```

Basically, we are mapping a `lambda` over the given string. This `lambda` takes the ordinal position of `x`, adds 1 (modulo 256), and then projects it back to the character value. This is an example of a homomorphism.

In general, string homomorphisms are string-to-string mappings. Given a string belonging to Σ^* (a "string over Σ^*"), a function h from domain Σ^* to codomain Γ^* (*i.e.*, strings over alphabet Σ to strings over alphabet Γ) is called a **string homomorphism** if it respects two conditions:

- $h(\varepsilon) = \varepsilon$. This means that the empty string must be mapped to the empty string.

- $h(xy) = h(x)h(y)$. This means that if you arbitrarily pull apart a string xy into two pieces x and y (*e.g.*, if xy = "Hello there", x could be "Hell" and y could be "o there"), apply the homomorphisms separately to x and y and put the results back through concatenation, you will get the same results as applying the homomorphism to the entire string. In our example,

  - h applied to "Hell" returns "Ifmm"
  - h applied to "o there" returns "p!uifsf"
  - Concatenating "Ifmm" and "p!uifsf" obtains "Ifmmp!uifsf"—the same as applying it to the non-pulled-apart "Hello there".

Crux of homomorphisms: The only reason why homomorphisms may not work in the manner illustrated with "Hell" and "o there" is if there are special-case rules. Suppose h were a strange function defined as follows:
- $h("Hel")$ = "ooo"
- $h("Hell")$ = "Ifmm"
- $h("l")$ = "z"
- Then, the concatenation of $h("Hel")$ and $h("l")$ gives "oooz" whereas $h("Hell")$ gives "Ifmm".

This is because homomorphisms are *string-to-string* mappings where the strings could be of length more than one. When defining such homomorphisms for strings of length above 1, one must take care not to "conflict" in mappings. We can observe this conflict above: "Hel" is mapped to "ooo" while "Hell" is mapped to "Ifmm" which does not respect the fact that mapping the first three characters is done disregarding how the mapping over the first four characters proceeds.

In contrast, mappings defined over strings of length 1 (*i.e.*, characters) are guaranteed to be homomorphisms, as in the homos function illustrated above. This is because there can't be such conflicts on substrings.[8] Here are some additional examples:

- Suppose you provide a keyboard with two keys (say 0 and 1) to someone, but are interested only in listening to the "tap" "tap" sound. You can then map both 0 and 1 to t (for tap). This is a homomorphism, and helps forget what is being typed (as if you are listening to the person from the adjacent room).
- Another handy homomorphism is when 0 is changed to a and 1 to b.

[8] Single characters don't have substrings other than themselves or ε; and generally one does not map ε to other than ε.

This is another simple homomorphism that maps strings over alphabet $\{0, 1\}$ to strings over alphabet $\{a, b\}$.

Exercise 3.5, Homomorphism

1. Is string reversal a homomorphism? Explain your answer.
2. Assume $\Sigma\{a\ldots z\}$. Define a function f that maps a to d, b to e, ..., w to z, and then x to a, y to b, z to c, etc. (every character is mapped two higher, in a modulo fashion). The same function also maps ab to c. Is f a homomorphism? Explain your answer.

$\square$

3.5.1 Taking Star Repeatedly

Theorem 3.5.1: For any language L, $L^{*^*} = L^*$.

Proof: Consider Definition 3 for Star on Page 27, repeated below for convenience:

$$L^* = \{x : x \in L^k \text{ for some } k \in Nat\}.$$

Using this definition, we can define L^{*^*}:

$$L^{*^*} = \{x : x \in L^{*^k} \text{ for some } k \in Nat\}.$$

So now, what is L^{*^k}?   Let us use $\bullet$ to denote concatenation. Then, it is an k-ary concatenation (for some $k \in Nat$) that can be expressed as $L^{*^k} = L_1 \bullet L_2 \bullet \ldots L_k$
where *each L_i* in this concatenation expression is $\{x : x \in L^m \text{ for some } m \in Nat\}$

Proof that every $x \in L^*$ is also in L^{*^*}:   A string x is in L^* if x is in L^k for some $k \in Nat$. Such an x is also in L^{*^*} because we can take the m in the definition of L_1 to be k and the rest of the m's in the definitions of L_2 through L_m to be 0. Then we will satisfy the equality $k = k + 0 + 0 + \ldots + 0$.

Proof that every $x \in L^{*^*}$ is also in L^*:   Now let $x \in L^{*^*}$ by picking $m = m_1$, $m = m_2$, etc, up to $m = m_k$ in the definitions of $L_1, L_2, \ldots, L_m$. Now we see that we can choose a k equal to $m_1 + m_2 + \ldots + m_k$, and for this k, $x \in L^k$.[9] $\square$

[9] See supplementary notes at https://bit.ly/Automata_Jove under StarStar for a more detailed proof.

3.6 Enumerating Strings in a Language

There are two popular ways of listing the strings in a language, namely

- according to the *lexicographic* order, and
- according to the *numeric* order.

Lexicographic order follows standard dictionary order. In it, pig comes before poblano, but pig comes after pago.

Despite its simplicity, there is a huge problem with the lexicographic listing order when applied to many infinite sets: *one may never get to list certain strings at all!* As an example, if you are asked to list the contents of $\{a,b\}^*$ lexicographically, you would go

$$\varepsilon, a, aa, aaa, \ldots .$$

You see, you'll never list b, bb, etc!

Here is where the notion of **enumeration** enters into the picture. Basically, enumeration is the same as listing, but with the guarantee of getting to every string after a finite amount of other strings. Here is *an* enumeration of $\{a,b\}^*$:

$\varepsilon,$

$a, b,$

$aa, ab, ba, bb,$

$aaa, aab, aba, abb, baa, bab, bba, bbb,$

$aaaa, \ldots$

In **numeric ordering**, we exhaust each length group before we go to the next length group. Within each length group, we follow the lexicographic ordering. Thus, pig comes before poblano. However, pago also comes after pig in numeric order. This is because we must list pig and all length-group 3 strings before we come to length-group 4 where pago belongs.

Formal definition of Lexicographic

Here is a formal definition of lexicographic order.

> Formally, two strings s and t are in lexicographic order $s \leq t$ if $s[i] == t[i]$ holds up to a point, and we find $s[i+1] \leq t[i+1]$. (We don't care what holds beyond $i+1$.)

We can write a Python function to help us understand lexicographic ordering. For two strings s and t, let predicate lexlt ("lexicographically less than") be defined as follows:

```
def lexlt(s, t):
    if (s==""):
        return True
    if (t==""):
        return False
    if (s[0] < t[0]):
        return True
    return (s[0] == t[0]) & lexlt((s[1::], t[1::]))
```

The following exercise helps you explore the use of lexlt.

Exercise 3.6, Lexicographic order

Given the following languages

```
L1 = {"abacus", "bandana", "pig", "cat", "dodo", "zulu", "physics"}
L2 = {"dog", "zebra", "zzxyz", "pimento"}
```

Define a function that lists all those pairs (a, b) from the Cartesian product of L1 and L2 such that a is lexicographically before b. Hint: form the Cartesian product using Python's built-in operator product, and then use lexlt to implement the filter operation over that list. □

Exercise 3.6, Numeric order

1. Generate the first ten strings over the alphabet $\{0, 1\}$ in numeric order.
2. Arrange the above collection of strings in lexicographic order. Produce a printout showing the numeric and the lexicographic orders. □

Coding-up a numeric order generator

Figure 3.1 has one implementation of nthnumeric for an alphabet of size 2.

Figure 3.1: Code for nthnumeric for $|\Sigma| = 2$.

```python
from math import floor, log, pow
def nthnumeric(N, Sigma={'a','b'}):
    """Assume Sigma is a 2-sized list/set of chars
       (default {'a','b'}). Produce the Nth string in numeric
       order, where N >= 0.
       Idea : Given N, get b = floor(log_2(N+1)) - need that
       many places; what to fill in the places is the binary
       code for N - (2^b - 1) with 0 as Sigma[0] and 1 as Sigma[1].
    """
    if (type(Sigma)==set):
        S = list(Sigma)
    else:
        assert(type(Sigma)==list
        ), "Expected to be given set/list for arg2 of nthnumeric."
        S = Sigma
    assert(len(Sigma)==2
        ),"Expected to be given a Sigma of length 2."
    if(N==0):
        return ''
    else:
        width = floor(log(N+1, 2))
        tofill = int(N - pow(2, width) + 1)
        relevant_binstr = bin(tofill)[2::] # strip the 0b
                                           # in the leading string
        len_to_makeup = width - len(relevant_binstr)
        return (S[0]*len_to_makeup +
                shomo(relevant_binstr,
                    lambda x: S[1] if x=='1' else S[0]))
```

Here are some tests using nthnumeric for the alphabet {a,b}:

```python
>>> nthnumeric(0)
''
>>> nthnumeric(1)
'a'
>>> nthnumeric(5)
'ba'
>>> [nthnumeric(i) for i in range(16)]
['', 'a', 'b', 'aa', 'ab', 'ba', 'bb', 'aaa',
 'aab', 'aba', 'abb', 'baa', 'bab', 'bba', 'bbb', 'aaaa']
```

Part II: Machines

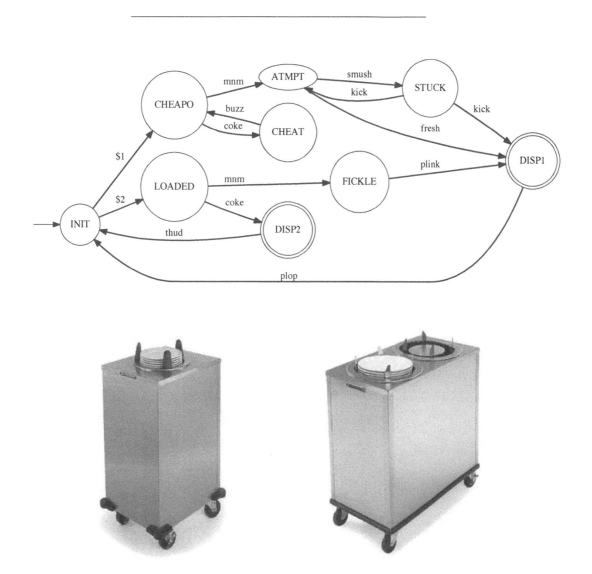

(Images Courtesy of Mission Restaurant, Supply Inc.)

The diagram at the top is a hypothetical **finite-state automaton**. It takes '$1' or '$2' and delivers an M&M ('mnm') or 'coke.' A 'smush'ed mnm package is "liberated" (nondeterministically) by a 'kick' input, and 'plop's to the output tray when successful. Attempts to underpay are rejected with a 'buzz.' We don't distinguish between inputs (e.g., 'kick') and outputs (e.g., 'plink') in this example.

You can obtain a **pushdown automaton** by adding one stack (left plate-stacker, above) to a finite automaton. You can obtain a **Turing machine** by adding one infinite tape (Figure 1.2). An infinite tape can be simulated (details in Chapter 13) using two stacks, as shown in the right-hand side plate-stacker.

4

Basics of DFA

> **Chapter Gist:** *DFA are centrally important to computing theory and practice (§4.1). We first present them through examples (§4.2) and then formally (§4.3). We define the language of a DFA (§4.4) and DFA acceptance (§4.5). The (so called) Pumping Lemma is presented (§4.6), with how to use this lemma to prove a language to be non-regular (§4.7). We discuss common pitfalls in the use of the Pumping Lemma (§4.8). One can ameliorate proof effort by subjecting the given languages to regularity-preserving transformations (§4.9).*

4.1 *DFA Everywhere*

Finite-state machines are crucially important to theoretical computer science, as well as computing practice. They are employed within compilers for checking the syntax of programs, and extracting primitive quantities such as numbers and strings (collectively called *tokens*) for further processing. Such finite-state machines are called *lexers* or *scanners*, and their correctness is crucial to the overall correctness of any compiler.

Finite-state machines are the ones that cycle traffic lights through various stages, sensing the arrival of vehicles and detecting the pressing of pedestrian crossing buttons. They are even "embedded" inside a human pilot's head, governing her actions. Actions of airplane control panels are also governed by finite-state machine components. These two finite-state machines (one governing the pilot's thought processes and the other underlying the control panel) must mesh. If any discord arises between these finite-state machines, the pilot may end up exerting the wrong control action at the wrong moment, leading to potential disasters. This state of a pilot's mind is called *mode confusion*. It is the responsibility of airplane control panel designers to minimize the chances that such mode confusion arises.

[1] An example of an infinitary bad behavior is the following: suppose you try to summon an elevator by pressing its call button; will the elevator actually come to your service, or infinitely-often go past your floor without stopping for you?

[2] The kick given to the vending machine illustrated at the beginning of Part II has a non-deterministic effect, as the machine can go to one of two states following it: ATMPT or DISP1.

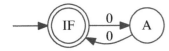

Figure 4.1: A DFA over alphabet {0} to recognize an even number of 0's. Notice that we do not need to keep a count of 0's, but only their parity (whether or not the count is even), and so two states suffice.

```
{'Q': {'A', 'IF'},
 'Sigma': {'0'},
 'Delta': {('A', '0'): 'IF',
           ('IF', '0'): 'A'},
 'q0': 'IF',
 'F': {'IF'}}
```

Figure 4.2: DFA of Figure 4.1 in Python

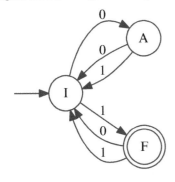

Figure 4.3: This DFA *seems to* accept all strings ending in a 1. Does it? See Exercise 2 and strengthen the condition on the language accepted.

Finite-state machines are involved in deep content inspection hardware deployed over the internet to look for malware in transit. Finite-state machines in such hardware units look for specific patterns within the bit-streams in flight. Suspicious packets are set aside for further scanning by anti-virus software.

Real-world finite-state machines must be modeled and designed correctly, and they must also be capable of processing information at an adequately fast rate. If finite-state machines that handle front-end information processing within the internet turn out to be too slow, hackers can perform *denial of service* attacks. Such malicious users simply bombard these front-end finite-state machines with meaningless requests, thus effectively choking them. This can prevent their services from becoming available to well-intentioned users.

The finite-state machines that one encounters in life are often variations of the basic type studied in this book; they include: (1) Mealy machines that consume inputs and produce outputs at each transition; (2) communicating finite-state machines that exchange messages; (3) Unified Modeling Language diagrams used in software engineering; (4) semantic rules to specify program behaviors; and (5) Büchi automata that model infinitary behaviors.[1] Deterministic finite automata (DFA) studied in this chapter and non-deterministic finite automata[2] studied in later chapters capture the essence of these practical machine types.

4.2 Elements of a DFA

Figure 4.1 describes our first example of a DFA that is assumed to have an alphabet {0}. This figure presents a directed graph with two types of nodes, one with a single circle and another with a double circle. These nodes are called **states**, with the double circled states called **final states**, with the others being **non-final states**. States are connected via labeled edges to other states. These arrows are state **transitions** or "jumps" the DFA can make. There is always one state (called the *initial* state) that has an arrow coming "from the air" poking at it; such an arrow is trying to say "start here!" In our example, it is the state named IF. Initial states can also be final states, as is the case here.

We will adhere to certain naming conventions for states (with only some rare exceptions). We will name all initial-only states with names that begin with an "I" or an "i." We will name all final-only states with names that begin with an "F" or an "f." States that are initial and final will be given names starting with "IF."

We now present another example DFA in Figure 4.3. This DFA has distinct initial and final states named I and F. In general, a DFA may not have a final state at all. It may also have many final states; in fact, all its states could be final states! A DFA always has a single initial state.

DFA are used to *describe* (or specify) languages of interest. More specifically, a DFA describes exactly the language (set of strings) L where each string $w \in L$ describes a path from the initial state to one of the final states (*i.e.*, the ith edge in such a path will be labeled by the ith symbol of string w). Such a string w is said to be **accepted** by the DFA. The set of strings accepted by a DFA is the language **recognized** by it.

In a typical DFA, there will be multiple paths (even an infinite number of paths) that start from the initial state and end in one of the final states. For a DFA without a final state, no such path exists, and so the language of such a DFA is $\emptyset$, the empty set. A DFA with an initial state that is also final has a path of length 0 to its final state, and so its language includes ε. For example, the DFA in Figure 4.1 describes the language $\varepsilon, 00, 0000, \ldots$, or in general $(00)^i$ for i being even.

The DFA of Figure 4.3 *seems to* accept all strings that end with a 1. Does it? See Exercise 4.2.2 below.

| **Exercise 4.2, DFA basics** |

1. Draw a DFA to recognize the set of strings over $\{0,1\}$ that have an even number of 0's and *any number of* 1s. (Difference with the DFA in Figure 4.1: that DFA does not have 1 in its alphabet.)

2. Accurately describe the language of the DFA of Figure 4.3. Does there exist a 2-state DFA with this language? ☐

4.3 Formal Structure of DFA

Formally, a *deterministic finite-state automaton D* is described by five items presented as a tuple, $(Q, \Sigma, \delta, q_0, F)$ (see a Python rendering of this five-tuple in Figures 4.2 and 4.4), where:

- Q is a *finite nonempty* set of states,
- Σ is a *finite nonempty* alphabet,
- $\delta : Q \times \Sigma \to Q$ is a *total* transition function,
- $q_0 \in Q$ is the initial state, and
- $F \subseteq Q$ is a *finite, possibly empty* set of final (or *accepting*) states. These are shown as double-circled nodes in the graph of a DFA.

The transition function δ is also sometimes depicted as a state table (one example appears in Figure 4.5).

In mathematics, all functions are "total" in that each function maps its entire domain into its *range*, which is a subset of its codomain. We stress the notion of total functions here, because it is often convenient to define a DFA partially and then "totalize it"—i.e., fill all its uninteresting mappings automatically.

To illustrate this point, let us change the DFA in Figure 4.1 by expanding its alphabet from $\{0\}$ to $\{0,1\}$ (adding one more key "1" to the DFA's Python Delta function). The language to be recognized must still be all even-length 0's *with no 1's allowed*! (Contrast this with the DFA of Exer-

```
{'Q': {'A', 'F', 'I'},
 'Sigma': {'0', '1'},
 'Delta': {('A', '0'): 'I',
           ('A', '1'): 'I',
           ('F', '0'): 'I',
           ('F', '1'): 'I',
           ('I', '0'): 'A',
           ('I', '1'): 'F'},
 'q0': 'I',
 'F': {'F'}}
```

Figure 4.4: DFA of Figure 4.3 in Python

State	Input	Next State
I	0	A
I	1	F
A	0	I
A	1	I
F	0	I
F	1	I

Figure 4.5: State table for the DFA of Figure 4.3

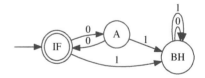

Figure 4.6: The DFA of Figure 4.1 whose alphabet has been expanded to now also have 1; we then "totalize" this DFA by sending it crashing to the "BH" (blackhole) state whenever a 1 is input.

[3] Black-hole states can also be called Roach-Motel or RM states. You can check in to such a motel; you can't check out! Often they are also called *sink* states or *empty* states—the latter term arising from the fact that such states have an *empty* language, in the sense that if one were to start from these states, there would be no paths that lead to a final state.

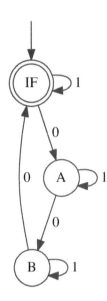

Figure 4.7: A DFA to recognize strings over {0,1} with the number of 0s being a multiple of 3

```
{ 'Q': {'A', 'IF', 'B'},
 'Sigma': {'0', '1'},
 'Delta': { ('IF', '0'): 'A',
            ('IF', '1'): 'IF',
            ('A', '0'): 'B',
            ('A', '1'): 'A',
            ('B', '0'): 'IF',
            ('B', '1'): 'B' },
 'q0': 'IF',
 'F': {'IF'}   }
```

Figure 4.8: DFA of Figure 4.7 in Python

cise 1 where we allowed *any number of* 1's.) The question now is "what will happen if someone types the 1 key? Well, we want the DFA to crash and go into a "black hole state" (BH) from which it never gets out.[3] We accomplish totalization through the construction in Figure 4.6 where all user-unspecified moves are "sent to the black hole."

A DFA can be viewed as a "goto-based" program. The states of a DFA are labels, and the transitions are "goto" statements executed when specific inputs are read. You may say "Eeew, goto?! My instructor said not to use goto!" Well, DFA programs are low-level programs—much like assembly language programs. At this level, gotos are *the primary mechanism* for control flow.

4.4 The Language of a DFA

We now define some of the important notions underlying DFA:

> The language of a DFA is the set of all strings accepted by it. We say that the DFA *recognizes* this language. When a DFA recognizes language *L*, it *accepts* all the strings within *L*. Notice that the word **accept** is reserved for (single) strings and the word **recognize** is reserved for languages (which are sets of strings). **A regular language over alphabet Σ is the language of some DFA with this alphabet.**

Note: Classically, regular languages were defined inductively through regular expressions (or regular operators)—namely, union, concatenation, and star. It was Kleene's theorem that established the connection between regular sets and finite automata.

4.4.1 DFA as String Classifers

The main purpose of almost all "machines" we define in this book is to partition Σ^*, the universal language over Σ, into *accepted* strings and *rejected* strings. Viewed this way, each DFA is a *string classifier*. Viewing a fresh example – that given in Figure 4.7 (with Python encoding in Figure 4.8), this is a DFA that accepts all strings that contain multiples of 000 interspersed with an arbitrary number of 1's. It rejects all other strings. Using set comprehension, this language can be specified as

$$L_{3Z} = \{w \mid w \in \{0,1\}^* \text{ and } (\#_0(w)) \% 3 = 0\}$$

where $\#_0$ is a function that returns the number of zeros in its argument string, and % denotes Python's modulus (mod) operator.

4.4.2   Basics of Designing a DFA

In this chapter, we present some simple approaches to building a DFA. Chapter 5 covers these ideas in more depth.

One way of designing a DFA is by treating it as a program that has only two constructs: (1) test the input symbol to match one of the symbols in the alphabet, (2) upon a match, *go to* a target state. Here is how the DFA of Figure 4.7 could have been designed:

- Write down a few strings that are accepted and some that are rejected. For example:
    - Accept ε, 000, and in fact, any repetition of 000's interspersed with an arbitrary number of 1's.
    - Reject all other strings.
- Since ε is accepted, make the initial state also a final state.
- Name each state according to our naming convention.
- Make sure that each state responds to all the symbols in the alphabet.

4.5   Formal Definition of DFA Language Acceptance

To define the notion of DFA acceptance we define three functions:

1. A function δ that, given a DFA D, a state q, and an input symbol $a \in \Sigma$ determines what the next state is. That is, $\delta(D,q,a)$ is the next state in Q attained by marching D on symbol a when D is at state q.

2. A function $\hat{\delta}$ that, given a DFA D, a state q, and an input string w, tells which next state the DFA falls into after processing every character within w (if any). If w is of the form ax, this function is defined by

$$\hat{\delta}(D,q,ax) = \hat{\delta}(D,\delta(D,q,a),x)$$

and if $w = \varepsilon$, it is defined by

$$\hat{\delta}(D,q,\varepsilon) = q.$$

3. A predicate $accepts(D,q,w)$ that is true exactly when $\hat{\delta}(D,q,w)$ falls within F (the set of final states of D).

There are two natural cases in $\hat{\delta}$ (Figure 4.9) reflected within run_dfa_h:

- (Case 1 when w is empty): return q, the current state.

- (Case 2 when w is non-empty): return the result of applying run_dfa_h with respect to DFA D on string w[1:] (the rest of the string), when the DFA starts from the state step_dfa(D, q, w[0]). This is the recursive case of run_dfa_h(D, w, q).

Thus, a DFA accepts w exactly when $\hat{\delta}(D,q_0,w) \in F$.

```python
def step_dfa(D, q, a):
    """Run DFA D from state q on character a.
    """
    assert(a in D["Sigma"])
    assert(q in D["Q"])
    return D["Delta"][(q,a)]


def run_dfa(D, w):
    """In : D (consistent DFA)
          w (string over D's sigma, including "")
       Out: next state of D["q0"] via string w.
    """
    curstate = D["q0"]
    if w=="":
        return curstate
    else:
        return run_dfa_h(D, w[1:], step_dfa(D,curstate,w[0]))


def run_dfa_h(D, w, q):
    """Helper for run_dfa. Compute the next state attained
       by w running on D starting from state q.
    """
    if w=="":
        return q
    else:
        return run_dfa_h(D, w[1:], step_dfa(D, q, w[0]))


def accepts_dfa(D, w):
    """ Checks for DFA acceptance. Input : DFA D, string w.
        Output : Boolean (True|False).
    """
    return run_dfa(D, w) in D["F"]
```

Figure 4.9: Functions to step a DFA on a single character q via δ (function step_dfa), run it on a string w via $\hat{\delta}$ (function run_dfa), and check for acceptance via $\hat{\delta}(q_0, x) \in F$ (function accepts).

Reading Suggestions for Code

This book provides all the code to work with DFA using Jove (except the Jupyter notebooks underlying Jove often have more detailed codes). Here is the general approach we recommend to understand the code in this book:

> • Always strive to understand the code enough to drive it—much like you would understand a car enough to drive it. (There will always be occasions where we will push you to understand more, but we also hope you are naturally curious and would love to learn how to write/improve such code.)
> • All important functions for each machine-type are provided in files named `Def_..` (for instance, `Def_DFA.ipynb`, `Def_NFA.ipynb`, etc.). A list of these functions is included at the end of each `Def_..` file, and one can always find out more about any function `foo` by typing `help(foo)` in Python.

4.6 "Lasso" Shape of DFA and the Pumping Lemma

DFA often look like a "lasso": they start at a state, optionally go forward a few steps, and then curl around (see Figure 4.1). In Figure 4.10, a DFA that accepts a single a or three or more a's, the lasso shape is more apparent. In Figure 4.7 (repeated in Figure 4.11 for your convenience), the alphabet isn't singleton (it is $\Sigma = \{0, 1\}$), and so the DFA must respond to both 0 and 1 at every state. Nevertheless, the DFA must "curl around" (revisit an already existing state); *i.e.*, it can't go on sprouting new states and transitions leading to them. So while not exactly a lasso, DFA still have cyclic paths in them. This fact is exploited in the so called **Pumping Lemma**—the topic of this section.

Notice also that when you take a string w accepted by a DFA that is at least as long as the number of states in the DFA, it will imply that there exists a piece of w that can be repeated to yield another string that is also in the language of the DFA. For example, 0100 is in the language of the DFA of Figure 4.11 (we named this language L_{3Z}). We can view w as the concatenation of three strings x, y, z where y is non-empty. We can now see that $xy^i z$ is also in L_{3Z} for $i \geq 0$.

For instance, let us start out with the observation that $0100 \in L_{3Z}$. Now, 0100 is longer than the number of states of this DFA. We can now view 0100 as xyz where $x = 0$, $y = 1$, and $z = 00$. Now we can see that $xy^i z \in L_{3Z}$ for $i \geq 0$. That is,

• For $i = 0$, $000 \in L_{3Z}$. This avoids taking the 1-loop at state A.
• For $i = 1$, $0100 \in L_{3Z}$. This is how we started out! This is where we take the 1-loop at A exactly once.

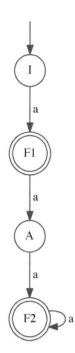

Figure 4.10: A "lasso" shape of a DFA over $\Sigma = \{a\}$

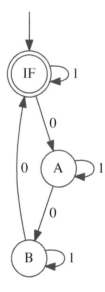

Figure 4.11: The DFA of Figure 4.7 repeated for easy reference

- For $i = 2$, we have $01100 \in L_{3Z}$. This is where we take the 1-loop at A twice.

This property is called the *Pumping Lemma*, and is fully described in §4.6.1.

Exercise 4.6, DFA Lasso

1. Argue that DFA over singleton alphabets must always have a lasso shape similar to that in Figure 4.10: after going forward a few steps, the DFA must transition back to one of the earlier states.

2. Argue that for a DFA D that recognizes language L, there are an infinite number of other DFA that also recognize L. □

4.6.1   General Statement of the Pumping Lemma

Recall that for any language L that wants to call itself a *regular language*, there must exist a DFA D_L of N states whose language is L. Furthermore, for any string $w \in L(D_L)$ such that w is *at least* of length N, it must describe a lasso (a "pump") in D_L. We must be able to take this "pump" as many times as we wish, or skip the pump entirely—*these too must be legal paths in the DFA.*

Figure 4.12: The Pumping Lemma illustrated on an N-state DFA. The input string of length $M \geq N$ revisits state s_p, "embossing" a state visitation number v_p the first time, and v_{p+k} the second time. Thus, $|y| = k$ in this case, and is greater than 0.

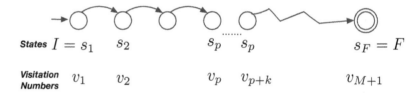

Input of length M which is at least N

States $I = s_1$ s_2 s_p s_p $s_F = F$

Visitation Numbers v_1 v_2 v_p v_{p+k} v_{M+1}

Figure 4.12 illustrates all these statements:

- This figure depicts the journey of a string w through the N-state DFA D_L starting from the initial state of s_1, ending with a final state which, for convenience, we call s_F. Thus, according to this diagram, some number of states s_1, s_2, etc. were touched during string w's traversal.
- Notice that this journey touches some state s_p for the first time, and afterwards, touches state s_p again. This is the lasso (or pump, or loop).
- To make these concepts more specific as well as visual, we imagine *embossing visitation numbers* v_1, v_2, ..., v_{M+1} when we visit states.[4] Given that we are taking a string of length $M \geq N$, we can emboss v_{M+1} such visitation numbers (one per step along the path taken by the string w of length M).
- Notice that we embossed v_p when we first encountered s_p. Then, after

[4] We have a rubber-stamping device, and go "click," stamping a state's visitation number on it when we arrive at it.

taking k more steps, we visited state s_p again, now embossing it with v_{p+k}. *Any state that has been embossed twice carries a pump.* Also, there can be multiple such pumps *anywhere along the journey.* However, in order to state a simple and crisp argument, we choose to focus on the **first pump** carried by state s_p between visitation numbers v_p and v_{p+k}. This means that we executed a lasso of length $k > 0$ at this state.

- Now, if this journey exists in this DFA, two other types of journeys also exist:
  - One journey where, after embossing v_p onto s_p, we don't take that path which revisits s_p. We take the "other path" that takes us toward state s_F directly. This is called *pumping down*. This journey is associated with a different string than w—*one which takes the pump away from w.*
  - Another journey where, after embossing v_p onto s_p, we take that pump again (and perhaps again, and again, ...) and after some time, we take the path toward s_F. This is called *pumping up*, and is associated with a different string than w—*one that splices in copies of the pump into w.*
- Thus, if the normal single revisitation occurs (*i.e.*, embossing v_p and then v_{p+k} onto s_p), then the pump-down journey (one such) and pump-up journeys (as many as we wish) also exist in the same DFA D_L. This DFA is forced to admit all these strings also into its language!

Theorem 4.6.1 : Pumping Lemma for Regular Languages.
- **IF** L is regular, then that **IMPLIES**
- There is a natural number N such that
- for any string $w \in L$ where w is at least of length N,
- we must be able to read out w as xyz
- where y is non-empty (x and z could be empty)
- and xy is confined to the first N steps of w
- and furthermore, for all $i \geq 0$, $xy^i z \in L$ must be true.

Proof Sketch: One can trace out the state-repeating path of w leading to a final state. The *first* loop would be found within the first N symbols of w.[5] The x is the prefix that first visits the repeating state, y is the non-empty path that second visits the repeating state, and z is the remainder of w. In such a situation, one may either skip the y loop or take it more than once before journeying towards the final state.   □

The reason that a string w must curl around follows from the *pigeonhole principle*,[6] as follows: if there are $M \geq N$ transitions (arrows), there must be some state s_p that repeats, and gets "re-embossed."[7] We call the string that takes us for the first time to s_p as x, the string that is involved in visiting s_p a second time as y, and the remainder of the string taking

[5] Here, N stands for the number of states of the DFA.

[6] The pigeonhole principle is discussed in https://en.wikipedia.org/wiki/Pigeonhole_principle, and asserts that with $N > 0$ pigeonholes housing $N + 1$ pigeons, there must be at least one pigeonhole that houses more than one pigeon.

[7] To explain the $M = N$ case, view states as being similar to duck fingers (or digits) and transitions that connect states being like webs between the digits. Thus, in an N-state DFA, there is at most a journey of length $N - 1$ that does not call for states to repeat. The moment the journey has length N, some state must repeat.

[8] There are versions of the Pumping Lemma that are of the form "$Regular(L) \Leftrightarrow Condition$" that we don't address in this book. You can read about Jaffe's version of the *if and only if* Pumping Lemma and its proof at https://dl.acm.org/citation.cfm?id=990528.

us to s_F as z. *Note that this forces xy to be situated within the first N steps,* as it is the first pump that we choose to focus on.

This allows us to conclude that xy^0z (the *pump down* case) also takes us to s_F (*i.e.,* $xy^0z \in L$, or simplifying, $xz \in L$). Likewise, $xyyz \in L$ (pump up once), $xyyyz \in L$ (pump up twice), etc., are also true. In short, $xy^iz \in L$ for all i.

When solving problems, it is strongly encouraged that you focus first on picking the y element that describes the pump. The choice of x and z will then be almost automatic: x is anything that precedes y, and z is "the rest of the string" after xy. There are no constraints on x and z themselves (they could be empty).

4.7 Proving a Language to Be Non-Regular

The Pumping Lemma that we've stated in §4.6.1 has a *one-way* implication.[8] It can be written as:

$$Regular(L) \Rightarrow Cond(L)$$

where we call $Cond$ **the pumping condition**, and it is the condition (stated over multiple lines) coming after the implication "**IMPLIES**," above. Given all this, we offer this strong advice:

> Never use the Pumping Lemma that we have stated above for *proving* something is regular (because it is only a "one way" implication). Use it only to *disprove* that something is regular.

We will be using the **contrapositive form** of the Pumping Lemma to prove that a given language is not regular:

$$\neg Cond(L) \Rightarrow \neg Reg(L).$$

We will achieve our goal by showing $Cond(L)$ is false (or, equivalently, $\neg Cond(L)$ is true). In more detail, here is how we proceed. Suppose that we are handed a suspicious language L_s and are asked to prove that this language is not regular. Before we apply the Pumping Lemma, we will spend some time and see if strings in L_s can be recognized by a finite memory device. For example, as illustrated in Figure 4.1, we will check whether we can summarize the information to be processed in terms of *an even number of bits seen so far* (instead of actually keeping counts). When we become reasonably convinced that L_s *is non-regular*, we will embark on showing that $Cond(L)$ is false.

> **Note:** We have tried teaching the Pumping Lemma as a proof by contradiction. This approach lets you assume that the language is regular, and derive a contradiction. However, given that most

of the languages we consider in our exercises are non-regular, it is better (and more direct) to begin with the assumption that the pumping condition (namely $Cond(L)$) is false. This makes the proof much more *goal-directed* right from the beginning. This approach also turns into a "recipe" that can be taught more reliably, in our experience.

Example Use of the Pumping Lemma: Let us proceed methodically to see an actual example involving the Pumping Lemma.[9] Suppose you walk into your office one fine morning and find a note left on your desk by someone claiming that the language L_{01} is regular, where

$$L_{01} = \{ 0^i 1^i : i \geq 0 \}$$

Your intuition tells you that unless we can *count* the number of 0's, we cannot later check it against the number of 1's.[10] So now you are doubly sure that what you must do is *show that $Cond(L)$ is false*, thus showing $\neg Reg(L)$. Let us rewrite the Pumping Condition $Cond(L)$ once, and then keep its negated form handy.
$Cond(L) =$

- There is $N \in Nat$ such that
- for any string $w \in L$ where w is at least of length N,
- there exists a split of w as xyz,
- where y is non-empty (x and z could be empty),
- and the length of xy is $\leq N$
- and furthermore, for all $i \geq 0$, $xy^i z \in L$ must be true.

$\neg Cond(L) =$

- For any $N \in Nat$,
- for some string $w \in L$ where w is at least of length N,
- for all splits of w as xyz,
- where y is non-empty (x and z could be empty),
- and the length of xy is $\leq N$
- there exists i such that $xy^i z \notin L$.

4.7.1 Why All Splits of x, y, z?

By showing $\neg Cond(L)$, we are showing that the pumping condition *cannot hold* for any **DFA** that someone might propose. Given that all regular languages are required to imply (abide by) the pumping condition, this will then prove that *no DFA whatsoever* can exist. Given that we are interested in ruling out *any DFA* at all, we cannot assume a particular location for the pump y. This forces us to consider all splits. □

[9] See supplementary material at https://bit.ly/Automata_Jove under PainlessPL for a rigorous presentation of the Pumping Lemma, as well as more intuitions.

[10] Also, checking the date on the calendar, you see "April 1, 2017," making you think about the note even more...

Showing that L_{01} is not regular: To show that L_{01} is not regular, let us make $\neg Cond(L_{01})$ (below) true.

> - For any $N \in Nat$,
> - for some string $w \in L_{01}$ where w is at least of length N,
> - for all splits of w as xyz (view $w = xyz$),
> - where y is non-empty (x and z could be empty),
> - and the length of xy is $\leq N$
> - there exists i such that $xy^i z \notin L_{01}$.

Here is how we can achieve this goal:
- Pick $w = 0^N 1^N$ which is in L_{01}.
- Now, split w into three pieces x, y and z (i.e., $w = xyz$) where xy is of length $\leq N$. This forces all possible y's to consist only of 0's.
- Now,
  - pumping down results in a string with fewer 0's than 1's
  - and pumping up results in a string with more 0's than 1's
- Thus there exists $i = 0$ (pumping down) or $i \geq 2$ (pumping up) where the new string is not in L_{01}.

This makes the pumping condition false, and hence $\neg Reg(L_{01})$ is established.

Caveat: There are cases when one applies the Pumping Lemma systematically to a given *non-regular* language, and still fails to derive that $Cond(L)$ is false. Can we then conclude that the language is regular? The answer is **no**! There are cases where the Pumping Lemma we introduce is not powerful enough, and therefore we cannot derive that $Cond(L)$ is false. §4.8 has some examples.

Exercise 4.7.1, Regular or not?
---

1. Show that the language of evenly matched braces is not regular. This language

$$L_{br} = \{\{^i\}^i : i \geq 0\}$$

2. Show that

$$L_{01} = \{0^i 1^j : i,j \geq 0\}$$

   is regular (you show something regular by building a DFA).

3. Show that

$$L_{cat} = \{0^i : i \geq 0\}\{1^i : i \geq 0\}$$

   is regular. Notice that L_{cat} is the concatenation of two languages.
   $\square$

4.8   Grossly Abusing the Pumping Lemma

For someone wielding a hammer, every problem looks like a nail. The Pumping Lemma we presented to you is your hammer, but not every non-regular language is your nail. That is, **only if you draw a contradiction can you claim that a language is non-regular.** To make matters worse, for some *non-regular* languages, *you may never derive a contradiction.*[11]

4.8.1 *Inability to Prove with this Pumping Lemma*

(This is from Sipser's book "Introduction to the Theory of Computation"): Consider the language

$$L_{if} = \{a^i b^j c^k : i,j,k \geq 0 \wedge \text{if } i = 3 \text{ then } j = k\}$$

In this language, strings such as *aaabbcc, aaabbbccc, aaabbbbbbcccccc, aaa, aabbbccc* and *abbcc* are allowed to exist, because *whenever* there are three *a*'s, the number of *b*'s and *c*'s are the same (even 0 *b*'s and 0 *c*'s are considered).

However, strings such as *aaabcc,aaabbbcc,aaabbbbbbccbbccc,aaab* are **not** allowed to exist. This is because we can't have unequal *b*'s and *c*'s when there are three *a*'s.

Now, if we claim that L_{if} is regular,

- There must be an N-state DFA for it.
- It is easy to see that L_{if} is infinite. Thus, there are strings of any finite length in it.
- Thus if we choose a string w length $\geq N$, there must be one DFA-state repeating ("lasso").
- Choose a string $w = aaa\, b^N\, c^N$ (blanks inserted for clarity). This is a string of length $2N + 3$, which is fine for our purposes: it definitely has many places that it has loops ("lassos").
- We don't know where, while accepting the string, our DFA "curls" into a lasso. In fact, when we chug along and consume the three *a*'s and then embark on consuming the N *b*'s, we would be curling around (as we would have traversed at least N symbols).
  - If it curls while processing the *b*'s, we can "pump" the *b*'s to increase or decrease the *b*-count, making it unequal to the *c*-count. This pump takes us outside the language L_{if}. So the pumping condition *Cond* is being falsified for this *xyz* split.
  - If it curls around while straddling some *a*'s and some *b*'s, we can pump and produce "mangled" patterns where some *a*'s appear, then some *b*'s, then some *a*'s, then some *b*'s, which takes us outside L_{if}. Even here, the pumping condition *Cond* is being falsified.
  - If it curls around while straddling some *b*'s and some *c*'s, we can pump and produce "mangled" patterns where some *b*'s appear, then some *c*'s, then some *b*'s, then some *c*'s, which takes us outside L_{if}. Even here, the pumping condition *Cond* is being falsified.[12]
  - However, if it curls around within the *a*'s, we can pump this loop

[11] This is because what we gave you was a rubber hammer, *i.e.*, a form of the Pumping Lemma that sometimes (often enough) works and helps prove the pumping condition to be false. However, we could easily have given you a wooden hammer ("more general Pumping Lemma;" see Exercise 2), or even a vanadium-steel hammer (the *complete Pumping Lemma* of which several kinds exist; either Jaffe's Pumping Lemma mentioned on Page 50 or Stanat and Weiss's Pumping Lemma [43]). With the complete Pumping Lemma in hand, you can "knock off" any language.

[12] We don't need to consider the case of processing just the N *c*'s, as this falls outside the scope of the *first N positions*.

to create more a's or less a's. **This however does not take the so-pumped w outside L_{if}.** Thus we fail to prove for all xyz splits!

- The non-existence of a DFA for L_{if} must be argued *no matter what this DFA is.* This is what forces us to consider all xyz splits. Short of it, we cannot prove that L_{if} is not regular.[13]

Exercise 3 asks you to prove that L_{if} is not regular using the more general Pumping Lemma you define in Exercise 2.

4.9   Regularity-Preserving Transformations Aid Proofs

In Chapter 7 we will demonstrate that the reverse of a DFA can be expressed as a language-equivalent NFA. By the end of Chapter 10, we will have shown that DFA, NFA, and RE are all equally powerful; that is, all and only regular languages are expressible using them. Thus, the reverse of a DFA has a regular language, as well.

But now, suppose we take L_{if} and assume it is regular. Then clearly, the reverse of L_{if} (call it L_{ifrev}) must be regular. But suppose we *successfully pump* and show that L_{ifrev} is not regular. Then, *we will have shown that L_{if} is also non-regular.*

What is L_{ifrev}? It is

$$L_{ifrev} = \{c^k b^j a^i : i,j,k \ge 0 \wedge \text{if } i = 3 \text{ then } j = k\}$$

Now, this language is interesting. Suppose this language is regular. That then means that this language has an N-state DFA whose language exactly matches L_{ifrev}. But now if we consider the string

$$c^N b^N a^3$$

We can observe these:

- We can definitely claim that there **is** a loop within c^N
- So we can pump it
- But pumping this loop gives us a string that is *outside L_{ifrev}*
- Thus, this can't be a regular language, because we will end up accepting strings even outside of the specification of L_{ifrev}. Thus,
  - Either a DFA that exactly filters out strings from L_{ifrev} cannot exist, or
  - The closest a DFA exists for L_{ifrev} is one that accepts a language other than L_{ifrev}.
- Thus, a DFA for exactly L_{if} cannot exist.

Exercise 4.9, Regularity Preserving

1. Consider the language

   $$L_{ifabc} = \{a^i b^j c^k d^l : i,j,k,l \ge 0 \wedge \text{if } i = 3 \text{ then } j = k \text{ else } k = l\}$$

   (a) Show that this language is not regular. Note again that our Pumping Lemma does not work directly on L_{ifabc}.

[13] By way of analogy, if one tries proving $x \ne x^2$ for all x, one cannot stop with $x = 2$ or $x = 3$ where this inequality holds, and conclude that $\forall x : (x \ne x^2)$. One has to consider all x. But for $x = 0$ and $x = 1$, we cannot make the inequality hold. This tells us that this inequality does not hold for all x. We get the same kind of a "proof" if we were to conveniently pick x, y, z for which "the proof works." It still does not rule out *all possible DFA.*

(b) However, if you reverse L_{ifabc} to obtain $L_{ifabcrev}$, you can indeed argue through the cases. Please try this and report your experience.

2. On Page 49, we stated

> However, in order to state a simple and crisp argument, we choose to focus on the **first pump** carried by state s_p between visitation numbers v_p and v_{p+k}.

One can define a *more general Pumping Lemma* that allows you to pick an *xyz* split of *any segment of length N* of the given string w. The reason we avoid introducing this Pumping Lemma is for simplicity of exposition. Try to state this more general Pumping Lemma by situating the *xyz* split after an arbitrary initial segment h ("head") and allowing for an arbitrary final segment t ("tail").

Thus, we will have $w = hmt$ where m ("middle") is a segment of length N, and furthermore, m is split into *xyz* in all possible ways.

3. Prove using the more general Pumping Lemma of Exercise 2 that L_{if} is not regular.

4. In order to reliably use the Pumping Lemma, one must define it in predicate logic. Below, we define the Pumping Lemma in this fashion where one can clearly see where the Pumping condition *Cond* lies.

$Reg(L) \Rightarrow$
$\quad \exists N \in Nat:$
$\quad \forall w \in L : [\, |w| \geq N$
$\qquad \Rightarrow$
$\qquad\quad \exists x, y, z \in \Sigma^* :$
$\qquad\qquad\quad w = xyz$
$\qquad \wedge \qquad |xy| \leq N$
$\qquad \wedge \qquad y \neq \varepsilon$
$\qquad \wedge \qquad \forall i \geq 0 : xy^i z \in L \;]$.

State the negated condition in predicate logic, and then relate it to the recipe stated in English on Page 52 (the bulleted list under "Showing that L_{01} is not regular").

5. If you are given a "lineup" of languages, can you pick out those which are regular and those which are probably not? It is good to check your ability to do so: Here are some of the languages given in the *Pumping Lemma tutor* of the JFLAP tool.[14] For those that are regular, develop a DFA. For those that are not regular, write a proof showing that to be the case.

(a) $L_1 = \{0^i 1^i : i \geq 0\}$

(b) $L_2 = \{w \in \{a, b\}^* : \#_a(w) < \#_b(w)\}$

(c) $L_3 = \{(ab)^n a^k : n > k, k \geq 0\}$

(d) $L_4 = \{a^n b^k c^{n+k} : n, k \geq 0\}$

[14] http://www.jflap.org

(e) $L_5 = \{a^n b^l c^k : n > 5,\, l > 3,\, k \le l\}$

(f) $L_6 = \{a^n : even(n)\}$

(g) $L_7 = \{a^n b^k : odd(n) \text{ or } even(k)\}$

(h) $L_8 = \{bba(ba)^n a^{n-1} : n \ge 0\}$

(i) $L_9 = \{b^5 w : w \in \{a, b\}^*,\, 2\#_a(w) = 3\#_b(w)\}$

(j) $L_{10} = \{b^5 w : w \in \{a, b\}^*,\, (2\#_a(w) + 5\#_b(w)) \bmod 3 = 0\}$

(k) $L_{11} = \{b^k (ab)^n (ba)^n : k \ge 4,\, n \ge 1\}$

(l) $L_{12} = \{(ab)^{2n} : n \ge 1\}$

(m) $L_{13} = \{a^i b^j c^k : \text{if } (i = 3) \text{ then } (j = k)\}$ $\square$

5

Designing DFA

> **Chapter Gist:** *DFA design is really low level programming; hence, subtle mistakes are quite likely. It is important to avoid them through best practices, clearly understanding the language for which a DFA is to be developed, and double-checking our DFA designs (§5.1). We must check that the DFA accepts all strings in L, and none outside of L (§5.1.2). We illustrate these ideas in action (§5.2) and present a markdown language that makes us treat DFA design like assembly language programming through the use of clear state-names and comments to describe states and transitions (§5.3).*

5.1 Understanding the Language to Be Realized

Languages must be ideally specified using any combination of these approaches: (1) A precise English description; (2) a clear set builder (comprehension) description; (3) through a sufficient number of *positive examples* (strings in L) and *negative examples* (strings outside L); and (4) descriptions that take different perspectives.

Set comprehensions can be difficult to arrive at, and can be very difficult to understand. Examples are foolproof, but seldom complete. All these methods are prone to mistakes, but if two methods agree, we can be surer (if not, we can debug more easily).

5.1.1 The Language of Equal Changes

Suppose you are asked to design a DFA over the alphabet $\Sigma = \{0,1\}$ for language L_{eqc} where every string in L_{eqc} undergoes an equal number of 0-to-1 and 1-to-0 changes.

English: The above sentence happens to be a technically precise English description.

Set Builder (Comprehension): The following comprehension is a good first attempt:

$$L_{eqc} = \{\, s \,:\, s \in \{0,1\}^* \text{ has the same number of } 0 \to 1 \text{ and } 1 \to 0 \text{ changes}\,\}$$

The predicate within this set comprehension is written out in English to enhance readability.[1] Let us now expand the predicate a little:

- Without loss of generality, assume that the first character is 0. Suppose a series of 0's follow.
- As soon as a band of one or more 1's appear, we see a $0 \to 1$ change occur. We must "restore order" by ensuring that a band of one or more 0's comes later.

Examples:

1. *Positive:* 010, 10101, etc.
2. *Negative:* 0111, 10100, etc.

The reasoning thus far is not systematic, as we have inadvertently left out ε! It is always recommended that you perform a *numeric order* enumeration of all strings from Σ^* and admit all such strings up to a certain convenient length.[2] Here are those "small instances," by adding which we have a more well-rounded set of examples:

1. ε vacuously satisfies the predicate by having no symbol at all!
2. Likewise, a single 0 and a single 1 also have an equal number (meaning *zero*) changes.

A completely different way of specifying L_{eqc}: After writing this many specifications, one often has a "light-bulb moment":

> Hey, this language is nothing but all strings over $\{0,1\}$ that begin and end with the same symbol. *[Not quite!]*

Well, this specification does take a different approach (and is nearly right; see Exercise 5.1.1.1).

Exercise 5.1.1, equal-change DFA

1. What is missing from this alternate definition?
2. Design a DFA for L_{eqc} and argue that it correctly includes *all* positive examples and correctly excludes *all* negative examples. □

5.1.2 *Best Practices to Correct DFA Design and Verification*

It is worth repeating that DFA-programming is very similar to assembly-language programming, and as such we must thoroughly document DFA descriptions. Just leaving behind the drawing of a DFA without any documentation is insufficient (except for the simplest of DFA). We recommend a few additional tips to produce well-documented DFA:

- *Make sure that the state names are mnemonic.* Thus instead of calling a state "Foo," you might name it F010 to reveal the purpose of the state:

[1] A mathematical rendering would include defining a predicate eq10(s) to capture $1 \to 0$ changes in s and a similar predicate eq01(s). Then the predicate would read $(\text{eq10}(s) \wedge \text{eq01}(s))$.

[2] As Bob Kurshan, famous formal methods researcher, once told me: "If you cannot get the small sizes right, how do you expect all large cases to be handled correctly?"

– it is a final state (begins with "F")

– it is a state that is entered immediately after seeing a "010".

- *Having named states mnemonically, ensure that the transitions preserve the mnemonic significance.* Thus,

  – F010 upon seeing a '0' may transition to S001, where 'S' says "non-final" and "001" is a right-shift of "010" with the newly arriving '0' prepended.[3]

- *Use a markdown notation that permits line-by-line comments.* We have developed a convenient markdown language for this purpose. §5.3 introduces this notation and provides a detailed example.

Minimal DFA are *unique* for a given regular language—this is the *Myhill-Nerode* theorem, which is discussed in Chapter 6. Using this theorem, one can design DFA D_1 and D_2 taking two different perspectives. If they match state-for-state and transition-for-transition,[4] one can be quite sure that D_1 and D_2 have the same language. In fact, the DFA isomorphism test supported by function iso_dfa produces an error trace that helps you debug, in case two DFA happen to be non-isomorphic.[5]

5.2 Examples of Designing DFA

5.2.1 The Language of Blocks of 3

Suppose you are asked to design a DFA over the alphabet $\Sigma = \{0, 1\}$ for language L_{b3} described using set comprehension (and some English) as follows:

$$L_{b3} = \{ x : \text{Every contiguous block of 3 bits in } x \text{ must have exactly two 1s} \}$$

Here are positive examples:
1. ε is included, as it does not have a block of 3 contiguous bits.[6]
2. $0, 1, 00, 01, 10, 11$ are included for the same reason (vacuous cases again).[7]
3. The actual (non-vacuous) cases of strings included are 011, 101, 1011, etc.

Here are negative examples:
1. Notice that 110011 is not included because not every contiguous block of three bits satisfies the condition:
   (a) you have to imagine a window of size 3 being dragged over this string, as illustrated:

   ⌐110⌐011, 1⌐100⌐11, 11⌐001⌐1, 110⌐011⌐

   (b) when the window contains 110 or 011, the condition is met, but when the window contains 100 or 001, the condition is violated.

---

[3] It might also transition to F001, if demanded by the logic of the problem. Here, we are simply giving generic examples.

[4] One can place the printouts on top of each other and "hold them to light" to find that they match up. The only possible difference will be in the state names. Please use Appendix B to locate and read the definition of iso_dfa.

[5] Such error-trace generation is very reminiscent of how model-checkers (central tools in Formal verification) give user-feedback. Also, NFA are generally easier to specify than DFA, so in Chapter 7, you will learn how to convert an NFA to a DFA automatically. You will learn a similar method in Chapter 9, but starting from a regular expression (which are again very often much easier to arrive at).

[6] Well, we are learning fast about vacuous cases!

[7] You should ask "where is my vacuum cleaner?!" You should try and pick up all these vacuous cases early!

State	to	New State
S	$\xrightarrow{0}$	S0
S	$\xrightarrow{1}$	S
S0	$\xrightarrow{0}$	S0
S0	$\xrightarrow{1}$	S01
S01	$\xrightarrow{0}$	S010
S01	$\xrightarrow{1}$	S
S010	$\xrightarrow{0}$	S0
S010	$\xrightarrow{1}$	S0101
S0101	$\xrightarrow{0}$	S010
S0101	$\xrightarrow{1}$	S

Figure 5.1: Pseudo-code for a DFA for "ends in 0101"

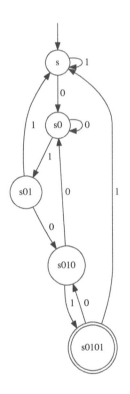

Figure 5.2: A DFA for "ends in 0101"

1. Design a DFA for L_{b3}, describing its design clearly in a few sentences.
2. Now design a DFA for the complement of L_{b3}. Was this easier to design? Justify your answer briefly. (Note: Often, the complement of a condition is easier to specify.)
3. What is the approach to obtaining the desired DFA from the complement DFA?   □?

5.2.2 DFA for "Ends with 0101"

In Figure 5.1, we design a DFA over $\Sigma = \{0,1\}$ that recognizes the language of all strings in $\{0,1\}^*$ that end in 0101. Here are the recommended steps in designing this language (call it L_{E0101}):

- Make sure that the language appears to be regular, *i.e.*, passes the cursory finite-memory property. In this case, it appears that we need to remember only the last four input bits seen so far.
- List a few strings in the language and a few that aren't, to gain further insights:
  - $0101 \in L_{E0101}$; also $010101 \in L_{E0101}$
  - $0 \notin L_{E0101}$; also $101 \notin L_{E0101}$
- Use the state name as a carrier of the history of the last few *relevant* bits (sometimes four bits, but often much shorter). We do not need to keep all four bits all the time. For instance, upon seeing a 0 from state S0, we jump to state S0 and not S00, as seeing two 0's is tantamount to seeing only one 0 (as far as accepting a 0101 goes).
- Likewise, from state S0101, if we see a 0, we go back to S010, while if we see a 1, we fall back to state S (no advantage gained with respect to ending in 0101). Thus, a state name S1011 is not introduced.
- We start with state S. For each state name we introduce, we consider the 0 input case first, and then the 1 input case (as in Figure 5.1). This way, we won't accidentally forget a case (the δ function must be specified for each state and input).
- We introduce state names on demand; in this DFA, we get away explicitly representing only five (of the total) 16 possible states. (The number 16 comes from the number of arrangements of four bits.) When no more state names are deemed necessary, we stop our construction. We obtain the DFA in Figure 5.2.

5.2.3 DFA for "MSB/LSB-first Binary Number is Divisible by 3"

Positionally weighted binary words have a most significant bit (MSB) and a least significant bit (LSB). Let us define what we mean by the "MSB-First" scenario. The scenario presented by this design challenge is that bits are revealed to us from the MSB toward the LSB. Thus, we might see a progress of bits as follows, given that you typed in 1, then 0, then 0, then 1, then 0, and finally 1.

$$\varepsilon \to \underline{1} \to 1\underline{0} \to 10\underline{0} \to 100\underline{1} \to 1001\underline{0} \to 10010\underline{1} \to \ldots$$

We interpret the number described so far in unsigned binary (treating ε as 0). For this sequence, we obtain these "numbers so far"

$$0 \to 1 \to 2 \to 4 \to 9 \to 18 \to 37\ldots$$

and clearly, these numbers are evenly divisible by 3 (or not) as shown by the Boolean truth values

$$1 \to 0 \to 0 \to 0 \to 1 \to 1 \to 0\ldots$$

The task for you is to build a DFA that is in a final state exactly when the sequence seen so far represents an "evenly divisible by 3" situation as per these Boolean truth values. Thus, after seeing 100, the DFA must be in a non-final state and after seeing 1001, it must be in a final state. Similar to the DFA for "Ends with 0101," a moment's reflection reveals that what we have to record is the *remainder* and not the whole number, thus convincing us that a DFA is possible. The following recurrence will aid you in constructing the state evolution: a "number so far," *i.e.*, N, will become $2N + b$. Also, N starts out at 0.

$$S_N \xrightarrow{\ b\ } S_{2N+b}$$

Here, b is the new incoming LSB. When that bit arrives, it pushes the bits seen so far to the left by one position.

Since only the modulus (remainder after division, denoted by % in Python) is to be remembered, we can write the following transition recipe:

$$S_{N\%3} \xrightarrow{\ b\ } S_{(2N+b)\%3}$$

The general approach we are following is to generate states for the possible input bits recursively until no new states are generated. More specifically: (1) do the mod calculations "in your head" and arrive at the state names, or (better approach) (2) exploit properties of mod given in Figure 5.3 and arrive at the state names.

DFA FOR THE "LSB-FIRST" SCENARIO: The case for LSB-First arrival is even more interesting, and relies upon more ideas to be turned into a DFA-based language recognition problem. Let us say that you *still type in the same bits in the same order*; *i.e.*, 1, then 0, then 0, then 1, then 0, and finally 1. However, the interpretation differs as shown below:

$$\varepsilon \to \underline{1} \to \underline{0}1 \to \underline{0}01 \to \underline{1}001 \to \underline{0}1001 \to \underline{1}01001 \to \ldots$$

$$\begin{aligned} (a+b)\%N &= (a\%N + b\%N)\%N \\ (a \cdot b)\%N &= (a\%N \cdot b\%N)\%N \end{aligned}$$

Figure 5.3: Rules of the mod operator, %. These rules allow you to reduce "the number so far," namely N, to "the number after a calculation, modulo N." In our general rule stated in the main body of the text, observe that N becomes $(2N + b)\%3$. The rules for the mod operator, %, tells us that $(2N + b)\%3$ can be simplified to $((2N)\%3 + b\%3)\%3$. But since b is a bit, $b\%3$ is nothing but b. Now, working on $(2N)\%3$, we can simplify it to $((2\%3) \cdot (N\%3))\%3$. Notice that $(2\%3)$ is simply 2. Thus we can indeed simplify $(2N + b)\%3$ to $(2 \cdot (N\%3) + b)\%3$. Now, given that we are maintaining $(N\%3)$ in the state name, we can directly work with the number coming after "S," and process it using the equation $(2 \cdot (N\%3) + b)\%3$. This is what we have been doing "in our head."

State	to	New State
I	$\xrightarrow{0}$	S0
I	$\xrightarrow{1}$	S1
S0	$\xrightarrow{0}$	S00
S0	$\xrightarrow{1}$	S01
S1	$\xrightarrow{0}$	S10
S1	$\xrightarrow{1}$	S11
S00	$\xrightarrow{0}$	S000
S00	$\xrightarrow{1}$	S001
S01	$\xrightarrow{0}$	S010
S01	$\xrightarrow{1}$	S011
S10	$\xrightarrow{0}$	F100
S10	$\xrightarrow{1}$	F101
S11	$\xrightarrow{0}$	F110
S11	$\xrightarrow{1}$	F111
S000	$\xrightarrow{0}$	S000
S000	$\xrightarrow{1}$	S001
S001	$\xrightarrow{0}$	S010
S001	$\xrightarrow{1}$	S011
S010	$\xrightarrow{0}$	F100
S010	$\xrightarrow{1}$	F101
S011	$\xrightarrow{0}$	F110
S011	$\xrightarrow{1}$	F111
F100	$\xrightarrow{0}$	S000
F100	$\xrightarrow{1}$	S001
F101	$\xrightarrow{0}$	S010
F101	$\xrightarrow{1}$	S011
F110	$\xrightarrow{0}$	F100
F110	$\xrightarrow{1}$	F101
F111	$\xrightarrow{0}$	F110
F111	$\xrightarrow{1}$	F111

Figure 5.4: Pseudo-code for a DFA that recognizes the language "third-last is a 1."

[8] Proof idea: Suppose we don't maintain all the bit combinations shown here up to (and including) all 3-bit combinations. Then, for the next N bit arrivals, we will not have enough information in our hands to know whether to be in a final state or not.

Each new bit lands left-most, thus attaining growing weight as we enter more. The "number so far" stays the same. The recurrence must now keep the weight growing:

$$S_{(2^k, N)} \xrightarrow{b} S_{(2^{k+1}, N + 2^k \cdot b)}$$

In other words, we keep an ordered pair $(2^k, N)$ within the state. When a bit b arrives, it is weighted by 2^k and added to N. The weight itself is now adjusted up to 2^{k+1}, essentially preparing for the arrival of the next b bit (that must be weighted higher). This can also be finite-state encoded, as we do not need to know the exact values of $N + 2^k \cdot b$—only whether it is evenly divisible by 3. See Exercise 5.2.4.2 on Page 62.

5.2.4   DFA for "Third-last bit is a 1"

Suppose we have to design a DFA which is in a final state whenever the third-last bit you typed is a 1. The concept of the *last bit* is clear; second last is otherwise known as penultimate, and third-last is the one before that. Suppose you typed in bits as follows: 1, then 0, then 0, then 1, then 0, then 0 and finally 1. We now mark the third-last in the sequence below with an underline:

$$\varepsilon \to 1 \to 10 \to 1\underline{0}0 \to 1\underline{0}01 \to 10\underline{0}10 \to 100\underline{1}00 \to 1001\underline{0}01 \to \ldots$$

The DFA we design must be in a final state whenever the underline is under a 1. Figure 5.4 shows the design of this DFA following the state history idea; here, I is the initial state, and any state with name beginning with F is a final state. Notice however an exponential blow-up in the number of states needed (see Figure 5.5). In particular, the DFA for "third-last bit is 1" maintains 15 states. If we were to build a DFA for fourth-last is a 1, we can show that we will need to maintain 31 states.[8] This example shows that DFAs can be exponentially sized—a big nuisance that will send us on a quest for better automata representations in the next chapter.

Exercise 5.2.4, DFA exp blowup

1. Prove that the exponential blow-up is unavoidable for the DFA implementing the language "Nth last bit is a 1" ($L_{Nthlast1}$). This language is

$$L_{Nthlast1} = \{x1y : x \in \{0,1\}^* \wedge y \in \{0,1\}^{(N-1)}\}$$

Here, we refer to N as the "look-back" of this language.

2. Follow the approach described on Page 61 and design a DFA that enters a final state exactly when the magnitude of the number seen so far (arriving LSB-first) is evenly divisible by 5.   □

Figure 5.5: DFA drawing for 'second-last is 1' (above), and for comparison (to show the exponential growth), we include 'third-last is 1' also (below). Mind you, these are *minimal* DFA!

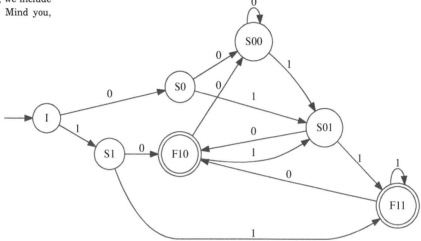

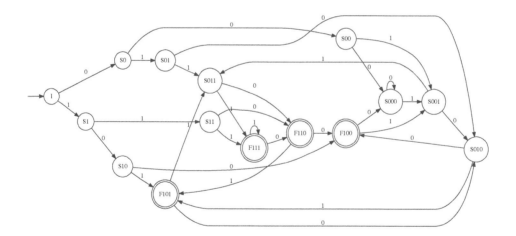

5.3 Automd: A Markdown Language for All Machines

We have designed for your convenience a language called Automd that can accept a simple textual input file and produce beautiful DFA drawings, plus formal descriptions of the kind shown in Figure 4.2 automatically![9] This way, you can step and run DFA as soon as you finish entering things into your markdown.

We will provide a fresh example – "second-last is a 1" – and its complete creation + execution within Jove.

5.3.1 Markdown for DFA

Figure 5.6 shows the markdown input for the language "second-last is 1" while Figure 5.5 presents the drawing automatically generated. Even this DFA has an exponential number of states with respect to 2, but is half as big to easily illustrate our basic ideas.

Exercise 5.3.1, DFA Jove design
---

1. Enter the 'third-last is 1' language and carry out all the steps illustrated in this section for the 'second-last is 1' language.

2. For all the DFA you have been asked to design in this chapter, enter them using the markdown syntax of Jove. Then run them under Jove by emulating the examples provided in file Drive_DFA.ipynb. Demonstrate that the functions step_dfa, run_dfa and accepts_dfa work as described in Chapter 4. In your tests, feed as input the first 10 strings as per the numeric order (defined in Chapter 3) by using function nthnumeric(N).

   Here is how I carried out the same tests for the 'second-last is 1' DFA presented in Figure 5.5:

```
tests = [ nthnumeric(i, ['0','1']) for i in range(12) ]
for t in tests:
    if accepts_dfa(secondLastIs1, t):
        print("This DFA accepts ", t)
    else:
        print("This DFA rejects ", t)


# Test Results

This DFA rejects
This DFA rejects  0
This DFA rejects  1
This DFA rejects  00
This DFA rejects  01
This DFA accepts  10
This DFA accepts  11
This DFA rejects  000
```

[9] The function **md2mc** in Jove (§B.1.5) is the one that converts the markdown syntax for DFA, NFA, PDA or TMs into the internal representation (dict) used by Jove (we call the internal representation "machine" or "mc"; hence the name "md2mc").

Figure 5.6: DFA Markdown input for "second-last is 1." Notice that merely by typing the contents of this figure into Jove, we can generate a DFA (it could also be read by our tools from a file). While the information in this description is essentially the same as expressed in Figure 5.4, we do allow documentation strings (comments) that begin at !! and end at the end-of-line. The particular command we have implemented to achieve this markdown processing is called md2mc(...), *i.e.*, "markdown to machine" where "machine" is one of DFA, NFA, PDA, or TM. We will later introduce all details of how this language is parsed and how the various types of machines are built. A study of our markdown parser implemented within the md2mc command will be another practical example of how we use the theory introduced in this book to build practical tools (in this case, a mini-compiler).

```
!!-----------------------------------------------------------------
!! This DFA looks for patterns of the form ....1.
!! i.e., the second-last (counting from the end-point) is a 1
!!
!! DFA find such patterns "very stressful to handle",
!! as they are kept guessing in the form  of 'are we there yet?',
!! 'are we seeing the second-last' ?
!! They must keep all the failure options at hand. Even after
!! a 'fleeting glimpse' of the second-last, more inputs can
!! come barreling in to make that "lucky 1" a non-second-last.
!!
!! We take 7 states in the DFA solution.
!!-----------------------------------------------------------------

DFA

!!-----------------------------------------------------------------
!! State : In -> ToState !! Comment
!!-----------------------------------------------------------------


I   : 0 -> S0  !! Enter at init state I
I   : 1 -> S1  !! Record bit seen in state letter
                !! i.e., S0 means "state after seeing a 0"

S0  : 0 -> S00 !! continue recording input seen
S0  : 1 -> S01 !! in state-letter. This is a problem-specific
                !! way of compressing the input seen so far.

S1  : 0 -> F10 !! We now have a "second last" available!
S1  : 1 -> F11 !! Both F10 and F11 are "F" (final)

S00 : 0 -> S00 !! History of things seen is still 00
S00 : 1 -> S01 !! Remember 01 in the state

S01 : 0 -> F10 !! We again have a second-last of 1
S01 : 1 -> F11 !! We are in F11 because of 11 being last seen

F10 : 0 -> S00 !! The second-last 1 gets pushed out
F10 : 1 -> S01 !! The second-last 1 gets pushed out here too

F11 : 0 -> F10 !! Still we have a second-last 1
F11 : 1 -> F11 !! Stay in F11, as last two seen are 11


!!-----------------------------------------------------------------
```

```
This DFA rejects  001
This DFA accepts  010
This DFA accepts  011
This DFA rejects  100
```

☐

6

Operations on DFA

Chapter Gist: *It is often easier to build a DFA for the complement of the desired language, and then apply the complementation algorithm to obtain the desired DFA (§6.1). Sometimes we can express the desired language L as $L = L_1 \cup L_2$ where the DFA for L_1 and L_2 are easier to obtain; at other times, $L = L_1 \cap L_2$ would be the recommended decomposition. (§6.2). DFA minimization and the Myhill-Nerode theorem about the uniqueness of minimal DFA are key results (§6.4), and we also present the key fixpoint algorithm used during minimization (§6.4.2). With these operations, we design a number of DFA (§6.5), checking minimal DFA obtained through two perspectives for agreement via isomorphism (§6.5.1). DeMorgan's law (§6.5.2) often eases DFA construction.*

6.1 Complementation of DFA

Complementation is achieved by swapping final and non-final states. We must however totalize the given DFA before we begin our work. This is because the 'black-hole' states of the original DFA (states that you can enter but not leave, and are non-accepting states) will now become 'white-hole' states (states that you can enter but not leave, and are accepting states).[1] The Jove code for the complementation algorithm is provided in Figure 6.1.

6.2 Union and Intersection of DFA

The formal constructions for **DFA union** and **DFA intersection** are given below, and the code for union/intersection is in Figure 6.2.

- Given DFAs $D_1 = (Q_1, \Sigma, \delta_1, q_0^1, F_1)$ and $D_2 = (Q_2, \Sigma, \delta_2, q_0^2, F_2)$
- Output $(Q_1 \times Q_2, \Sigma, \delta, (q_0^1, q_0^2), F)$
- Here, $\delta((q_1, q_2), a) = (\delta_1(q_1, a), \delta_2(q_2, a))$
- The final states are defined as follows:

```
def comp_dfa(D):
    """In : D (DFA)
       Out: D's complement
       Method: Swap final/non-final
               states
    """
    Dt = totalize_dfa(D)
    return mk_dfa(Dt["Q"],
                  Dt["Sigma"],
                  Dt["Delta"],
                  Dt["q0"],
                  Dt["Q"]-D["F"])
```

Figure 6.1: DFA Complementation algorithm

[1] White-hole states can also be called 'heavenly states'; you enter, but do not leave, but have a great time, as it is a final or accepting-state.

  – For union, $F = (F_1 \times Q_2) \cup (Q_1 \times F_2)$
  – For intersection, $F = F_1 \times F_2$.

Notice how we tie together the states of the DFA into a pair and step each part of the pair as per the δ function of the appropriate DFA.[2] The final states for union are defined as ordered pairs where at least one of the DFA is in a final state; final states for intersection are defined as ordered pairs where *both* DFA are in a final state. Jove provides more error-checking and also handles the case of the DFA alphabets being different.

The following code pretty much implements the math behind the union operation. Notice the call pruneUnreach that removes unreachable configurations. Its need will become apparent if you carry out Exercise 6.2 using Jove.

[2] Ganging two DFA is like treating them as two work-horses and tying them to a common yoke.

Figure 6.2: Union algorithm (Intersection is similar, and is left as an exercise; it is just a one-line change!)

```
def union_dfa(D1in, D2in):
    """In : D1in
            D2in
       Out: DFA for language union of D1in, D2in
    """

    D1 = totalize_dfa(D1in)
    D2 = totalize_dfa(D2in)
    # The states can be anything in the Cartesian product
    Q   = set(product(D1["Q"], D2["Q"]))
    # Accept if one of the DFAs accepts
    F   = (set(product(D1["F"], D2["Q"])) |
              set(product(D1["Q"], D2["F"])))
    # Start a lock-step march from the respective q0
    q0  = (D1["q0"], D2["q0"])
    # The transition function attempts to march both
    # DFAs in lock-step per their own transition functions
    Delta = { ((q1,q2),ch) : (q1p, q2p)
                for q1 in D1["Q"] for q1p in D1["Q"]
                for q2 in D2["Q"] for q2p in D2["Q"]
                for ch in D1["Sigma"]
                if D1["Delta"][(q1,ch)] == q1p and
                   D2["Delta"][(q2,ch)] == q2p }
    return pruneUnreach(
        mk_dfa(Q, D1["Sigma"], Delta, q0, F))
```

Exercise 6.2, DFA Jove, ∪,∩

1. Take the DFA of Figure 4.7, calling it D. Obtain its complement, calling it Dc. Obtain the union of D and Dc without the pruneUnreach call. Print the resulting DFA. Is this still a DFA? (Hint: DFA are allowed to have disconnected states. However, these are annoying and many algorithms do not allow disconnected states; hence we

prune unreachable states using the algorithm in Figure 6.3.)

2. Describe the algorithm implemented by pruneUnreach in Jove. Its code is in Figure 6.3.   □

Figure 6.3: Algorithm for pruneUnreach

```python
def pruneUnreach(D):
    """In : D (consistent DFA)
       Out: Consistent DFA.
       Given a consistent (and total) DFA D, returns a new
       (consistent) DFA with unreachable states in D removed.
       Transitions from each unreachable state are also removed.
       Reachable states are those that can be reached in
       |D["Q"]| - 1 steps or less.
    """
    Nsteps  = len(D["Q"]) - 1 # Search this far
    Frontier = set({D["q0"]}) # BFS frontier
    AccumF  = Frontier        # Used to accumulate Frontier changes
    for n in range(Nsteps):
        for q in Frontier:
            for ch in D["Sigma"]:
                AccumF = AccumF | set({step_dfa(D, q, ch)})
        Frontier = AccumF
    newQ     = Frontier
    newF     = D["F"] & Frontier
    newDelta = dict({ ((q,ch),qp)
                      for ((q,ch),qp) in fn_trans(D["Delta"])
                      if q in Frontier })
    return mk_dfa(Frontier, D["Sigma"], newDelta, D["q0"], newF)
```

Figure 6.4: Algorithm for DFA Language Equivalence

```python
def langeq_dfa(D1, D2, gen_counterex=False):
    """Check whether D1, D2 are lang. eqlt. gen_counterex is a flag
       that triggers the printing of a counter-example. Two DFAs are
       language-equivalent if they accept the same set of strings.
       We determine this through a joint depth-first walk."""
    if D1["Sigma"] != D2["Sigma"]:
        print("Error: the DFA cannot be compared...")
    else:
        (eqStatus,
         cex_path) = h_langeq_dfa
                        (D1["q0"], D1, D2["q0"], D2, Visited=[])
        if not eqStatus:
            if gen_counterex:
                print("The DFA are NOT language equivalent!")
                print("Path leading to counterexample is: ")
                print(cex_path)
        return eqStatus # True or False


def same_status(q1, D1, q2, D2):
    """Check if q1,q2 are accepting/non-accepting wrt D1/D2."""
    return (q1 in D1["F"]) == (q2 in D2["F"])


def h_langeq_dfa(q1, D1, q2, D2, Visited):
    """If (q1,q2) is in Visited, no screw-up so far, so continue.
       Else if they agree in status, recursively check for all reachable
       configurations (a DFS in recursion). Else, return (False, Visited)
       Visited is the counter-example."""
    if (q1,q2) in Visited:
        return (True, Visited)
    else:
        extVisited = Visited + [(q1,q2)]
        if not same_status(q1,D1,q2,D2):
            return (False, extVisited)
        else:
            l_nxt_status = list (
            map(lambda symb:
                h_langeq_dfa(D1["Delta"][(q1,symb)], D1,
                             D2["Delta"][(q2,symb)], D2,
                             extVisited), D1["Sigma"])   )
            l_rejects = list(filter(lambda x: x[0]==False, l_nxt_status))
            if l_rejects==[]:
                return (True, extVisited)
            else: # This is the first offending (status,cex)
                return l_rejects[0]
```

6.3 Language Equivalence and Isomorphism

Figure 6.4 defines function langeq_dfa that checks for DFA language equivalence, while Figure 6.5 defines function iso_dfa that checks for DFA isomorphism (both functions assume alphabet agreement).

A *natural* way to check whether two DFAs are language equivalent is to lock-step march them, feeding all sequences of inputs (achieved by a concurrent depth-first search). So long as the DFAs "track each other" — meaning that upon seeing every string possible, they both are in an accepting state or in a non-accepting state — they are language equivalent. This process will converge. Can you argue why? (Hint: In the limit, we will have visited all the states in the Cartesian product of the states of the DFA.) Let us walk through the code:

- After checking whether the alphabets agree, call helper h_langeq_dfa which returns the language equality status eqStatus and a possible counter-example path cex_path. If there is a mismatch (*i.e.*, a string is accepted by one DFA and the same string is rejected by the second DFA), then cex_path is a list of pairs of states of DFAs D1 and D2 ending in a state pair (q1,q2) where the predicate same_status emerges false (one of these states is accepting and the other is not).

- h_langeq_dfa keeps a list of already visited state pairs (Visited); if the current state pair is in there, then there is no point chasing down this pair of DFA states. Just return True.

- Otherwise add (q1,q2) to the visited set. If (q1,q2) don't have the same status, return False.

- Otherwise, obtain the list of statuses of the next states attainable from (q1,q2). This is achieved by recursively mapping h_langeq_dfa on the next states attained (see the expression D["Delta"][(q1,symb)] for instance).

- Then employ the filter function to check for any lurking mismatches (False) in the list of statuses. If a mismatch is found, return the reject status plus the counterexample path.

The use of higher-order programming using map and filter makes the code elegant to read.[3]

To summarize this section,

> Two DFA are isomorphic if and only if they are language equivalent and have the same number of states.

```
def iso_dfa(D1,D2):
    """Given consistent + total DFA
       D1 and D2, they are isomorphic
       * If they have the same number
         of states,
       * and are language-equivalent
    """
    return(len(D1["Q"]) == len(D2["Q"])
           and
           langeq_dfa(D1, D2))
```

Figure 6.5: Algorithm for DFA Isomorphism

[3] In all the code written for pedagogy, one emphasizes clarity over efficiency, even though higher-order functions can indeed be efficient in modern implementations.

6.4 DFA Minimization and Myhill-Nerode Theorem

We **minimize a DFA** based on a simple idea. Suppose you are given a magical stethoscope called "the language stethoscope." You can touch any state x of a DFA $D = (Q, \Sigma, \delta, q_0, F)$ with it, and you will "hear" the language of state x. This language is defined to be the set of strings that take x to a state within F via $\hat{\delta}$.[4] Now if you touch this stethoscope on q_0, you will "hear" the language of the DFA.[5] Given this setup, two states p and q are indistinguishable if they have the same language. (Exercise: identify all indistinguishable pairs of states in Figure 6.7.)

A DFA is then said to be minimal if no two distinct states q_i and q_j are indistinguishable. The algorithm we are to present implements this idea using dynamic programming. Here is an important property of minimal DFA:

> **Theorem 6.4:** If D_1 and D_2 are two minimal DFA for the same regular language L, then D_1 and D_2 are isomorphic.

For a proof, see [26].

Intuitively, we can print D_1 and D_2, hold one printout on top of the other, and observe a state-for-state and transition-for-transition match. More formally, D_1 and D_2 are isomorphic (as tested by the Jove predicate `iso_dfa`) if:

- We can put the set of states $Q(D_1)$ and the set of states $Q(D_2)$ into a bijective (1-1, onto) mapping under some function f such that a pair of states coupled by this bijection are either both final states or both non-final states.
- $q_0(D_1)$ and $q_0(D_2)$ are linked by this bijection f.
- For every pair of states q_1 and q_2 that are linked by this bijection f, for all symbols $c \in \Sigma$, $\delta_1(q_1, c)$ and $\delta_2(q_2, c)$ are also linked under this bijection.

6.4.1 Fully Worked-out Example of DFA Minimization

Consider the example in Figure 6.7 (automatically generated from the markdown in Figure 6.6). We have to test all pairs of states in the DFA using our language stethoscope. There are $\binom{6}{2} = 15$ state pairs so we will employ a data structure, a 'frame,' that allows us to easily compare them. The algorithm, going frame by frame, is presented in Figure 6.7 (bottom).

- Frame-0 (marked Initial): Our frame design allows us to "clash" all combinations of states taken two at a time. We put a -1 against each pair of states to denote that they have not been found distinguishable yet. Thus, in Frame-0, all combinations carry a -1.

[4] A game of pretense in which you pretend that x is the DFA's initial state.

[5] Its heart...

```
bloated_dfa = md2mc('''
DFA
IS1 : a -> FS2
IS1 : b -> FS3
FS2 : a -> S4
FS2 : b -> S5
FS3 : a -> S5
FS3 : b -> S4
S4  : a | b -> FS6
S5  : a | b -> FS6
FS6 : a | b -> FS6''')
dotObj_dfa(bloated_dfa)
```

Figure 6.6: Markdown for a bloated DFA

- Frame-1 (marked 0-distinguishable): We now put a 0 where a pair of states is 0-distinguishable. This means the states are distinguishable after consuming a string of length 0 (*i.e.*, ε) – that is, one state is a final state and the other is a non-final state. Thus, all pairs in the Cartesian product of the set of states {IS1,S4,S5} and {FS2,FS3,FS6} are marked 0-distinguishable, since exactly one of the members of these pairs is a final state and the other is a non-final. Thus, we see nine entries having "0" in them.

- Frame-2 (marked 1-distinguishable): We now put a 1 where a pair of states is 1-distinguishable. This means the states are distinguishable after consuming a string of length 1 (a single symbol). This is only possible if one state transitions to a final state and the other transitions to a non-final state *after consuming a symbol*. State pairs (FS6,FS2) and (FS6,FS3) are of this kind.

  Take (FS6,FS2) as an example. While both FS6 and FS2 are final states (and hence are 0-indistinguishable), after consuming an 'a', FS6 stays at FS6, while FS2 goes to S4. We already know that (FS6,S4) are 0-distinguishable (we can say that *'a' helps pull apart FS6 and FS2*).

  - *Just one* way to force distinguishability suffices. In our example, even 'b' helps pull apart (FS6,FS2).
  - (FS6,FS3) get pulled apart similarly.
  - **General rule:** Let us say that (pre_p, pre_q) are of unknown distinguishability status. Suppose $p = \delta(D, pre_p, c)$, $q = \delta(D, pre_q, c)$, and we already know that (p, q) are k-distinguishable. Then we can declare that (pre_p, pre_q) are $k+1$-distinguishable.

- Applying our general rule, we mark state pairs (S5,IS1) and (S4,IS1) to be 2-distinguishable. Let us explain this for the (S5,IS1) case (the (S4,IS1) case is left as an exercise).

  Notice that after consuming an 'a', S5 ends up at FS6 while IS1 ends up at FS2, and (FS6, FS2) are already marked 1-distinguishable. Thus, we mark (S5, IS1) as being 2-distinguishable.

- Another way to see things is this. After seeing an 'aa', S5 ends up at FS6 while IS1 ends up at S4. Since FS6 and S4 are 0-distinguishable, S5 and IS1 are 2-distinguishable.[6]

- Thus from Frame-3 which captures 2-distinguishability, we try to generate Frame-4 (which captures 3-distinguishability). Alas, no more such distinctions can be established. In other words, Frame-4, if built, would emerge identical to Frame-3. (We denote this by writing "Frame-3 = Frame-4" on top of the 2-distinguishability case.) Thus the algorithm terminates. At this point, all the state pairs that could not be distinguished in any manner (*i.e.* still carry a "-1" against them) are considered to be equivalent ($\equiv$). Now suppose there are states s_1, s_2 and s_3 such that $s_1 \equiv s_2$ at this stage, and $s_3 \equiv s_3$ at this stage, we can claim that $s_1 \equiv s_3$. That is, $\{s_1, s_2, s_3\}$ belong to one equivalence class.[7]

[6] We avoid going from Frame 3 back to Frame 1, as we will then be forced to consider all strings of length 2, which are quite numerous. We always end up considering **single symbols** and manage to extend the distinguishability relation with respect to a prior frame.

[7] Equivalence classes are reviewed in Appendix A.

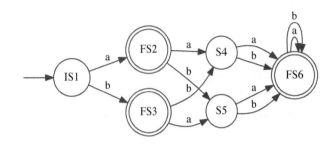

Figure 6.7: Schematic for the bloated DFA of Figure 6.6 (top), its minimized version (middle), and the frames generated during DFA minimization (bottom) are all shown.

Brief summary of minimization: Consider the pair (S5, IS1) which starts out at -1, and becomes 2 in Frame-3. This is because S5 ends up at FS6 after consuming an 'a' while IS1 ends up at FS2 after an 'a'. Given that (FS6, FS2) has been determined to be 1-distinguishable, (S5, IS1) becomes 2-distinguishable.

State pairs that remain at -1 are indistinguishable, and must be merged into equivalence classes. One such equivalence class, namely (FS2,FS3), results in state FS2_FS3. Another equivalence class, namely (S4,S5), results in state S4_S5. The minimized machine now has these state names.

*

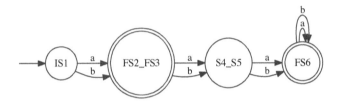

```
Frame-0                 Frame-1                 Frame-2                 Frame-3 = Frame-4
(Initial)               (0-distinguishable)     (1-distinguishable)     (2-distinguishable)
-------------------     -------------------     -------------------     --------------------

FS2  -1                 FS2   0                 FS2   0                 FS2   0

FS3  -1  -1             FS3   0  -1             FS3   0  -1             FS3   0  -1

S4   -1  -1  -1         S4   -1   0   0         S4   -1   0   0         S4    2   0   0

S5   -1  -1  -1  -1     S5   -1   0   0  -1     S5   -1   0   0  -1     S5    2   0   0  -1

FS6  -1  -1  -1  -1  -1 FS6   0  -1  -1   0   0 FS6   0   1   1   0   0 FS6   0   1   1   0   0

     IS1 FS2 FS3 S4 S5       IS1 FS2 FS3 S4 S5       IS1 FS2 FS3 S4 S5       IS1 FS2 FS3 S4 S5
```

- At this point, we need to *merge equivalence classes*, as we now detail.
  - We established in the end that (FS2,FS3) and (S4,S5) are indistinguishable: they are still left with the initial "-1". Thus
    * (FS2,FS3) forms a single equivalence class.
    * (S4,S5) forms another equivalence class.
    * Whenever two equivalence classes overlap, we can merge them. Thus, if (FS2,FS3) and say (FS3,Y) were to be two equivalence classes for some state Y, we would have ended up creating a "super equivalence class" (FS2,FS3,Y). But there is no such state "Y". So, in this example, the process of merging equivalence classes stops.[8]
    * Thus we can introduce a new state, say "FS2_FS3" modeling the equivalence class (FS2,FS3) and yet another new state S4_S5 modeling (S4,S5).

  Figure 6.7 (Page 74) presents the results of the minimization (our algorithm employs a simple heuristic to rename equivalence classes).

6.4.2 Salient Code Excerpts

We now present our algorithm which is a fixed-point computation. The DFA minimization algorithm is included in Jove in module `Def_DFA.ipynb` and illustrated on interesting examples in `Drive_DFA.ipynb`.

- We set a `changed` flag `False` upon entry to the loop, setting it `True` and break out whenever a change occurs.
- The hash-table maintains state-pairs as keys and the values are the distinguishability distances.
- You can see that `ns0` and `ns1` are the next states attained after each character `c` is applied to the current states `s0` and `s1`.
- If the states are the same, obviously they are indistinguishable; so we continue.
- When an entry is at -1, we set the distinguishability distance based on the distance of the (`ns0,ns1`) pair *plus* 1.
- We check whether (`ns1,ns0`) is in the hash-table; if so, it serves as a proxy for (`ns0,ns1`). This step is needed because we store only the lower triangle of the matrix.
- That is it! When the hash-table ceases to change, we exit and return `ht`.

We call this a fixed-point computation because that is the standard name for applying a function f to something, and then applying f again, and so on, till nothing changes.[9] For instance, 1 is a fixed-point of the factorial (fac) function because $fac(1) = 1$; so is 2.

[8] Each equivalence class is like a drop of water, and when two equivalence classes touch, they coalesce. An equivalence class containing a pair (a,b) "touches" another equivalence class containing (b,c), and the super-equivalence class now includes (a,b), (b,c), and (a,c).

[9] If you xerox your face on a copier, and then xerox the xerox, and xerox the xerox of the xerox, ..., very soon the image will fuzz up to a point—becoming one fixed-point of the copier.

```python
def fixptDist(D, ht):
    """In : D (consistent DFA)
            ht (hash-table of distinguishability pair distances)
       Out: ht that has attained a fixpoint in distinguishability.
       Determine the min. distinguishability distances, going frame
       by frame. Fixpoint attained when ht ceases to change.
    """
    changed = True
    while changed:
        changed = False
        for kv in ht.items():
            s0 = kv[0][0]
            s1 = kv[0][1]
            for c in D["Sigma"]:
                ns0 = D["Delta"][(s0,c)]
                ns1 = D["Delta"][(s1,c)]
                #
                # Distinguishable state pairs carry
                # "distinguishability distance" in the ht
                if ns0 == ns1:
                    continue
                if (ns0, ns1) in ht:
                    # s0,s1 are distinguishable
                    if ht[(s0,s1)] == -1 and ht[(ns0, ns1)] >= 0:
                        # acquire one more than the
                        # dist. number of (ns0,ns1)
                        ht[(s0,s1)] = ht[(ns0, ns1)] + 1
                        changed = True
                        break
                else:
                    # ht stores only (ns0,ns1);
                    # so check the other way
                    if (ns1, ns0) in ht:
                        if ht[(s0,s1)] == -1 and ht[(ns1, ns0)] >= 0:
                            ht[(s0,s1)] = ht[(ns1, ns0)] + 1
                            changed = True
                            break
                    else:
                        print("ht doesn't cover all reqd state combos.")
    return ht
```

6.5 Examples of Language Design and Manipulation

We now present examples that we highly recommend that you experiment with using Jove.

6.5.1   Use of Union, Minimization, and Language Equivalence

Here are the specific commands we can use to conduct these experiments:

- Create a DFA for L_1, a language over $\{0, 1\}$ that has an odd number of 1s. The following Jove command for our markdown language accomplishes this construction:

```
dfaOdd1s = md2mc('''
DFA
I : 0 -> I
I : 1 -> F
F : 0 -> F
F : 1 -> I
''')
```

- Define the DFA for our second language of interest, say L_2 (also over $\{0, 1\}$) which is "ends in 0101:"

```
ends0101 = md2mc('''
DFA
I     : 0 -> S0
I     : 1 -> I
S0    : 0 -> S0
S0    : 1 -> S01
S01   : 0 -> S010
S01   : 1 -> I
S010  : 0 -> S0
S010  : 1 -> F0101
F0101 : 0 -> S010
F0101 : 1 -> I
''')
```

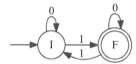

Figure 6.8: DFA for L_1, which corresponds to dfaOdd1s

- Obtain the union of $L = (L_1 \cup L_2)$ using
  odd1sORends0101 = union_dfa(dfaOdd1s,ends0101)
- Minimize L to obtain L_{min}
  Minodd1sORends0101 = min_dfa(odd1sORends0101)
- Check if the machines for L and L_{min}, are isomorphic, using
  iso_dfa(odd1sORends0101, Minodd1sORends0101). The answer is
  False
- Check if the machines for L and L_{min} are language equivalent using
  langeq_dfa(odd1sORends0101, Minodd1sORends0101). The answer
  is True. Thus, the minimization did reduce the DFA size. But the
  unminimized machine also has the same language.

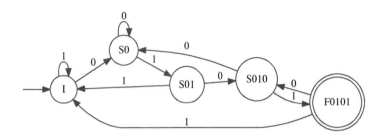

Figure 6.9: DFA for L_2, which corresponds to ends0101

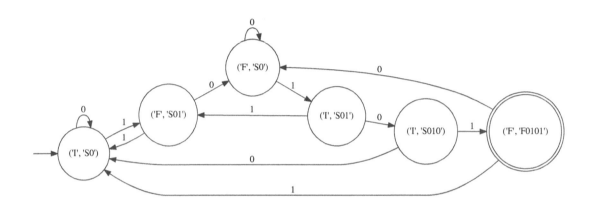

Figure 6.10: Minimized Intersection for "odd 1s" *and* "ends in 0101"

6.5.2 Use of DeMorgan's Law

Next, suppose we are interested in obtaining the intersection of L_1 and L_2. We first obtain it directly, and then later obtain it also using DeMorgan's law, as shown below.:

- Obtain the intersection, *i.e.*, $L_1 \cap L_2$, directly using
  `odd1sANDends0101 = intersect_dfa(dfaOdd1s,ends0101)`
- Minimize the intersection using
  `Minodd1sANDends0101 = min_dfa(odd1sANDends0101)`
- Complement the first DFA, *i.e.* obtain $\overline{L_1}$, via
  `CdfaOdd1s = comp_dfa(dfaOdd1s)`
- Complement the second DFA, *i.e.* obtain $\overline{L_2}$, via
  `Cends0101 = comp_dfa(ends0101)`
- Obtain the complement of their union, *i.e.*, obtain

$$\overline{(\overline{L_1} \cup \overline{L_2})}$$

using

```
C_CdfaOdd1sORCends0101 =
comp_dfa(union_dfa(CdfaOdd1s, Cends0101))
```

- Minimize it using

```
MinC_CdfaOdd1sORCends0101 = min_dfa(C_CdfaOdd1sORCends0101)
```

- Check for isomorphism using

`iso_dfa(MinC_CdfaOdd1sORCends0101, Minodd1sANDends0101)` and the answer is True.[10]

This demonstrates that DeMorgan's law indeed works for DFA, and this can be checked through isomorphism because *minimal DFA for a given regular language are isomorphic*. It should be quite apparent that without these tools, it would be quite tedious and error-prone to obtain a DFA such as in Figure 6.10—the intersection of the DFAs in Figure 6.8 and 6.9. We will have more occasions to present cases where the complement of a language is much easier (and more reliable) to specify than the desired language itself.

[10] One can obtain the drawings of these DFA, place them one on top of the other, and "hold them to light." The state names and the transition names must then match up.

| Exercise 6.5.2, DFA, DeMorgan's Law |

1. Argue that if `langeq_dfa(D1,D2)` holds but `iso_dfa(D1,D2)` does not hold, then the bijection mentioned under the Myhill-Nerode theorem does not exist.

2. Attempt to directly design a DFA that accepts exactly the strings that contain an odd number of 1s and end in 0101. Proceed by trying to write a markdown description directly or draw the DFA on paper and convert it to a markdown.

3. Step through Figure 6.10 (call it DFA D) and make sure that the language is indeed the intersection of these two languages. Write down three positive examples handled by D and three negative examples avoided by D. Now, using our tools, check for the existence of a negative string in D with respect to "ends in 0101" as follows:
   - Obtain the complement of "ends in 0101".
   - Intersect with D and make sure that the intersection is empty.
   - As extra practice, minimize this intersection; what must it emerge as (a specific kind of DFA; describe that in a sentence)?

4. Design a DFA for recognizing the language of all strings over $\Sigma = \{a,b\}$ that contain an odd number of a's (call it D_{oa}). Next obtain D_{eb}, a DFA that recognizes strings with an even number of b's. From D_{oa} and D_{eb}, show how to obtain a DFA for $D_{ea} \cup D_{ob}$ using DeMorgan's Law. Show all the steps using Jove.

5. Minimize the DFA "blimp" described in the markdown below, by hand. Next, enter it into Jove and obtain the minimal DFA. Confirm that these minimal DFA are isomorphic (if not, locate and correct errors). □

```
blimp = md2mc('''
DFA
I1 : a -> F2
I1 : b -> F3
F2 : a -> S8
F2 : b -> S5
F3 : a -> S7
F3 : b -> S4
S4 : a | b -> F6
S5 : a | b -> F6
F6 : a | b -> F6
S7 : a | b -> F6
S8 : a -> F6
S8 : b -> F9
F9 : a -> F9
F9 : b -> F6
''')
```

7

Nondeterministic Finite Automata

> **Chapter Gist:** *We begin by introducing the key concept of nondeterminism through nondeterministic finite automata (NFA, §7.1). NFA can be exponentially more succinct than DFA. One can also view NFA as modeling parallel search in which each forked behavior pursues one search option. We formally define NFA (§7.2) and present how the language of an NFA can be intuitively understood (§7.3). This intuition is made crisp by the idea of Eclosure (§7.4). Subset construction, the centrally important algorithm to convert an NFA to a DFA then follows (§7.5). A clever DFA minimization algorithm due to Brzozowski is to simply reverse the given DFA, determinize it, then reverse it, and determinize it again (§7.6)! A complete illustration of this "unbelievable algorithm" follows (§7.7).*

7.1 Overview of NFA

Figure 7.1 is an NFA. Two things stand out: (1) On an input of '1', control can move from state I back to itself or to state S0. (2) There are states without any moves out of them, such as F. In fact, with NFA, *there is no requirement that we equip each state with one move per symbol in its alphabet.* Here is how to read this diagram, and "execute" its steps.

- When the NFA is "started", we place one **token** at state I, indicating where the control-flow is.
- When a '0' input is supplied, the token goes back to I.
- When a '1' is supplied, the token *splits* (also referred to as "forks" or "clones") into two, with one copy ending up at I and the other at S0. Now, one token continues execution from I while the other executes from S0.
- The ability to fork models the act of *guessing*:
  - The token that stays back at I on input '1' is saying *hmm, this '1' isn't the third-last character in the string.*

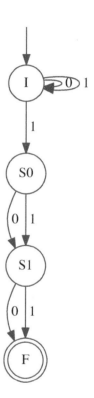

Figure 7.1: An NFA to recognize "third-last is 1". **A regular expression (RE) describing these NFA is** (0+1)*1(0+1)(0+1), as we shall see in Chapter 8. If you revisit this chapter after reading Chapter 8, you will gain added insights.

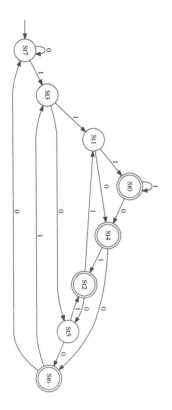

Figure 7.2: DFA for the third-last bit being a 1

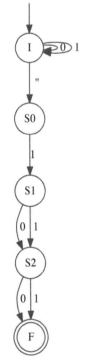

Figure 7.3: A variant of the NFA of Figure 7.1 with an Epsilon (ε). Note: We use " to denote ε in our PDF diagrams and our Python code.

– The token that goes to state S0 says *yes, this is indeed the "lucky 1": the third-last character in the string.*

• When a token reaches state F, that forked copy *accepts*—and hence the whole NFA is deemed to have accepted the input so far.

– Thus, if a '111' is typed, you will find one token at state F, another at state S1, yet another at S0, and finally one token at I. This "laggard" of a token at I is the most pessimistic of tokens, saying *no, I've not yet seen the third-last 1.*

– When a '1111' is typed, the token sitting at F has nowhere to go; it "falls out of the diagram." But the token at S1 now lands at F, thus accepting the string so far (there is a third-last 1 now).

• Thus, if *any token* reaches a final state, the NFA *accepts the input that has propelled that token so far.*

Reasons for why NFA are Succinct: *One can notice that the pattern recognized by this NFA quite clearly stands out;* in fact, its "tail" leading to F focuses on the "third last is 1" pattern of interest. Everything else (the self-loop at "I") is *waiting for this pattern to eventually appear.* In contrast, the DFA that recognizes this language in Figure 7.2 (also shown in Figure 5.4) seems to "go all over the place."

If we generalize our "third-last is a 1" language to "Nth last is a 1," any DFA implementing it will grow exponentially with N (refer to Exercise Exercise 5.2.4 on Page 62, where N was defined to be the look-back). However, the NFA for this language will be *linearly sized* as a function of the look-back—essentially placing the point at which the "magical 1-transition" is taken N steps before the final state F. This example illustrates that NFA are in general much more compact than DFA.

We will demonstrate in this chapter that every NFA can be converted to a language-equivalent DFA. Moreover, every DFA is also an NFA. Thus, NFA do not end up specifying anything other than regular languages, showing that "they are equally powerful." With NFA, *users are the winners*: they have less to write, and still end up implying a much bigger DFA that they did not have to specify laboriously. Since DFA are a special case of NFA, it is also clear that an NFA need not be larger than the equivalent minimal DFA.

Is the use of ε in an NFA essential? To discuss this point, consider Figure 7.3 which is also an NFA. Its first transition is labeled with ' ' which is how we type ε in ASCII syntax. Labeling a transition with ε allows the NFA to take the transition **without consuming an input**. Note that *an NFA's alphabet does not include ε*, as detailed in §7.2.

The question of whether or not to employ an ε in an NFA is up to the "programmer" (designer) of an NFA. The use of ε is strictly not necessary, but often makes the design simpler and/or clearer. In our current example, we can cleanly partition the two cases, *without altering the remaining inputs*:

- Stay back at I.
- Commit to the case of seeing a '1' in state S0.

7.2 Formal Description of NFA

Let Σ_ε stand for $(\Sigma \cup \{\varepsilon\})$. An NFA N is a structure $(Q, \Sigma, \delta, Q_0, F)$, where:

- Q is a *finite non-empty* set of states (as with DFA);
- Σ is a *finite non-empty* alphabet (as with DFA);
- $\delta : Q \times \Sigma_\varepsilon \to \mathscr{P}(Q)$, is a transition function. An NFA's δ function takes a state in Q and a symbol or ε and returns a *set of states* (which is a member of $\mathscr{P}(Q)$, the *Powerset* of Q). See Figure 7.4 for the state transition table 'δ' for the example NFA.
- $Q_0 \subseteq Q$ is a *set of initial states*; and
- $F \subseteq Q$, is a *finite, possibly empty* set of final states.

Let us briefly contrast NFA and DFA:

- NFA allow ε transitions from any state to a *set* of next states.[1]
- NFA allow one character $c \in \Sigma$ to transition to a *set* of next states.
- The set of next states targeted can have a cardinality of one or more.
- An NFA transition leading to an empty set of states is never explicitly drawn out. For instance, a '0' move from state S0 in Figure 7.3 is not drawn. During NFA to DFA conversion, such "missing" moves are automatically considered to be landing in an empty set of states.

Notice that NFA can start from a *set* of states and not just a single state.[2] Figure 7.4 provides the transition function δ for our example NFA of Figure 7.3, as well as our Python encoding thereof. As with DFA, we allow a simple markdown language in Jove that allows NFA to be specified much more conveniently. Here is all you have to type into Jove to produce this NFA's transition function given in Figure 7.4:

```
thirdlast1_b = md2mc('''
NFA
!!-----------------------------------------------------------
!! This NFA looks for patterns of the form ....1..
!!-----------------------------------------------------------
I : 0 | 1 -> I  !! On input 0 or 1, stay at I
I : ''     -> S0 !! Commit to state S0 nondeterministically
S0: 1      -> S1 !! This is speculated to be the third-last 1
S1: 0 | 1 -> S2 !! The second-last character could be 0 or 1
S2: 0 | 1 -> F  !! The last character could be 0 or 1
''')
dotObj_nfa(thirdlast1_b) # Draws the NFA thirdlast1_b
```

[1] Note that ε self-loops are useless.

[2] A majority of the books in this area model NFA with a single initial state q_0, while the book "Automata and Computability" by Dexter Kozen requires NFA to have a *set* of initial states Q_0. We follow Kozen's convention for one crucial reason: **it ensures that the DFA minimization algorithm by Brzozowski yields minimal** DFA (discussed later in this chapter). Most of our examples may have a singleton set of initial states, however.

State	Next state upon inputs		
	0	1	ε
I	$\{I\}$	$\{I\}$	$\{S0\}$
S0	$\{\}$	$\{S1\}$	$\{\}$
S1	$\{S2\}$	$\{S2\}$	$\{\}$
S2	$\{F\}$	$\{F\}$	$\{\}$
F	$\{\}$	$\{\}$	$\{\}$

```
{'Q'    : {'F', 'I',
          'S0','S1','S2'},
 'Sigma': {'0', '1'},
 'Delta':
 {('I', '0')  : {'I'},
  ('I', '1')  : {'I'},
  ('I', '')   : {'S0'},
  ('S0', '1') : {'S1'},
  ('S1', '0') : {'S2'},
  ('S1', '1') : {'S2'},
  ('S2', '0') : {'F'},
  ('S2', '1') : {'F'}},
  'q0': {'I'},
  'F' : {'F'}}
```

Figure 7.4: The transition table δ for the NFA of Figure 7.3. Our Python encoding for NFA is also shown in this figure.

7.3   Language of an NFA: Example Driven

We first describe informally how the NFA in Figure 7.5 (reproduced from Figure 7.1 to avoid page-flipping) works on test input 100 and then 1000. We then highlight the essential differences with how the NFA of Figure 7.3 works.[3]

[3] See supplementary material at https://bit.ly/Automata_Jove under NonDet-InCS for additional intuitions about NFA, including how it helps during formal verification through model checking.

7.3.1   Simulations of the NFA of Figure 7.5

Imagine there is a *token* of control flow at state I. Let us feed our NFA the input '100'. Let us also keep track of where all the tokens are after each input symbol arrives.

The first 1 sends one copy of the token back to I while another copy is sent to S0. Thus, the tokens are now present in the set of states {I,S0}. The next input symbol of 0 moves the token in state I back to the set of states {I} while it moves the token at S0 along to the set of states {S1}. Thus, the tokens are now at the set of states {I,S1}. The next 0 results in the tokens being at {I,F}. Given that the entire string has been consumed *and* one token has reached F (the intersection of {I,F} with the final set of states is non-empty), we conclude that the NFA has accepted '100'. The sequence of transitions between *sets of states* is the following:

$$\{I\} \xrightarrow{1} \{I,S0\} \xrightarrow{0} \{I,S1\} \xrightarrow{0} \{I,F\}$$

When we simulate this NFA on input '1000', an additional 0 comes in to confront the state {I,F}. In this situation, the token at F "falls off the diagram" (*i.e.*, the transition takes us to $\emptyset$, the empty set of states), while the token at I transitions to {I}. Thus, the machine ends up in state {I}$\cup\emptyset$, which is {I}. Since the intersection of {I} with the set of final states of the NFA is empty, (*i.e.*, "no tokens have reached a final state"), the input 1000 is rejected. The full sequence of transitions between *sets of states* is shown below:

$$\{I\} \xrightarrow{1} \{I,S0\} \xrightarrow{0} \{I,S1\} \xrightarrow{0} \{I,F\} \xrightarrow{0} \{I\}$$

Let us look at one more example: an input of '111':

$$\{I\} \xrightarrow{1} \{I,S0\} \xrightarrow{1} \{I,S0,S1\} \xrightarrow{1} \{I,S0,S1,F\}$$

The set of states reached after consuming this input is {I,S0,S1,F}. Observation: NFA Consider Multiple Scenarios in Parallel: From §7.3.1, it must be clear that NFA states are designed to be members of the powerset of Q. When an NFA receives an input symbol, it "hedges its bet" on a number of states, hoping that one of them will be able to accept the remainder of the (unseen) input. Once one understands this process of designing an NFA, one can indeed arrive at compact solutions to many

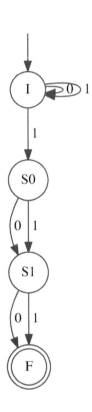

Figure 7.5: NFA reproduced from Figure 7.1

a language design challenge. This approach also tells us how to convert NFA to DFA: we essentially design a DFA whose states are members of the powerset of NFA states.

7.3.2 Simulations of the NFA of Figure 7.3

The main feature of this NFA (reproduced in Figure 7.6) is the ε move. This is a "silent" move that can be taken any time. Thus, as soon as we put a token at I, a copy "oozes" into state S0. This activity of oozing is called *Eclosure* or ε-closure.[4] Thus, the sequence of states that this NFA will traverse includes an Eclosure step at every point. The simulation sequence for input '100' is the following:

$$\{I,S0\} \xrightarrow{1} \{I,S0,S1\} \xrightarrow{0} \{I,S0,S2\} \xrightarrow{0} \{I,S0,F\}$$

Even though the initial state is merely I, the Eclosure step will put a token at {I,S0}. Now comes the interesting detail:

- The token at {I} will, upon seeing a 1, go to the set of states {I}. It will immediately Eclose itself into {I,S0}.
- The token at {S0} will, upon processing a 1, go to the set of states {S1}. Here, the Eclosure step doesn't add any states because there aren't any ' ' (ε) edges going out of S1.
- Thus, we take the union of {I,S0} and {S1} to obtain the set of states {I,S0,S1} in our simulation.

In a nutshell, here is how the simulation proceeds.

- Begin the NFA's simulation at a set of states S obtained by taking the Eclosure of the initial state of the NFA.
- At each step of the simulation, we take the current set of states S that the NFA is situated in. The simulation seeks to discover the set of states the NFA will be in after processing a symbol $x \in \Sigma$. For this, we consider each state $s \in S$ and advance s with respect to x to obtain a set of states s_x. (Observe that s_x will equal $\delta(s,x)$.) We take the Eclosure of s_x, denoted by $Eclosure(s_x)$. We do this for each state $s \in S$ for input x, and take a set union of the $Eclosure(s_x)$ states. The resulting set of states is where the NFA will be in after processing input x at its beginning set of states S.

7.4 Language of an NFA: via Eclosure

In §7.3.1 and §7.3.2, we provided many scenarios of NFA accepting as well as rejecting specific inputs. We illustrated why strings '100' and '111' were accepted, while '1000' was rejected. Do these scenarios fully define the notion of the language of an NFA? *Of course they don't*—these are just a few of the many possible inputs!

Do you recall how, in §4.5, we defined the notion of a DFA accepting a

[4] Imagine high voltage being applied to state I, with the ε-labeled (") edges serving as diodes that conduct one way. Then Eclosure puts tokens at all states that one can touch and get electrocuted. The high voltage leaks from I to S0. But a token that is merely at S0 does not leak back into I (the diode-like behavior we alluded to). Now if you have multiple ε edges that form a path, then the high voltage can traverse *as far as possible along these paths*.

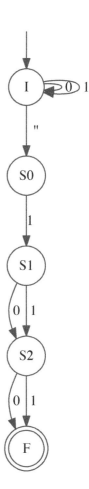

Figure 7.6: NFA reproduced from Figure 7.3

string x? We seek a similar rigorous (mathematical) definition for NFA. Unfortunately, the corresponding definition for an NFA is not going to be that straightforward, and you may wish to skip over to the next section during your first read; but *do come back and read these definitions*.[5]

7.4.1 Defining Eclosure

Let us first formally define **Eclosure** as a function that, when applied to a state q, returns the set of states that can be reached from q by traversing only ε edges.[6] Then,

- Let us introduce some notation, suggesting transitive closure along ε edges. Our notation suggests treating $\overset{\varepsilon}{\to}$ as a post-fix operator. Since we are applying $*$ to this operator, it suggests "zero or more applications."

$$Eclosure(q) = q \overset{\varepsilon}{\to}{}^{*}.$$

- Now, let us define $Eclosure(Q)$ for a set of states Q by taking the set unions of $Eclosure(q)$ for every $q \in Q$:

$$Eclosure(Q) = \cup_{q \in Q} Eclosure(q).$$

7.4.2 Definition of δ and $\hat{\delta}$

Now we embark on defining two functions: (1) the one-step next state function δ that moves a state (or a set of states) upon seeing a single character $a \in \Sigma$, and (2) the multi-step next state function $\hat{\delta}$ that moves a state (or a set of states) upon seeing a string $x \in \Sigma^*$.

- We make δ and $\hat{\delta}$ work over a *set* of states Q by basically applying them at every $q \in Q$ and taking the union of the results:

  $\delta(Q,a) = \cup_{q \in Q} \delta(q,a)$

  $\hat{\delta}(Q,x) = \cup_{q \in Q} \hat{\delta}(q,x)$

- Now we define $\hat{\delta}(q,x)$, the **string transfer function** for state q and string x inductively as follows:

  $\hat{\delta}(q,\varepsilon) = Eclosure(q).$

  For $a \in \Sigma$ and $x \in \Sigma^*$,

  $\hat{\delta}(q,ax) = \cup_{q_1 \in Q1} \hat{\delta}(q_1,x)$, where we have

  $Q1 = Eclosure(\delta(Eclosure(q),a))$

- That is, Eclose q, then apply δ and $Eclose$ again.
- The **language of an NFA** N is

  $\mathscr{L}(N) = \{w : \hat{\delta}(q_0,w) \cap F \neq \varnothing\}.$

  This means the following:

  – Discover all strings w such that $\hat{\delta}(q_0,w)$ has a non-empty intersection with F.

– All such strings are **accepted** by the given NFA.

7.5 NFA to DFA Conversion through Subset Construction

The NFA to DFA conversion algorithm follows pretty much all the steps in defining the language of an NFA given in §7.4. We now demonstrate the specific steps with the help of the NFA in Figure 7.7 (top):

- Obtain the Eclosure of the initial state I, resulting in the set of states {I,A,B,C,G}.
- Fire '0' from each of the aforesaid states, and Eclose the resulting sets of states to obtain {E,A,B,C,G} as follows:
  - States I, A, B, and G yield the empty set ($\emptyset$) of states;
  - State C yields the set of states {E}. We then perform Eclosure on {E} obtain {E,A,B,C,G}. Now, taking the set union with $\emptyset$ obtained in the earlier step still leaves us with {E,A,B,C,G}.
  - State {E,A,B,C,G} is a new (as yet unexpanded) state of the DFA.
- Repeat, by firing '1' from {I,A,B,C,G} to obtain {D,A,B,C,G,F}. Again this is a new state.
- Since these are new states, repeat the '0' and '1' firings from {E,A,B,C,G} and {D,A,B,C,G,F}. They yield only states previously generated, but introduce new transitions. This results in the DFA of Figure 7.7 (bottom).

Algorithm for Subset Construction:

- Input: An NFA $N = (Q, \Sigma, \delta, Q_0, F)$
- Output: A language-equivalent DFA D
- Method: **Subset Construction**
  - Add the Eclosure of the initial state of the NFA as an unexpanded state of the DFA D being built. This would also be the **initial state of the DFA being built.**
    Repeat
    Choose a state S of D that has not been expanded
    Expand(S)
    Until there are no more unexpanded states in D
  - **Expand(S):**
    Mark S as expanded;
    If $S \cap F \neq \emptyset$, **record S to be a final state of the DFA**
    For each symbol c in Σ
      For each state $s \in S$ do
        Let $s_c = \delta(s,c)$;
        Let $S_c = Eclosure((\cup_{s \in S} s_c))$;
        If S_c isn't present in D, add it as an unexpanded state;
        Add $S \xrightarrow{c} S_c$ to D's transition

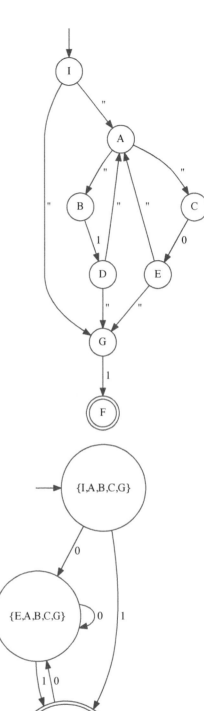

Figure 7.7: (Top) An NFA for "ends in 1"; (Bottom) DFA obtained through subset construction.

The states of the resulting DFA are subsets of the given NFA's states. While it is possible to generate every such subset – *i.e.*, the DFA states could be the powerset of the NFA states – in practice, this rarely happens. It is, however, guaranteed to happen for the NFA of Figures 7.1 and 7.3. Thus, these NFA, when converted to a DFA, end up yielding an exponential number of states with respect to the given NFA states.

> **Theorem 7.5:** A language L is regular if and only if L has an NFA recognizing it.

Proof, (Only if): A language is regular if there is a DFA describing it. Now every DFA is also an NFA.

Proof, (If): If a language is described by an NFA, we can obtain a DFA that is language-equivalent through the subset construction algorithm. Hence the language is also regular.[7]

[7] The fact that subset construction yields a language-equivalent DFA of course merits a proof. This can be argued at a high level by tracing the accepting paths in the DFA generated out of an NFA and relating it to corresponding paths in the NFA.

Exercise 7.5, NFA to DFA

1. Design an NFA for the language of all strings over alphabet $\{0, 1\}$ that contain a "0101" somewhere. Use Jove's markdown syntax to capture your design. Next,

   (a) Simulate this NFA using the `accepts_nfa` function on the first 20 strings in numeric order, showing that it works.

   (b) Next, apply subset conversion by hand, and convert this NFA to an equivalent DFA. Capture this DFA using Jove's markdown.

   (c) Now, invoke the `nfa2dfa` function that automatically converts your NFA to a DFA.

   (d) Show that these two DFA are language-equivalent using the `langeq_dfa` function.

2. Consider the "blocks of 3" language described in §5.2.1. Suppose you want to arrive at an NFA for the complement of this language, which means you have to negate this sentence.

   > For each block of words B
   > IF B has 3 symbols
   > THEN B has exactly two 1s.

   This negation is

   > There exists a block of words B such that
   > B has 3 symbols
   > AND B violates the condition "has exactly two 1s".

   (a) Argue that the above negation is correct.

   (b) Describe the NFA for this negated language using Jove's markdown. Convert this NFA to a DFA, and finally complement it and minimize it, thus producing a DFA for the "blocks of 3" language.

   (c) For testing your constructions, design an NFA for the set of strings that must *not* be in this DFA's language, and then show (using Jove) that your DFA does not contain any of these strings.

3. Suppose we are given a DFA $D = (Q, \Sigma, \delta, Q_0, F)$, and we are asked to design an NFA N such that it recognizes the middle-third of every string recognized by D. More formally, we want the language of the NFA N, $\mathscr{L}(N)$, to be

$$\mathscr{L}(N) = \{y \, : \, \exists x, y, z \in \Sigma^* : len(x) = len(y) = len(z) \wedge xyz \in \mathscr{L}(D)\}$$

Of course, for every string in $\mathscr{L}(D)$ whose length is not divisible by 3, there will be no corresponding "middle-third" string contributed to N.

Build the NFA using the mk_nfa call provided within Def_NFA, and then produce the minimized DFA using Jove's suite of functions.

Hint:

The idea is to imagine keeping a nondeterministically chosen milepost a third of the way and another nondeterministically chosen milepost two-thirds of the way. Model the NFA state as Q^5 (the Q being the set of states of the DFA), *i.e.*, $Q \times Q \times Q \times Q \times Q$. Let the initial state be $(q_0, q_1, q_1, q_2, q_2)$. It is as if we placed one mile-post token at some state q_1, another at q_2, and let a token from q_0 seek the one at q_1, let another token at q_1 seek q_2, and a final token at q_2 seek q_f, a state in F. We move three tokens with "equal velocities," and if/when they manage to hit a "magical configuration," we have found our middle-third string.

More specifically, the idea is then to consider the following types of state transitions (notice that the second and fourth components of the state tuple stay put, serving as fixed mileposts):

$(q_0, q_1, q_1, q_2, q_2) \rightarrow$

$(q_{01}, q_1, q_{12}, q_2, q_{21}) \rightarrow$

$(q_{02}, q_1, q_{12}, q_2, q_{22}) \rightarrow$

$(q_{03}, q_1, q_{13}, q_2, q_{23}) \rightarrow$

$\cdots$

$(q_1, q_1, q_2, q_2, q_f)$

In this scenario, the last state is our "magical configuration."

Putting these ideas to work, we write down the following components of our NFA:

$Q^N = Q^5$

$\Sigma^N = \Sigma$

$Q_0^N = \{(q_0, q_1, q_1, q_2, q_2) \, : \, q_1, q_2 \in Q\}$ (*i.e.*, start the NFA from the set of these initial states).

$F^N = \{(q_1, q_1, q_2, q_2, q_f) \, : \, q_f \in F\}$.

$$\delta^N((q_a, q_1, q_b, q_2, q_c), y) =$$
$$\{(\delta(q_a, c_1), q_1, \delta(q_b, y), q_2, \delta(q_c, c_2)) : c_1, c_2 \in \Sigma\}$$

One can see how only the middle token "listens to y"; the other two tokens chaotically wander. They seek to find the "magical configuration" in all possible ways. *Try animating this on a piece of paper using colored beads in place of tokens.*   □

7.6   Brzozowski's DFA Minimization Algorithm

In 1962, Janusz Brzozowski came up with an elegant DFA minimization algorithm [13]. Its Jove code will take your breath away:

```
def min_dfa_brz(D):
    """Minimize a DFA as per Brzozowski's algorithm.
    """

    return nfa2dfa(rev_dfa(nfa2dfa(rev_dfa(D))))
```

That is it! To minimize a DFA:

- Reverse the DFA to obtain an NFA. Reversing a DFA captures the reversal of the language of the original DFA. To reverse the DFA:
  - Take the final states of the original DFA as the initial *set of states* of an NFA,
  - Take the initial state of the original DFA as the only final state of the NFA, and
  - Reverse all the transitions.
- Then determinize the resulting NFA to obtain a DFA. Here, *determinize* means perform the subset construction (nfa2dfa). Brzozowski proved that at this point, we have a minimal DFA for the reverse of the original DFA. *But we want a minimal DFA for the given DFA.* What do we do? Simple! Do it again!
- So, reverse the DFA just obtained (min DFA of the reverse) and determinize it again!

7.6.1   Reversal of a DFA Yields an NFA

Reversal of a DFA also has an elegant piece of Jove code:

```
def rev_dfa(D):
    """In : D = a partially consistent DFA w/o unreachable states.
       Out: A consistent NFA whose lang. is D's lang. reversed.
    """
    NDict = { (q,ch) : inSets(D,q,ch)
                for q in D["Q"]
                for ch in D["Sigma"] }
    return mk_nfa(D["Q"], D["Sigma"], NDict, D["F"], {D["q0"]})


def inSets(D,trg,ch):
    """Return all those states from which there are moves
       to a given target state trg through character ch.
       ---
       In : D   = partially consistent dfa,
            trg = a target state in D["q"]
            ch  = a member of D["Sigma"]
       Out: a set of states. { q s.t. Delta[q,ch] == trg }
    """
    return { q for q in D["Q"] if D["Delta"][(q,ch)] == trg }
```

That is it! Here is how the code works:

- We form a dictionary called NDict that forms the NFA's transition function directly!
- The NFA transition function takes a (q,ch) and jumps to all the states that *have sent a transition into* q. This is computed by inSets as follows
  - We view each state of the DFA as a "target" (trg), and determine the set (call it Shooters) of all those DFA states q from which "arrows labeled by character ch are being shot" to hit trg. When we form an NFA by reversing the DFA, arrows turn around! It is now the *target* that now emits arrows labeled by ch hitting the set of Shooters.

Exercise 7.6.1, Brzozowski's DFA minimization

1. Apply Brzozowski's minimization algorithm by hand to the DFA of Figure 6.6.

2. Can you argue that just carrying out the reversal followed by determinization obtains the minimal DFA of the original DFA's reverse? (It is worth studying Brzozowski's proof here.) Just get the general idea of this proof, if not the whole proof.

3. Using Jove, run the two minimization algorithms on "blimp" (the topmost DFA in Figure 7.8), showing that the minimal DFA obtained are isomorphic.

□

7.7 A Complete Illustration of Brzozowski's Minimization

Let's start with "blimp," a bloated DFA introduced in Exercise 5 on Page 80. The full minimization sequence using Brzozowski's method is illustrated in Figure 7.8 and explained in its caption.

Figure 7.8: (1) We start with a DFA called "blimp," the first (topmost) diagram here. The steps are as follows: (2) We first reverse blimp and obtain the second diagram called rblimp. Notice that the final state of rblimp is I1 (blimp's initial state). Also the set of initial states of rblimp is {F2,F3,F6,F9} which are the final states of blimp. (3) We then determinize rblimp to obtain a DFA which we call drblimp (the third diagram from the top). We can already see that the diagram has gotten quite minimized. (4) Next, we reverse dr-blimp to obtain rdrblimp which is the NFA (fourth from top). Again we see a reversal of flows, and an NFA with multiple initial states. (5) We finally determinize NFA rdrblimp to obtain the final DFA. Notice that this language is its own reverse (the third and fifth diagram are isomorphic). This is often not the case.

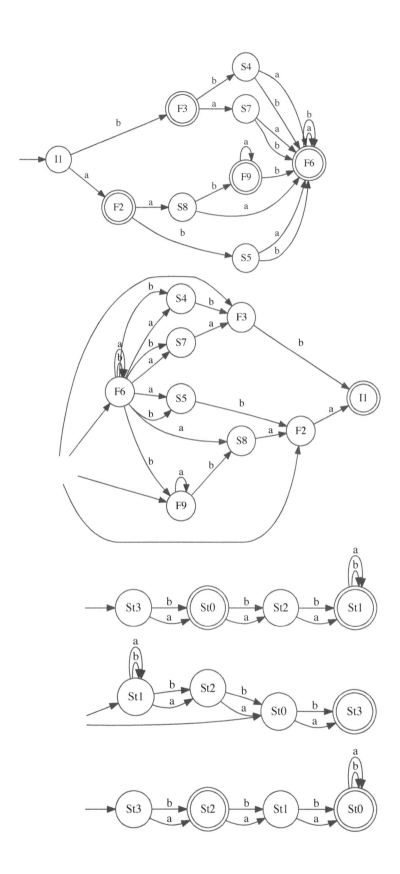

8

Regular Expressions and NFA

Chapter Gist: *In a sense, regular expressions provide the expression syntax for typing in NFA into a computer (§8.1). We present RE to NFA conversion through the interactive use of the* re2nfa *function (§8.2) and then solve a non-trivial language design problem (§8.3). Getting regular expressions wrong can open up security holes; one must employ good checkers (§8.4). We present a mini-compiler that parses REs and emits NFA (§8.5). We design an error-correcting DFA using two approaches and confirm its correctness through the isomorphism check of the generated minimal DFA (§8.6, §8.7). A cool application of regular expressions is to calculate what postage values one can attain using given stamp values (§8.8); a related problem is the McNugget number.*

8.1 Regular Expressions

User syntax	Mathematical Syntax	Language Denoted
"	ε	$\{\varepsilon\}$
1	1	$\{1\}$
a	a	$\{a\}$
aa	aa	$\{a\}\{a\} = \{aa\}$
a+b	$a+b$	$\{a\} \cup \{b\} = \{a,b\}$
(a+b)(a+c)	$(a+b)(a+c)$	$\{a,b\}\{a,c\} = \{aa,ac,ba,bc\}$
(ab)+(ac)	$(ab)+(ac)$	$\{ab\} \cup \{ac\}$
a*	a^*	$\{a\}^*$
nothing	$\emptyset$	$\{\}$

Regular expressions define an *expression* language to denote regular *languages* much like expressions over numbers such as 0, (3+4)*5, or (x+4)*y denote numbers. Regular expressions are either primitive ones (such as a string of length 1) or simple regular expressions that are glued together through union, concatenation, and star. In the table above, we

[1] We do not include other regular expression operators, as all regular languages are definable using just these three operators. Also there is one more primitive regular expression, namely $\emptyset$, that denotes the empty language. We do not provide a user-level syntax for $\emptyset$ till Chapter 10, where it is needed to specify derivative-based pattern matching (till then, we do not need this primitive regular expression).

present examples of the user-level syntax (for Jove users), mathematical syntax, and the regular language denoted. (Note: We suppress quotes around strings of length 1.)

We will now detail some of the entries of this table. Entry ' ' denotes the unit language $\{\varepsilon\}$. A single character a denotes the language $\{a\}$. Concatentation, plus (+), and star (*) are used to denote corresponding language operations. That is it![1]

The General Syntax for Regular Expressions (RE): REs can be defined over an alphabet Σ as follows:

1. ε is a RE denoting the regular language $\{\varepsilon\}$;

2. $a \in \Sigma$ is a RE denoting the regular language $\{a\}$;

3. if r is a RE, so is r^* as well as (r); the former denotes the regular language $(\mathscr{L}(r))^*$ and the latter[2] denotes $\mathscr{L}(r)$, the language of r;

[2] The reason why we introduce (r) as a separate category is to group regular expressions for readability, and also to prevent $*$ from binding too tightly. For example, $(a+b)*$ and $a+b*$ are not the same, as in the latter, $*$ binds to b, whereas in the former, $*$ applies to the whole RE.

4. if r_1 and r_2 are REs, so are $r_1 + r_2$, and $r_1 r_2$. These expressions denote $(\mathscr{L}(r_1)) \cup (\mathscr{L}(r_2))$ and $(\mathscr{L}(r_1))(\mathscr{L}(r_2))$ respectively.[3]

The core of this chapter deals with algorithms that convert regular expressions to NFA. We will also touch upon how regular expressions play a pivotal role in building compiler *scanners* (otherwise known as *lexical analyzers*). §8.4 covers some of the dangers of employing regular expressions without due precautions.

[3] Please note that concatenation $(r_1 r_2)$ has higher precedence than summation $(r_1 + r_2)$. Thus, you must read regular expression $ab + c$ as $(ab) + c$. Also, concatenation and summation are associative; thus, $abc = (ab)c = a(bc)$, and likewise, $a + b + c = (a + b) + c = a + (b + c)$.

8.2 RE to NFA Conversion: Examples, Algorithm Sketch

We now list each of the REs handled by Jove and the corresponding NFA generated. A sketch of the underlying conversion algorithm is then presented.

- The NFA for RE ε (' ') is obtained by typing:
  `dotObj_nfa(re2nfa(""))`, resulting in the following NFA that has language $\{\varepsilon\}$ (the unit language):

- The NFA for RE a is obtained by typing:
  `dotObj_nfa(re2nfa("a"))`, resulting in an NFA that has language $\{a\}$:

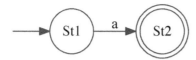

- The NFA for RE *ab* is obtained by typing:
  `dotObj_nfa(re2nfa("ab"))`, resulting in an NFA that has language {*ab*} (built by concatenating an NFA for *a* with an NFA for *b*.

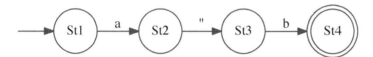

- The NFA for $(a + b)$ is obtained by typing:
  `dotObj_nfa(re2nfa("a+b"))`, resulting in an NFA that has language {*a*,*b*} (built by taking the union of an NFA for *a* with an NFA for *b*. Given that our NFA can start from a set of initial states Q_0, we can "leave the input states unconnected" as in the figure that follows.

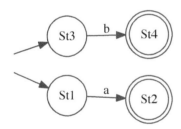

- In general, the algorithm for concatenation is: (1) put an epsilon-edge from every final state of the first NFA to every initial state of the second NFA, (2) make the final states of the first NFA non-final. This is illustrated by RE $(a + b)(a + b)$ whose NFA is shown below:

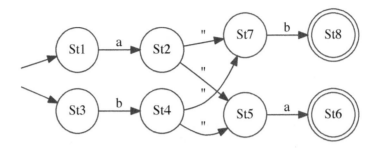

- The NFA for $a*$ is obtained by typing:
  dotObj_nfa(re2nfa("a*")), resulting in an NFA that has language $\{a\}^*$. This is obtained by taking an NFA for a and modifying it as follows: (1) introduce the NFA for a (between states St1 and St2 here), but make the final states of this NFA non-final; (2) introduce a new initial+final state (St3 here); (3) transition from the original final states of the NFA to this new initial+final state (St3); (4) transition from the new initial+final state (St3) to all the initial states of the NFA.

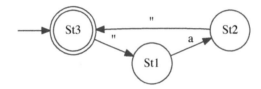

[4] Notice that we turn states St2 and St4 into non-final states. While the semantics of star would not be affected if we had left states St2 and St4 as final, further constructions (especially by hand) could turn erroneous. For instance, if we are asked to perform a subsequent concatenation such as (a+b)*c, we would be required to connect all final states of (a+b)* to the start state of c, *and render the final states of* (a+b)* *non-final.* After this construction, only the final state of the concatenated RE c (turned into its own NFA) would be a final state. However, if we had accidentally left behind St2 and St4 as final states, we would have an incorrect construction that allows acceptance even before seeing a c. So the act of limiting the set of final states to the new initial+final state—in this case St5—is to safeguard against this inadvertent mistake.

- To see how the star construction generalizes, let us obtain an NFA for the RE $(a + b)^*$. This is obtained as follows.
  dotObj_nfa(re2nfa("(a+b)*")), resulting in an NFA that has language $\{a, b\}^*$. We can see that the star procedure described above is followed: the NFA for $(a + b)$ has been inserted into the "star recipe" (involving states St1 through St4).[4]

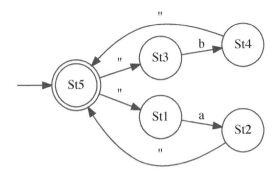

- Finally, to see how three operations—plus, concatenation, and star—come together, let us build an NFA for $(a+b)(a+b)^*$ by typing

  `dotObj_nfa(re2nfa("(a+b)(a+b)*"))`

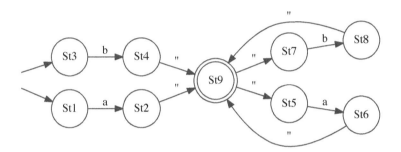

Exercise 8.2, NFA Operations

1. Write down two strings in the symmetric difference of the regular languages denoted by (0* 1 (1 0*)*) and (0* 1 (0* 1)*).

2. Build NFA for the following REs by hand (here, `''` is ϵ). *Do not hand-simplify the RE or the NFA obtained.*

   (a) ("+a)*(b+cd)*. Here is the recommended approach:
      - Build an NFA for ("+a) by building NFA for " and a, and applying the procedure for + (union).
      - Build an NFA for (b+cd).
      - Obtain their stars separately.
      - Apply the concatenation procedure on the resulting NFA.

3. Verify your overall construction by comparing the NFA you obtained against the original RE through Jove's conversion pipeline (convert them both to a minimal DFA and then check for isomorphism). □

8.3   A More Extensive Example

With the syntax of regular expressions at hand, a user can, with some
practice, begin *designing* regular languages systematically. As an example, suppose someone wants to develop a regular expression for the language over {0,1} that has either an odd number of 1's or exactly three 0's.
Here is how to design this RE, with steps indicated:

- *Focus on the salient pattern:*
  - One pattern is "odd 1's"
    * Odd 1's can be generated by adding '11's to a '1'.
    * Thus the basic pattern is `1(11)*`, as it describes a 1 followed by
      zero or more `(11)`s.
    * However, 0's can be present anywhere in this pattern.
    * Thus, this RE can be generalized to `0* 1 0* (1 0* 1 0*)*`
  - The other pattern is "exactly three 0s".
    * This is again easily obtained as `1* 0 1* 0 1* 0 1*`
- *Obtain the final pattern:*
  - Now that we have the constituents worked out, the full RE is
    `0* 1 0* (1 0* 1 0*)* + 1* 0 1* 0 1* 0 1*`
- *Test out using Jove:* Given we have the power of Jove, we can "by Jove"
  obtain an NFA and then a minimized DFA.
- The NFA is obtained through
  `dotObj_nfa(re2nfa("0*10*(10*10*)* + 1*01*01*01*"))`

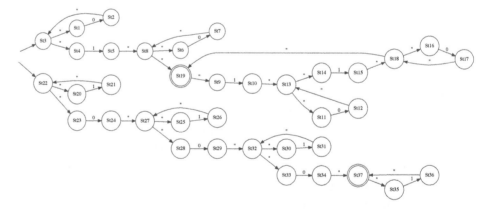

- Clearly, this NFA is meant for consumption by computers, and not
  humans. Yet, one can see RE idioms present in it. To obtain a human-
  readable machine, we turn this NFA into a minimal DFA which is
  clearly readable! One can trace paths leading to final states and study
  this rather fun shape (coming next) that encodes "odd 1's" or "three
  0's."
  `dotObj_dfa(min_dfa(nfa2dfa(re2nfa("0*10*(10*10*)*+1*01*01*01*"))))`

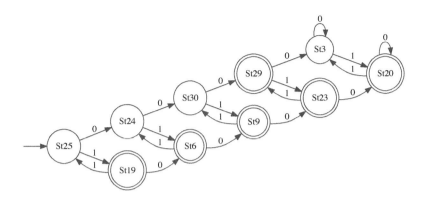

So far, we have described how to convert RE to NFA, and also NFA to DFA. In Chapter 9, we will describe an algorithm to convert NFA to RE. Given all this, we have the following theorem.

Theorem 8.3: The class of languages defined by DFA, NFA, and RE are all regular sets.

8.4   Regular Expressions: Ubiquitous, yet Error-Prone

Regular expressions pack a considerable punch in small syntactic confines. This makes them quite suitable for use in various settings—in parsing user web-forms, rejecting malformed command-line inputs, detecting malware within internet packets, etc. etc. Unfortunately, casually staring at regular expressions, or running a few test cases is not a sufficient check for some of the dangers that innocuous regular expressions might be hiding. Because of their cryptic nature, it can become impossible to fathom all the strings denoted by a regular expression; perhaps there are unintended strings lurking within the language defined by a regular expression. Simple typos can also change the meaning of regular expressions. For instance, a reader may want to puzzle over the exact difference between these two REs:

1. `(0* 1 (1 0*)*)`
2. `(0* 1 (0* 1))*`

Real-world regular expressions, as supported in real programming languages, allow even more rules of composition.

- They allow character-classes to be matched; for instance [a-z] means *any of the characters in the range* a *through* z.
- They allow negation; for instance, [^a-z] means *anything but* a character in the range a-z.
- They even allow fancy concepts such as *negative lookahead*. For instance, to express all the characters in the range A-Z except for S, P,

[5] http://www.pyregex.com/ is one such. I had to devise the Python version of this RE through much trial and error. It is not easy to perform the negation of more complex patterns!

[6] Mentioned at the webpage https://cwe.mitre.org/data/ definitions/185.html.

Q, and R, one has to write, in Python, a construct using negative lookahead:

```
(?![SPQR])[A-Z]
```

- However, in other settings such as Microsoft's .NET, such negations are supported more readily. There are online tools that can help debug Python regular expressions, and their use is highly recommended.[5]

When we require such complex regular expressions to be developed by ordinary users, dangerous bugs can be quite easily introduced. For example, consider a Perl program that employs a carelessly written RE matching test:[6]

```
$phone = GetPhoneNumber();
if ($phone =~ /\d+-\d+/) {
 # looks like it only has hyphens and digits
 system("lookup-phone $phone"); #$ Executes given string; oops !!
}
else { error("malformed number!"); }
```

The website goes on to say:

> An attacker could provide an argument such as: "; /bin/rm * ; echo 123-456". This would pass the check, since "123-456" is sufficient to match the "\d+-\d+" portion of the regular expression.

What really can happen is this: Perl's =~ operator asks if the pattern "some digits followed by a dash followed by more digits" (specified by Perl's regular-expression syntax)[7] appears somewhere in the user input. It does! However, because of the supplied input string, the exact command executed becomes

```
lookup-phone; /bin/rm * ; echo 123-456
```

[7] Here, the "back-slash followed by d" means "one digit" in Perl's RE syntax. Now, following something with a + means "one or more." It is like Kleene-star, except the latter specifies "zero or more."

which executes lookup-phone without any argument. Most likely, this would complain saying "no phone number given." The next piece is /bin/rm * which removes all of a user's files! Finally, it echoes 123-456 which is not terribly interesting to anyone (but was the "decoy" pattern present in the input that caused the pattern-match to succeed.) The website gives good advice on the use of REs:

> *Regular expressions can become error prone when defining a complex language even for those experienced in writing grammars. Determine if several smaller regular expressions simplify one large regular expression. Also, subject the regular expression to thorough testing techniques such as equivalence partitioning, boundary value analysis, and robustness. Even after testing and a reasonable confidence level is achieved, a*

regular expression may not be foolproof. If an exploit is allowed to slip through, then record the exploit and refactor the regular expression.

By focusing on a very simple regular expression syntax, we hope to drive the main conceptual points across much more easily. Practical languages such as Perl, Python, C# and Javascript all have regular expressions that are individually much more complex than the RE syntax we study, and (unfortunately) subtly different from each other. However, by acquiring a firm grounding on this topic through examples in Jove's RE syntax, you may have a leg up when it comes to studying real-world REs.

8.5 Anatomy of the RE to NFA Converter

It is clear that behind the scenes, function `re2nfa` is a miniature compiler. It dissects the inner structure of a given regular expression and first produces NFA for the elementary expressions, and then assembles them using the procedure described thus far. We will now go through salient excerpts from the module `Def_RE2NFA` (which the reader may kindly study and exercise).

- This mini-compiler consists of a *lexer* (the `lex()` call in Figure 8.1) and a *parser* (the `yacc()`). The lexer is automatically generated, and is a DFA that is implemented based on how the user specifies the structure of the so-called *tokens*—meaningful units of information for the parser.

- The tokens input by our parser are listed at the top. As one example, the construct `r'[a-zA-Z0-9]'` specifies that a single string processed as an RE symbol is
  - Any character in the `a-z` range
  - Any character in the `A-Z` range
  - Any digit in the `0-9` range

  In a sense, this is a taste of what real-world regular expressions look.

- The parsing rules are the subject of later chapters. But one can see the following useful facts:
  - We define the parsing rule `p_expression_plus` which is how REs are connected with a PLUS operator. We essentially obtain two little NFA via `t[1]` and `t[3]`, and "glue them together" using function `mk_plus_nfa`.
  - One can see that this function essentially follows the recipe given earlier for the + operator. Specifically,
    * The states are unioned
      ```
      Q = N1["Q"] | N2["Q"]
      ```
    * The alphabets are unioned
      ```
      Sigma = N1["Sigma"] | N2["Sigma"]
      ```
    * The starting states are unioned
      ```
      N1["Q0"] | N2["Q0"]
      ```

* The final state could be that of either NFA (hence unioned)
  `N1["F"] | N2["F"]`
* The transition relation is `delta_accum` which takes all the moves
  from both machines and unions them (the `delta_accum.update()`
  calls).

Figure 8.1: Salient Excerpts of the RE to
NFA Compiler

```
#-- The tokens that constitute an RE are these
tokens = ('EPS','STR','LPAREN','RPAREN','PLUS','STAR')
#-- The token definitions in terms of raw strings
t_PLUS    = r'\+'
t_STAR    = r'\*'
t_LPAREN  = r'\('
t_RPAREN  = r'\)'
t_EPS     = r'\'\'|\"\"'    # Epsilons are fed as '' or ""
t_STR     = r'[a-zA-Z0-9]'


#--- Here are the parsing rules for REs; each returns NFA as "code"
precedence = ( ('left','PLUS'), ('left','STAR'), )


def p_expression_plus(t):
    '''expression : expression PLUS catexp'''
    t[0] = mk_plus_nfa(t[1], t[3]) # Union of two NFAs returned


def mk_plus_nfa(N1, N2):
    """Given two NFAs, return their union."""
    delta_accum = dict({})
    delta_accum.update(N1["Delta"])
    delta_accum.update(N2["Delta"]) # Accumulate the transitions
    return mk_nfa(Q     = N1["Q"] | N2["Q"],
                  Sigma = N1["Sigma"] | N2["Sigma"],
                  Delta = delta_accum,
                  Q0    = N1["Q0"] | N2["Q0"],
                  F     = N1["F"] | N2["F"])
... other parsing rules omitted ...
def re2nfa(s, stno = 0):
    """Given a string s representing an RE and an optional
       state number stno (default 0), generate an NFA that
       is language equivalent to the RE"""
    ResetStNum()      # State name generator counter reset
    relexer = lex()   # Build the lexer
    reparser = yacc() # Build the parser
    myparsednfa = reparser.parse(s, lexer=relexer) # Feed lexer
    return myparsednfa
```

8.6 Example: Designing an Error-Correcting DFA

Given two bit-strings (over $\{0, 1\}$) of equal length, they are at a Hamming distance of k if there are exactly k positions where the strings differ. For example, 0101 and 1111 are at a Hamming distance of 2, as are 0101 and 0110. Two strings are *within* a Hamming distance of k if the Hamming distance between them is at most k. This notion is useful in defining "how corrupted" a string is compared to an "ideal string."

Now, given a regular language L, how can we design a DFA that recognizes all strings that are within a Hamming distance of k? We present two approaches to specify such languages, one using regular expressions, and the other using NFA. Following this, we will convert both descriptions to a minimal DFA and invoke the isomorphism test to see that the two independent thought processes lead to the same minimal DFA—thus greatly diminishing the chances of having made a mistake. For concreteness, our target language L is the following over $\Sigma = \{0, 1\}$, and $k = 2$:

$$L = \{x \mid x \text{ has an occurrence of } 0101 \text{ in it}\}$$

8.6.1 Error-correcting RE for "within Hamming Distance of 2"

Following the approach of RE design, we know that the pattern of interest is 0101 and the pattern before/after does not matter.

> The RE for L is therefore (0+1)*0101(0+1)*

To arrive at an RE for "within a Hamming distance of 2," we consider all possible ways of "dinging" the 0101 pattern: there must be $\binom{4}{2} = 6$ such patterns, where a ? represents a 0 or a 1.

- ??01
- ?1?1
- ?10?
- 0??1
- 0?0?
- 01??

We know how to model ?—simply use (0+1). Thus, the full RE for the "within 2 Hamming distance" is provided over multiple lines for clarity:

```
(0+1)*
( (0+1)(0+1)01
+ (0+1)1(0+1)1
+ (0+1)10(0+1)
+ 0(0+1)(0+1)1
+ 0(0+1)0(0+1)
+ 01(0+1)(0+1)
) (0+1)*
```

The minimal DFA for the above RE is obtained using these Jove commands:

```
h2_0101_re = ("(0+1)* ( (0+1)(0+1)01 +" +
                  " (0+1)1(0+1)1 +" +
                  " (0+1)10(0+1) +" +
                  " 0(0+1)(0+1)1 +" +
                  " 0(0+1)0(0+1) +" +
                  " 01(0+1)(0+1) )" +
              "(0+1)*")

minD_h2_0101_re = min_dfa(nfa2dfa(re2nfa(h2_0101_re)))


dotObj_dfa(minD_h2_0101_re)
```

The minimal DFA is given in Figure 8.2.

Figure 8.2: Minimal DFA generated through the RE-based approach

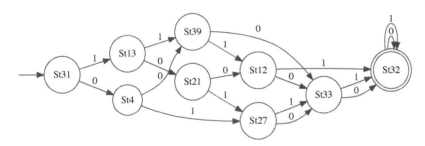

Exercise 8.6.1, RE, Error Correction

1. There are overlaps between the following two regular expressions used in defining h2_0101_re: (0+1)(0+1)01 and (0+1)1(0+1)1 (both their languages contain the string 0101). Argue that such overlaps do not matter.

2. Is the error-free occurrence of 0101 included in h2_0101_re? If so, how many times does it get included?

3. Given the regular expression language so far, can you think of a way to directly and compactly express the following regular language:

> The set of all bit-strings that *do not contain a 0101*

If not, state why this pattern is difficult to express. *Hint:* While one can easily complement DFA, there aren't easy methods for complementing an NFA or RE directly. □

8.6.2 NFA-based Design of "within Hamming Distance of 2"

Now that we have obtained one minimal DFA using RE, let us explore a design approach using NFA. The idea is simple: whenever we are "supposed" to make a move on a 0 (let us say) and a 1 comes, instead, we "silently" error-correct the 1 to a 0—but remember in the state name that we have suffered one "ding." We allow two "layers" of state names of this kind, after which we provide no more reprieve. We invoke the markup language of Jove and do the design directly in it (Figure 8.3) and obtain the NFA in Figure 8.4.

This results in an NFA which is much more tedious to obtain than the RE we designed on Page 107. However, human effort-wise, arriving at this NFA design is not that difficult (compared to manually deriving a DFA or a minimal DFA directly from the problem statement).

8.7 Minimal DFA and Isomorphism

We obtain the minimal DFA based on the NFA-based approach shown in Figure 8.5.

We finally administer the test `iso_dfa(minD_h2_0101_re, minD_h2_0101_nfa)`, which yields **True**.

This finishes the design using two distinct approaches and a confirmation that the results agree.

Figure 8.3: Design of an Error-Correcting NFA

```
h2_0101_nfa = md2mc('''NFA
!!------------------------------------------------------------
!! We are supposed to process (0+1)*0101(0+1)* with up to two
!! "dings" allowed.  Approach: Silently error-correct, but
!! remember each "ding" in a new state name. After two dings,
!! do not error-correct anymore
!!------------------------------------------------------------
!!-- Pattern for (0+1)* : no error-correction needed here :-)
I : 0 | 1 -> I


!!-- Now comes the opportunity to exit I via 0101
!!-- The state names are A,B,C,D with ding-count
!!-- Thus A0 is A with 0 dings, C2 is C with 2 dings


!!-- Ding-less traversal -- how lucky!
I  : 0 -> A0
A0 : 1 -> B0
B0 : 0 -> C0
C0 : 1 -> F
F  : 0 | 1 -> F  !!--  Phew, finally at F


!!-- First ding in any of these cases
I  : 1 -> A1
A0 : 0 -> B1
B0 : 1 -> C1
C0 : 0 -> F  !!-- ding-recording unnecessary; just goto F


!!-- Second ding in any of these cases
A1 : 0 -> B2
B1 : 1 -> C2
C1 : 0 -> F  !!-- ding-recording unnecessary; just goto F


!!-- No more dings allowed!
B2 : 0 -> C2
C2 : 1 -> F


!!-- Allow one-dingers to finish fine
A1 : 1 -> B1
B1 : 0 -> C1
C1 : 1 -> F
''')
minD_h2_0101_nfa = min_dfa(nfa2dfa(h2_0101_nfa))
dotObj_dfa(minD_h2_0101_nfa)
```

Figure 8.4:   NFA obtained through the markdown approach

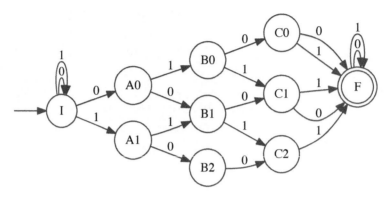

Figure 8.5:   Minimized DFA obtained through the NFA-based Approach

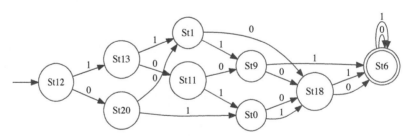

8.8   DFA Ultimate Periodicity to Solve the Postage Stamp Problem

Suppose you are given an unlimited supply of 3 cent and 5 cent postage stamps, and asked *what is the largest denomination of postage that you cannot make using these stamps?* By trial and error at first, you would proceed as follows: *I can make 3, but not 4 and 7. Then all denominations above 7 can be made.* So the answer is 7. Suppose you are asked to solve it for stamps that are 29 cents and 53 cents. Well, trial and error becomes tedious.

There is a formula to solve this problem discovered by a mathematician named Sylvester in 1884. For two relatively prime numbers p and q,[8] the largest natural number that cannot be expressed as a linear combination of the form $a \cdot p + b \cdot q$ for natural numbers a and b is called the **Frobenius number**,[9] and denoted by $Fr(p,q)$. It is given by the formula

$$Fr(p,q) = pq - p - q$$

Applying this formula to 3 and 5, we indeed get $3 \cdot 5 - 3 - 5 = 7$. Applying it to 29 and 53, we get 1455.[10]

Exercise 8.8, Sylvester's formula

1. Consider two numbers p and q that are not relatively prime, say 3 and 6 whose GCD is 3. Does there exist (in Nat) a largest number that cannot be expressed as a linear combination of 3 and 6?

2. What does your answer to the previous question tell you about why Sylvester's formula is applicable only to relatively prime numbers? Hint: if p and q have $gcd(p,q) > 1$, and say $p > q$, then what is the minimum "step size" in the series $p, p+q, p+2q, \ldots$?   □

8.8.1   Ultimately periodic sets and lengths of members of a regular language

The postage stamp problem closely relates to the notion of *ultimately periodic sets* of natural numbers.

> An ultimately periodic set S is a subset of Nat where there exists a bound $b \in Nat$ and a period $p \in Nat$ such that for all $x > b$, $x \in S$ if and only if $(x + p) \in S$.

Here are interesting examples of ultimately periodic sets:

- $S = \{3,5,8,9,10,11,12,13,\ldots\}$ is ultimately periodic.
  - We can choose $p = 1$ and $b = 7$ and find that

$$x \in S \iff (x+p) \in S$$

- $S = \{3,5,11\}$ is ultimately periodic. This is because we can choose $b =$

---

[8] Two natural numbers p and q are said to be relatively prime if their greatest common divisor is 1.

[9] There is also the related problem of the Chicken McNugget number, http://mathworld.wolfram.com/McNuggetNumber.html.

[10] For an excellent discussion of the number of amazing ramifications of the Frobenius number problem and its variants, as well as a proof of Sylvester's result, see the excellent talk slides *The Frobenius problem and its generalizations* by Jeffrey O. Shallit. https://cs.uwaterloo.ca/~shallit/talks.html. Also see Wikipedia for the "Coin Problem."

[11] As mentioned in Chapter 5, we do need a "vacuum cleaner" to avoid being trapped by such vacuous cases we might overlook.

11, and then there are no numbers above b in the set, and thus the ultimate periodicity condition is *vacuously* satisfied.[11]

The reason why ultimately periodic sets matter is this:

> **Theorem 8.8.1:** The set of **lengths** of strings in any regular language is ultimately periodic.

Proof sketch: If we take any DFA and convert all its symbols to the same symbol (thus obtaining an NFA) and then convert this NFA to a DFA and minimize the DFA, we will always obtain a single lasso that characterizes an ultimately periodic set. We will demonstrate this result in §8.8.5 using Jove.

8.8.2 Stamp Problem and Ultimate Periodicity via Jove

Automata theory – especially as mechanized by Jove – provides a very convenient medium through which to study these results. Let us revisit the question concerning 3 and 5 cent stamps raised at the beginning of §8.8. Suppose we represent a 3 cent stamp by the string 111, and a 5 cent stamp by 11111. Then a sequence of 3 and 5 cent stamps has *as many 1's as there are 1's in the regular expression* $(111 + 11111)^*$. Let us build a minimal DFA for this RE.[12] Given that this is an RE, a DFA must exist, and this DFA accepts all and only those strings that contain $3a + 5b$ 1's (we call this DFA "sylv_3_5" in honor of Sylvester's formula). See Figure 8.6 for this DFA:

[12] We do allow 0 cents of postage, thus justifying why we employ $(111 + 11111)^*$. If non-zero postage is desired, change this RE to $(111 + 11111)^+$—where the "Kleene-plus" operator denotes *one or more repetitions*.

```
sylv_3_5 = dotObj_dfa(min_dfa(nfa2dfa(re2nfa("(111+11111)*"))))
```

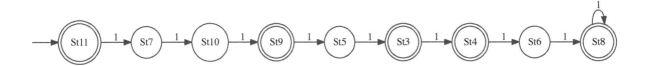

Figure 8.6: Sylvester's formula for $p, q = 3, 5$ "solved" via this minimal DFA that accepts all strings of length > 7. In general, given relatively prime numbers, such DFA will always have a **lasso shape**, with a "lasso" of size 1 at their last state (which will be final, and in fact, a *white-hole* or *heavenly* §6.1). The Frobenius number will then be the number of states minus 2, *i.e.* $|Q| - 2$. This is because there are $|Q| - 2$ transitions *before* the final transition that *first reaches* the final state.

Given the shape of this DFA, one can obtain $Fr(p, q)$ by taking the number of states of this DFA and subtracting 2 from it (subtract 1 to count the number of steps between the states, and subtract an additional 1 because only the penultimate step represents the last number that cannot be represented):

```
len(sylv_3_5["Q"]) - 1 - 1 = 7
```

8.8.3 Applying to numbers that are not relatively prime

Let us go ahead and obtain the minimal DFA for the regular expression $(111 + 111111)*$. This corresponds to all denominations that can be obtained using 3 and 6 cent stamps. Since their GCD is 3, there are an infinite number of "gaps" that cannot be realized (any even number not divisible by 3), and therefore Sylvester's formula does not apply. The minimal DFA is as shown in Figure 8.7.

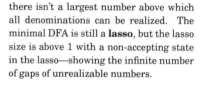

Figure 8.7: With 3 and 6 cent stamps, there isn't a largest number above which all denominations can be realized. The minimal DFA is still a **lasso**, but the lasso size is above 1 with a non-accepting state in the lasso—showing the infinite number of gaps of unrealizable numbers.

8.8.4 Solving for three stamps

If we have three relatively prime numbers (3, 5 and 7 for example), there isn't a (known) closed form formula that gives us the Frobenius number of three relatively prime numbers. However we can still use minimal DFA to solve such problems (see Figure 8.8):

```
stamp_3_5_7 = min_dfa(nfa2dfa(re2nfa("(111+11111+1111111)*")))
```

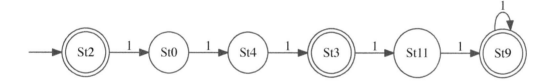

Figure 8.8: The Frobenius number for three relatively prime numbers can be calculated using the minimal automaton approach. Using stamps 3, 5, 7, all denominations above 4 can be realized (number of states minus 2)

8.8.5 Lengths of strings accepted by DFA

Consider the DFA "blimp" defined in Figure 7.8. In order to determine the lengths of strings of this DFA, first apply a homomorphism that changes all instances of a 'b' to an 'a' (see Figure 8.9, top):

```
blimpnfa = apply_h_dfa(blimp, lambda x:  'a')
```

This NFA can now be converted to a DFA and minimized. These steps preserve the lengths of strings in the language (Figure 8.9, bottom) demonstrating the theorem pertaining to the *lengths of strings in a regular language*.

```
dotObj_dfa(min_dfa(nfa2dfa(blimpnfa)))
```

Exercise 8.8.5, Postage Stamp

1. Determine the Frobenius number
   (a) For $p, q = 5, 11$ using Sylvester's formula
   (b) For $p, q = 5, 11$ as well as for $p, q, r = 5, 7, 11$ using the minimal automaton approach (use Jove).

2. Write a proof that the lengths of strings in any regular set form an ultimately periodic set.   □

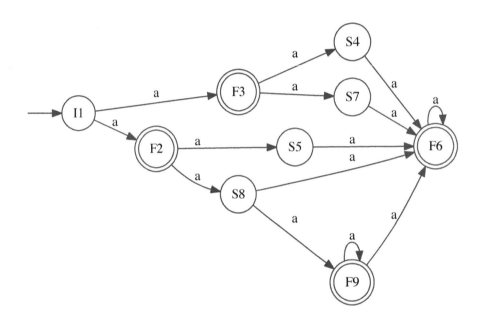

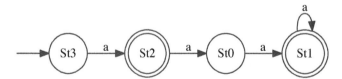

Figure 8.9: DFA "blimp" subject to a homomorphism that changes all b to an a. The minimized DFA (shown below) demonstrates Theorem 8.8.1 ("*Lengths of strings in a regular language*"). In this example, the lengths of strings in the DFA 'Blimp' are 1, 3, and all numbers above 3. Thus, the parameters b and p mentioned in the definition of ultimate periodicity take on values $b = 2$ and $p = 1$. While this is only a demonstration, it can be easily converted into a rigorous proof.

9

NFA to RE Conversion

Chapter Gist: *We describe how NFA can be converted to equivalent RE (§9.1), illustrating it on a pedagogical example that clearly shows how the RE size growth can be exponential (§9.2). A nontrivial NFA example with plenty of loops is then examined (§9.3). We check the conversion by converting the obtained RE back to an NFA and then a minimal DFA, and check it for isomorphism against the minimal DFA obtained from the given NFA (§9.4). These conversions demonstrate that DFA, NFA and RE are equivalent (§9.5).*

9.1   NFA to RE Conversion Algorithm

An NFA accepts a string $x \in \Sigma^*$ if there exists a path labeled by x from one of its initial states of the NFA to one of its final states.[1] Expressing each such path using a regular expression, and taking a union (sum) of these REs immediately gives us an algorithm to convert any NFA to a language-equivalent RE. The loops along each path will be expressed through a suitable Kleene-star-based RE. This procedure can be implemented by systematically building up the RE for each path as follows:

[1] All ε substrings within s can be ignored.

- Preprocess the given NFA by introducing a new initial state called Real_I and a new final state called Real_F.[2] Introduce ε-labeled transitions from Real_I to all the states of $Q0$, the initial set of states of the given NFA. Also introduce ε-labeled transitions from each state in F to Real_F, and make states in F non-final. The NFA so obtained is called a **Generalized NFA** or **GNFA**. Clearly, this GNFA ("G") has the same language that the given NFA ("N") has; it differs from N only in these respects:
  - Its initial set of states is {Real_I} and its set of final states is {Real_F}.
  - The transition relation of G includes all the members of N's transition relation plus the new ε-labeled edges mentioned above.

[2] These will serve as the states to "stand on" while state deletion (described momentarily) is in progress.

```
# Let this NFA be specified via
# our markdown as follows:
nfaExer = md2mc('''NFA
I1 : a -> X
I2 : b -> X
I3 : c -> X
X  : p | q -> X
X  : m -> F1
X  : n -> F2
''')
# First form the Dot Object...
DO_nfaExer \
 = dotObj_nfa(nfaExer)
# Check things by displaying
DO_nfaExer
# Form a GNFA out of the NFA
gnfaExer = mk_gnfa(nfaExer)
# Form a Dot Object
DO_gnfaExer \
 = dotObj_gnfa(gnfaExer)
# Check things by displaying
DO_gnfaExer


# Now invoke del_gnfa_states
# First argument is the GNFA
# of our exercise, gnfaExer
# The second arg (optional)
# is the deletion order of
# the states. If omitted,
# the tool picks the order
# (makes a HUGE difference).
(G, DO, RE) = \
del_gnfa_states(
 gnfaExer,
 DelList=["X", "I1",
         "I2","I3",
         "F1","F2"])
# Display DO[0] through DO[6]
# G is the final GNFA returned
# RE is the final RE compiled
```

Figure 9.1: NFA to RE example

- While there are states in the GNFA G other than Real_I or Real_F, pick one such state s, and delete it. However, account for all the paths supported by s by introducing substitute edges that bypass state s (whenever two parallel edges labeled with regular expressions r_1 and r_2 go from state A to state B, one should ideally replace this edge with a single edge labeled with $(r_1 + r_2)$).
- Do this till the GNFA only has two states left (which would be Real_I and Real_F). The desired regular expression would then be labeling the path connecting Real_I and Real_F.

We will present this algorithm implemented in Def_NFA2RE in this chapter on an example NFA. We also provide you with markdowns to permit experimentation using Jove. In one of the examples we provide (specifically the one in Figure 9.2), un-commenting a single line can quite dramatically increase the size of the RE generated. In fact, an exponential blow-up is waiting to occur during NFA to RE conversion. This is because in the worst case, there can be $O(2^N)$ paths in a graph of size N.

9.2 Illustration on Pedagogical Example

Consider the NFA in Figure 9.1 that can start from a set of three initial states, and we are asked to eliminate state X:

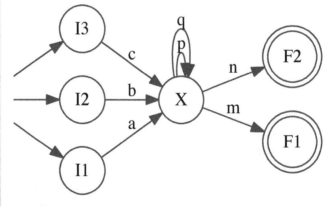

We first form a GNFA out of the given NFA using the procedure described earlier:

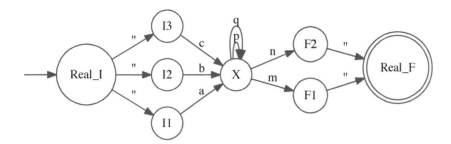

In preparing to eliminate X, we see that there are three paths going into X and two going out. Thus, six substitute paths must be introduced, going from I1, I2, and I3 to F1 and F2. We can achieve a specified order of elimination by invoking del_gnfa_states with the first argument being (of course) the GNFA to be processed, and a second argument called DelList which specifies the order of state elimination. We list state X first in this list, thus ensuring its elimination first:

We see two kinds of transformations taking place:

- The two alternatives, namely p and q get merged into one regular expression (p+q).
- Each of the six new paths now has an *incoming piece*, an *outgoing piece*, and a *self-piece*. For instance, I1 to F1 is labeled by (a (p+q)* m) where
  - a is the incoming piece, namely the I1 to X label,
  - m is the outgoing piece, namely the X to F1 label,
  - (p+q)* is the self-piece. This self-loop is, naturally, Kleene-starred, as it can be taken any number of times.

Referring to Figure 9.1, the result of eliminating state X is DO[1] shown below (the original NFA being DO[0]):

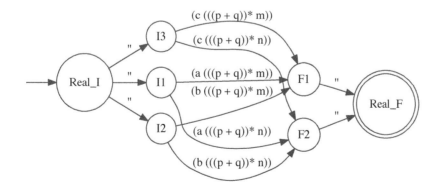

```
nfaEx = md2mc('''NFA
I :  '' -> B
I : a  -> A
!!-- reader: please try
!!-- uncommenting this
!!-- line: A : b  -> I
A : q  -> F
A : r  -> B
B : s  -> B
B : p  -> F
F : t  -> A
''')
DO_nfaEx = dotObj_nfa(nfaEx)
DO_nfaEx
```

Figure 9.2: NFA to be Converted to an RE

Exercise 9.2, NFA to RE

1. Try two other orders: (1) Delete all Ii states first, then all Fj states, and finally the X state. (2) Try the tool-default order. Contrast the size of the final REs obtained using these approaches to deleting the X state first. Which results in the smallest final RE and why?

2. Call the X state a "busy" state (many input, output and even self-loop edges). Heuristically, is it better to eliminate busy states first, or busy states last? Why?

3. How can we ensure that all the REs obtained through various elimination orders are language-equivalent? □

9.3 Illustration on Non-trivial Example

The example NFA we shall convert to an RE is described in Figure 9.2.

- The given NFA

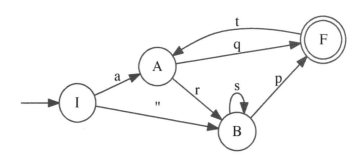

- The GNFA after introducing a new initial and final state is shown below

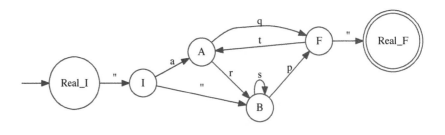

```
GNFA_nfaEx = mk_gnfa(nfaEx)
```

- We are going to employ a function del_gnfa_states (Figure 9.3) that not only does the deletion of states as described above, but also returns valuable intermediate information to look at. It also captures the regular expression generated from the given NFA (this RE can be fed back into our conversion pipeline as we shall illustrate).

NFA to RE Conversion Steps

- The GNFA is

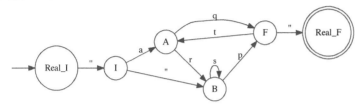

- After eliminating state F, we have the following GNFA:

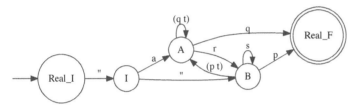

Notice that state F supports several paths through it, and here are the substitute paths introduced:

- A path from A to Real_F. This must now be labeled by the concatenation of q and ", which is q.
- A path from A back to itself, labeled by (q t), the concatenation of the edge-label q and edge-label t.
- A path from B to Real_F. This must now be labeled by the concatenation of p and ", which is p.
- A path from B to A. This must now be labeled by the concatenation of p and t, which is (p t).
- **In general, if the eliminated state has m inputs and n outputs, there will be $m \cdot n$ substitute paths**. In our case, we have $2 \cdot 2 = 4$ new paths.
- After Eliminating I, the GNFA changes in a simple way. There are only two paths supported by I, namely Real_I to B and Real_I to A. One can see the concatenation of the edge labels labeling these paths:
  - Real_I to B is labeled with the concatenation of " and ", which is "
  - Real_I to A is labeled with the concatenation of " and a which is 'a'

```
help(del_gnfa_states)
...
Return a triple
(Gfinal, dotObj_List,
final_re_str), where
Gfinal: The final GNFA
dotObj_List: Intermediate GNFA

(Gfinal, do_list, final_re)
= del_gnfa_states(GNFA_nfaEx)
**** Eliminating state F ****
**** Eliminating state I ****
**** Eliminating state B ****
**** Eliminating state A ****
```

Figure 9.3: The del_gnfa_states Function

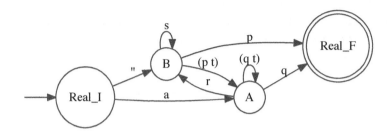

- B is a "very busy" state, and we are going to eliminate it now.
  - B has an input from state Real_I labeled by "
  - B has an input from A labeled by r
  - B has an output to state A labeled by (p t)
  - B has an output to state Real_F labeled by p
  - B has a self-loop labeled by s.
  - **The general rule to apply is this:**
    * If a state being eliminated has an input labeled by P, a self loop labeled by Q and an output labeled by R, then the substitute path to introduce is P(Q*)R.
  - Following this rule, we can see this instance:
    * There is an r label from A to B
    * There is a self-loop of s at B
    * There is an output from B via p to Real_F
    * Thus, there must now be a new path directly from A to Real_F labeled by r(s*) p.
  - We however notice that there *was* already an edge from A to Real_F labeled with q. When this happens, we have another rule:
    * **If two edges run from P to Q labeled by x and y, one can introduce a single edge from P to Q labeled by (x+y).**
    By applying this rule, the final label from A to Real_F is ((r(s*) p) + q)
  - The reader is invited to figure out all remaining edges introduced, and which rules were applied.
- Thus, after Eliminating B, we have this GNFA:

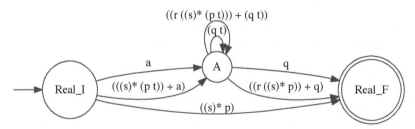

- Finally, after Eliminating A, we have the following GNFA:

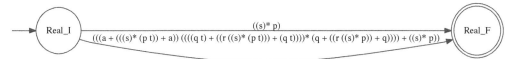

- We can see that the derived RE is a sum of the REs written on the two paths in this figure. It is given below:

```
(  ((s)* p)
 +
 ( (     (a + (((s)* (p t)) + a))
         ( (( (q t)  + ((r ((s)* (p t))) + (q t)) ))*
           (q + ((r ((s)* p)) + q))
           + ((s)* p)              ,        ))))
```

9.4 Checking the Conversion

We will provide salient excerpts from the code of del_gnfa_states. Given its complexity, one should constantly be vigilant against introducing bugs. A good way to check that things are correct should be familiar by now:

- Convert the generated RE back to a minimal DFA:

```
re_mindfa = min_dfa(nfa2dfa(re2nfa(final_re)))
```

This results in the following minimal DFA:

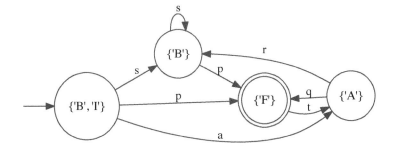

- Convert the original NFA to a min DFA:

```
dir_mindfa = min_dfa(nfa2dfa(nfaEx))
```

This results in the following minimal DFA:

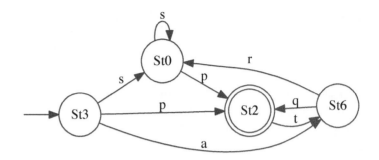

- Check that these DFA are isomorphic:

```
iso_dfa(re_mindfa,dir_mindfa)
True
```

Indeed, these DFA are isomorphic. This strongly minimizes the chance of our conversions being buggy.

9.5   DFA, NFA, and RE Are Equally Powerful

With this demonstration, not only did we illustrate practical uses of our tool, but we ended up showing that NFA, DFA, and RE are all equally powerful, and describe exactly the family of regular languages.
In more detail:

- RE can be converted to NFA (Chapter 8, `re2nfa`)
- NFA can be converted to DFA (Chapter 7, `nfa2dfa`)
- DFA can be minimized, and the minimal form is *unique* for a regular language:
  - Chapter 6 for the classical algorithm, `min_dfa`
  - Chapter 7 for Brzozowski's algorithm, `min_dfa_brz`
- In this chapter, we present two algorithms:
  - NFA can be converted to RE, with the whole conversion chain extracted for study:
    * Function `mk_gnfa` to preprocess NFA into "generalized NFA" before we delete NFA states one by one.
    * Function `del_gnfa_states` is the actual process of deleting NFA states one by one and producing substitute paths labeled by RE
    * Two support functions:
      · `mk_gnfa_from_D` (in case you want to turn a DFA into a GNFA before the conversion pipeline kicks in)

Exercise 9.5, nfa2re: RE Size

1. By hand, convert the NFA of Figure 7.7(a) into a regular expression, clearly discussing how each state is eliminated. The order of state elimination you must follow is I, A, B, C, D, E, G, F.

2. How large can the extracted REs get? Hint: Consider an NFA with a series of states A1, A2, ..., An and two paths between any Ai and Aj. How many total paths are there as a function of n?

3. Take the suggestion of uncommenting the single line in Figure 9.2 and redo the conversion using Jove. What is the size of the new RE? Explain the size increase.

4. Construct an NFA of N states such that converting it to an RE using our algorithm will result in an exponentially sized RE.   □

9.6   Implementation of NFA to RE

The implementation of NFA to RE conversion in Jove is in Def_NFA2RE, and salient excerpts from this Jupyter module are now described in Figures 9.4 and 9.5. The basic algorithm is to execute a double-nested for loop while there are states left to delete. In this double-nested for, every p and q states are considered with respect to the state qdel to be deleted, and the helper del_one_gnfa_state is called. There are *plenty* of corner-cases!

9.7   Closure Results Pertaining to Regular Languages

We now summarize all the closure results pertaining to NFA, DFA, and RE:

Theorem 9.7: NFA, DFA and RE are all closed under union, concatenation, star, complementation, reversal, and homomorphism. The general idea is that we have introduced conversions between NFA, DFA, and RE, and therefore one can perform any operation on one representation (say an NFA) and then convert it another representation (say an RE). Here is a slightly more detailed proof sketch:

1. DFA complementation was described in §6.1. Algorithms for union and intersection of DFA were described in §6.2.

2. The star of a DFA can be achieved the same way as star of REs and NFA were described in §8.2.

3. NFA and REs don't have a direct algorithm for intersection or complementation. In fact, swapping final and non-final states of an NFA does not result in the complementation of an NFA. You can easily verify it on an NFA that starts from an initial state and branches on an input a to two

```python
def del_gnfa_states(Gin, DelList=[]):
    """Given a GNFA Gin with no unreachable states, delete all states but Real_I or Real_F.
       Return a triple (Gfinal, dotObj_List, final_re_str)."""
    G = copy.deepcopy(Gin)              # Preserve the given GNFA
    if DelList==[]:                     # User hasn't provided a preferred order
        StatesLeft = list(G["Q"])       # ... so, form internal order of deletion from G["Q"]
    else:                               # User HAS provided a preferred order of deletion
        StatesLeft = DelList + ["Real_I", "Real_F"] # Tack-on these states to the user-given list
    dotObj_List = [ dotObj_gnfa(G) ] # Preserve the list of intermediate GNFAs for viewing later
    while len(StatesLeft) > 2:          # Continue so long as there is a state other than Real_I, Real_F
        (qdel, StatesLeft) = choose_state_to_del(G, StatesLeft) # Choose state to del. ; upd. StatesLeft
        print("**** Eliminating state " + qdel + " ****")
        New_Edges = dict()              # Brand new edges (ALL new paths supported by qdel) kept as a dict
        for p in StatesLeft:
            for q in StatesLeft:
                new_p_q_label = del_one_gnfa_state(G, p, qdel, q)  # Workhorse to delete ONE state
                if new_p_q_label != "NOPATH":                      # There is a p-qdel->q path
                    old_p_q_labels = Edges_Exist_Via(G, p, q)      # Exist p-qdel->q edges?
                    if old_p_q_labels != "NOEDGE":                 # There are.
                        combined_label = form_alt_RE( [new_p_q_label] + old_p_q_labels )
                        New_Edges.update( { (p, combined_label) : {q} } )
                    else:                                          # Only new_p_q_label needs to be added
                        New_Edges.update( { (p, new_p_q_label)  : {q} } )
        G["Q"] = set(StatesLeft)                                   # Fix G by adjusting its Q
        Surviving_Edges = []                                       # These edges don't get nuked
        for ((q,symb), States) in G["Delta"].items():
            if (q != qdel):                                        # (1) Removing all mappings out of qdel
                Surviving_Edges += [ ((q,symb), States - { qdel }) ] # (2) Remove from images
        G["Delta"] = dict( Surviving_Edges )
        G["Delta"].update( New_Edges )                             # Now bring in the brand new edges
        dotObj_List += [ dotObj_gnfa( gnfa_w_REStr(G) ) ]          # Convert REs to strings in dot objects
    G["Sigma"] = { edgelab for ((p,edgelab), q) in G["Delta"].items() } # Form the GNFA's ''Sigma''
    final_re     = form_alt_RE(Edges_Exist_Via(G, "Real_I", "Real_F"))  # RE from Real_I to Real_F
    final_re_str = RE2Str(final_re)                                # String form of RE
    Gfinal = {"Q": {"Real_I","Real_F"}, "Sigma":{final_re},"Delta":{("Real_I", final_re):{"Real_F"}},
              "Q0": { "Real_I" },        "F"    : { "Real_F" }}
    return (Gfinal, dotObj_List, final_re_str)
```

Figure 9.4: Implementation of del_gnfa_states.

```
def del_one_gnfa_state(G, p, qdel, q):
    """Delete state qdel if path p--qdel-->q exists. Return "NOPATH" if no such path.
       Else return new direct edge label p--new_label-->q (new_label will be a single RE)."""
    p_qdel_edges = Edges_Exist_Via(G, p, qdel)
    qdel_q_edges = Edges_Exist_Via(G, qdel, q)
    if (p_qdel_edges == "NOEDGE" or qdel_q_edges == "NOEDGE"):
        return "NOPATH"
    else:
        p_qdel_RE = form_alt_RE(p_qdel_edges)
        qdel_q_RE = form_alt_RE(qdel_q_edges)
        qdel_qdel_edges = Edges_Exist_Via(G, qdel, qdel)
        if qdel_qdel_edges == "NOEDGE":
            return form_concat_RE(p_qdel_RE, qdel_q_RE)
        else:
            qdel_qdel_RE = form_alt_RE(qdel_qdel_edges)
            return form_concat_RE(p_qdel_RE, form_concat_RE(form_kleene_RE(qdel_qdel_RE), qdel_q_RE))
```

Figure 9.5: Implementation of `del_one_gnfa_state`. This helper is rather clean. First, we check whether edges exist via `qdel`, starting from p and ending in q. If none, we return "NOPATH". Else, we grab all the p to `qdel` edges, fuse them via a `form_alt_RE` call. Likewise, we grab all the `qdel` to q edges, fuse them via a `form_alt_RE` call. Next, we check for a self-loop to be present. The result could be "NOEDGE" in which case we concatenate the previous two REs. Else we stick in a `form_kleene_RE` call in the middle to finish the story of $input(self)^*$ $output$.

states, say A (a non-final state) and F (a final state). In this case, swapping the final and non-final status of A and F still leaves a in the language of the NFA. However, you can convert them to a DFA, perform intersection or complementation, and re-obtain an NFA or an RE.

4. Reversal of a DFA was explained in §7.6.1. The same procedure also works for an NFA.

5. Concatenation of two DFA can be performed by treating them as NFA. The result will be an NFA which can be converted back to a DFA.

6. A homomorphism can be applied to a DFA, and this may result in an NFA. For NFA and RE, homomorphisms are straightforward to carry out.

10

Derivative-Based Regular Expression Matching

> **Chapter Gist:** *RE derivatives are based on the idea that REs themselves may be treated as states (§10.1). One can then systematically compute target REs as the result of applying the transition function δ to the source REs (§10.2.1). The derivation of a sequence of REs may be stopped as soon as one RE is found nullable (contains ε in its language, §10.2.2). The RE syntax can be extended to have negation; then one can pattern-match a given string against such an RE without first doing an NFA to DFA conversion (§10.3). The code presents yet another tiny RE to AST compiler and a pattern-matcher.*

10.1 Introduction to RE Derivatives

A finite automaton chugs along, consuming symbol after symbol from its input string till the string is exhausted. At this point when the automaton finds itself in a final state, the string is deemed accepted. In 1964 (a few years after DFA and NFA were introduced by Rabin and Scott), Brzozowski [14] came up with a similar idea except this time a regular expression (RegExp) chugs along, eating a string and also morphing itself into new regular expressions! A DFA does not need to morph after eating a character, as it can put itself in the "next state." A regular expression has no "internal state," and so it must morph into another regular expression, reflecting the equivalent of a new state.

The idea is simple. Consider Figure 10.1, a reproduction of Figure 4.1 for your convenience. After "eating a 0 in state A," the DFA goes to state B. Now consider the regular expression modeling this DFA: clearly, it is $(00)*$. We can rewrite this regular expression into $\varepsilon + 00(00)^*$ to explicate the two cases hidden within $(00)*$. Now, when presented with 0, the

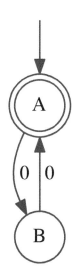

Figure 10.1: A DFA to recognize an even number of 0's. The regular expression modeling its language is $(00)^*$.

regular expression must abandon the ε case, and pursue the $00(00)^*$ case. This component of the regular expression "eats" a 0 and evolves itself into $0(00)^*$. We can depict the whole cyclic walk of RE $(00)^*$ as follows:

$$(00)^* = (\varepsilon + 00(00)^*) \xrightarrow{0} 0(00)^* \xrightarrow{0} (00)^*.$$

This was Brzozowski's insight! In a nutshell, the derivative of $(00)^*$ with respect to 0 is $0(00)^*$. In this chapter, we will employ the notation $E \xrightarrow{c} \ldots$ to denote "E derives via c."

In addition to their mathematical beauty, derivatives are capable of handling operators such as negation and intersection, which we did not include in our regular expression syntax of Chapter 8. With derivatives, handling these "troublesome" operators is easy: a regular expression harboring such operators simply evolves itself into new regular expressions (possibly harboring the same operators). Thus, we never have to "expand out" the negation. If we take the DFA route to handle negation (for instance), we would have to convert the regular expression into an NFA, convert the NFA into a DFA, and then complement the resulting DFA. You should recall from Chapter 4 that NFA to DFA conversion can often result in an exponentially sized DFA (with respect to the size of the NFA). The derivatives approach avoids this blow-up. Additionally, derivatives help introduce the idea of *rewrite rules* or *transition systems* (Figure 10.2) which are widely used for specifying software systems.

In this chapter, we shall introduce a derivative-based pattern matching routine that matches a given string against a given regular expression. In this test, think of the given string as "steering" the given regular expression by taking derivatives character by character. When all the characters are consumed, the regular expression must end up being *nullable, i.e.*, it must denote a language that contains ε. If so, the string is accepted; else, it is not.

In more detail, let us start from the regular expression r_0 and input string w_0, and ask the question whether r_0 matches string w_0. To check this, we start with $r = r_0$ and $w = w_0$ and execute the following steps:

1. If w is ε, then we have reached the end of the input string (no more steering is necessary). All we now need to check is whether the language itself contains ε, and if so, we can declare acceptance. Checking whether the language denoted by r contains ε is achieved by subjecting r to the **nullability test** that checks whether (the language of) regular expression r contains ε.

2. Otherwise, let $w = cw_1$ where $c \in \Sigma$ and w_1 is the "rest of w" that is yet to be consumed. We obtain the derivative of r with respect to c yielding r_1 (*i.e.*, $r \xrightarrow{c} r_1$). Here, r_1 models what r must morph itself into after seeing the input symbol c. So now the problem at hand (whether r matches w) reduces to whether r_1 matches w_1. To check this, we set $r = r_1$ and $w = w_1$, and repeat all the steps.

Note: we can stop and declare "no match" whenever r's derivative turns out to be $\emptyset$ (the empty RegExp).

The nullability test returns True if $\varepsilon \in r$ (*i.e.*, the language of r). This means that after a string w has been consumed, we are "left with" ε, coming "after the w" (so to speak), which is matched by a nullable RE r.

EXAMPLE-1:   The following derivation shows that $w = w_0 = 00$ is matched by $r = r_0 = (00)^*$:

$$(00)^* \xrightarrow{\ 0\ } 0(00)^* \xrightarrow{\ 0\ } (00)^*,$$

and since w is fully consumed and $r_2 = (00)^*$ is nullable, we accept $w = 00$. On the other hand, $w = 0$ is not matched by $(00)^*$:

$$(00)^* \xrightarrow{\ 0\ } 0(00)^*$$

and since w is fully consumed but $0(00)^*$ is not nullable, we reject 0.

EXAMPLE-2:   The following derivation shows that $w = ab$ is matched by $r = c + (a + b)(a + b)$:

$$(c + (a + b)(a + b)) \xrightarrow{\ a\ } (a + b) \xrightarrow{\ b\ } \varepsilon$$

and since w is fully consumed and (clearly) ε is nullable, we accept ab. On the other hand, cc is not matched by $c + (a + b)(a + b)$:

$$(c + (a + b)(a + b)) \xrightarrow{\ c\ } \varepsilon \xrightarrow{\ c\ } \emptyset$$

and clearly $\emptyset$ is not nullable. Notice that we "derive" $\emptyset$ from ε for any character c.

10.2   Definitions

10.2.1   Derivative Rules

We now define derivatives through a collection of nine rules, where the "Precondition" column lists the preconditions (could be empty), and the "Derivation" column lists the derivation being defined.

We now explain each case in greater detail:

(Rule 1) Derivation $c \xrightarrow{\ c\ } \varepsilon$: If we try to steer the regular expression c using character c, character c gets "eaten" by this regular expression, morphing it to ε ("all gone").

(Rule 2) Derivation $c \xrightarrow{\ d\ } \emptyset$: This is when regular expression c tries to "eat" incoming character d; the result is a "choke," as $d \neq c$. The regular expression morphs (gags) itself into $\emptyset$.

Rule	Precondition	Derivation	Explanation
1	None	$c \xrightarrow{c} \varepsilon$	RegExp c matches input c, leaving behind ε.
2	$c \neq d$	$c \xrightarrow{d} \emptyset$	RegExp c doesn't match input d; $\emptyset$ signifies blockage.
3	None	$\varepsilon \xrightarrow{c} \emptyset$	RegExp ε doesn't match character c; $\emptyset$ signifies blockage.
4	None	$\emptyset \xrightarrow{c} \emptyset$	RegExp $\emptyset$ does not match character c.
5	$E \xrightarrow{c} E_{1c}$	$E^* \xrightarrow{c} E_{1c} E^*$	Expand E^* as EE^*; obtain $(E_{1c} E^*)$ as the derivative.
6	$E \xrightarrow{c} E_{1c}$	$!E \xrightarrow{c} !E_{1c}$	Derivative of the non-negated form (with negation outside) suffices.
7	$E_1 \xrightarrow{c} E_{1c}$, $E_2 \xrightarrow{c} E_{2c}$	$(E_1 + E_2) \xrightarrow{c} E_{1c} + E_{2c}$	The derivation can proceed via either "arm" of the alternative.
8	$nullable(E_1)$, $E_1 \xrightarrow{c} E_{1c}$, $E_2 \xrightarrow{c} E_{2c}$	$(E_1 E_2) \xrightarrow{c} ((E_{1c} E_2) + E_{2c})$	If E_1 is nullable, then E_2 might consume c.
9	$\neg nullable(E_1)$, $E_1 \xrightarrow{c} E_{1c}$	$(E_1 E_2) \xrightarrow{c} (E_{1c} E_2)$	Since E_1 is not nullable, c must be expanded via E_1.

Figure 10.2: Derivative Rules

(Rule 3) Derivation $\varepsilon \xrightarrow{c} \emptyset$: Forcing regular expression ε to "eat" character c is futile; the regular expression morphs itself to $\emptyset$.

(Rule 4) Derivation $\emptyset \xrightarrow{c} \emptyset$: Forcing regular expression $\emptyset$ to "eat" character c is even more futile; the regular expression morphs itself to $\emptyset$, as it was to begin with.

(Rule 5) Derivation $E^* \xrightarrow{c} E_{1c}E^*$: Here, we are forcing E^* to consume c. Now we know that E^* hides two cases inside it. First, the star of anything contains ε. But since an actual character c has to be eaten, this case is not of interest. The other case is that any E^* can open itself as EE^*. This case is of interest! The first E that reveals itself now has to eat c. Suppose E derives E_{1c} by eating c. Then we know what E^* likes to morph itself into: $(E_{1c}E^*)$, *i.e.*, let the first E that shows up eat c and keep an E^* "in reserve."

(Rule 6) Derivation $!E \xrightarrow{c} !E_{1c}$: If we force regular expression $!E$ to eat character c, it is just easy to force E to eat c while keeping a negation operator outside. This works because when E becomes nullable, $!E$ won't be, and vice-versa. Thus we ignore the negation in calculating the derivative, but "keep the negation waiting outside" for the final nullability check. Since nullability is the "final stop" in determining acceptance, this approach works out fine. This can be established by induction on the length of a string $w \in \Sigma^*$ based on the notion of a string derivative $E \xrightarrow{w} E_{1w}$ (extending the notion of a single-character derivative we have defined).

For illustration, let us consider two examples:

Is $ac \in L(!(ab))$?

- Determine this answer through derivatives:
$$!(ab) \xrightarrow{a} !b \xrightarrow{c} !\emptyset$$

- Explanation:
    - We obtain $!(ab) \xrightarrow{a} !b$ through the derivative rule for negation (Rule 6) just now introduced.
    - We obtain $!b \xrightarrow{c} !\emptyset$ again through Rule 6.
- $\emptyset$ is not nullable (see §10.2.2).
- Thus $!\emptyset$ is nullable (see §10.2.2).
- Thus, $ac \in L(!(ab))$.

Is $ab \in L(!(ab))$?

- Determine this answer through derivatives:
$$!(ab) \xrightarrow{a} !b \xrightarrow{b} !\varepsilon$$

- Explanation:
  - We obtain $!(ab) \xrightarrow{a} !b$ through the derivative rule for negation (Rule 6) just now introduced.
  - We obtain $!b \xrightarrow{b} !\varepsilon$ again through Rule 6.
- ε is nullable (see §10.2.2).
- Thus $!\varepsilon$ is not nullable (see §10.2.2).
- Thus, $ab \notin L(!(ab))$.

(Rule 7) Derivation $(E_1 + E_2) \xrightarrow{c} E_{1c} + E_{2c}$*:* Given the task of regular expression $(E_1 + E_2)$ having to eat c, either expression may want to eat c, and thus they morph into E_{1c} and E_{2c} respectively. This explains the derivative for this case.

(Rule 8) Derivation $(E_1 E_2) \xrightarrow{c} ((E_{1c} E_2) + E_{2c})$*:* This is a situation of a concatenation expression $(E_1 E_2)$ wanting to eat c. We are considering the case that E_1 is nullable.

If E_1 is nullable in a concatenation $E_1 E_2$, then we know that E_1 might become ε, thus "exposing E_2 through itself." On the other hand, nullable only means "contains ε" (i.e., E_1 may very well harbor other non-ε matches). That is, E_1 might as well have a c derivative that reflects E_1 eating c.

We therefore have two cases: (1) E_1 "exposing E_2 through itself"[1] (by exercising the ε option), allowing E_2 to process c; and (2) also E_1 being able to process c (with E_2 waiting in reserve after E_{1c}). This explains the derivative obtained for this case.

(Rule 9) Derivation $(E_1 E_2) \xrightarrow{c} (E_{1c} E_2)$*:* This is a situation of a concatenation expression $(E_1 E_2)$ wanting to eat c. We are considering the case that E_1 is *not* nullable. We then have only one case: E_1 being able to process c (with E_2 being the regular expression available after E_{1c}). This explains the derivative obtained for this case.

Brzozowski proved that every regular expression has only a finite number of dissimilar derivatives.[2] This means that the process of obtaining derivatives always terminates.

10.2.2 Nullability Rules

The **nullability check** can easily be turned into a predicate test:

Nullability of c: Regular expression c is not nullable.

Nullability of $\emptyset$: Regular expression $\emptyset$ is not nullable.

Nullability of ε: Regular expression ε is nullable.

Nullability of E^*: Regular expression E^* is nullable.

---

[1] E_1 becomes transparent like clear glass, thus anyone staring at the concatenation $E_1 E_2$ sees E_2.

[2] Two regular expressions R_1 and R_2 are similar if one of them can be transformed to the other using three identities: Idempotence ($R + R = R$), commutativity ($R_1 + R_2 = R_2 + R_1$), and associativity (($R_1 + R_2$) + $R_3 = R_1 + (R_2 + R_3)$). Two regular expressions R_1 and R_2 are dissimilar if they are not similar.

Figure 10.3: A RegExp Scanner and Matcher. Exercise 1 asks you to extend this implementation to cover negation and conjunction.

```python
#!/usr/bin/env python
from rederivparse import *
def opr(E):      return E[0]
def arg1(E):     return E[1][0]
def arg2(E):     return E[1][1]
def arg(E):      return E[1]
def nullable(E):
    if (opr(E) == "str"):
        return False
    elif (opr(E) == "@"):   # '@' is how I represent Epsilon
        return True
    elif (opr(E) == "mty"): # 'mty' is how I represent an empty RE
        return False
    elif (opr(E) == "*"):
        return True
    elif (opr(E) == '+'):
        return nullable(arg1(E)) or nullable(arg2(E))
    elif (opr(E) == '.'):
        return nullable(arg1(E)) and nullable(arg2(E))
    else:
        return "???"
def dv(c, E):
    if (opr(E) == "str"):
        if (arg(E) == c):
            return ("@", "@") # Return derivative "@"; since dv must
        else:                 # return pairs, ("@","@") is returned
            return ("mty", "mty") # Ditto. Make pair ("mty","mty")
    elif (opr(E) == "@"):
        return ("mty", "mty")
    elif (opr(E) == "mty"):
        return ("mty", "mty")
    elif (opr(E) == "*"):
        return (".", (dv(c, arg(E)), E))
    elif (opr(E) == "!"):
        return ("!", dv(c, arg(E)))
    elif (opr(E) == '+'):
        return ("+", (dv(c, arg1(E)), dv(c, arg2(E))))
    elif (opr(E) == '.'):
        if nullable(arg1(E)):
            return \
            ("+", ( ('.', (dv(c,arg1(E)), arg2(E))), dv(c, arg2(E)) ))
        else:
            return ('.', (dv(c,arg1(E)), arg2(E)))
    else:
        return "???"
def matches(w, E):
    if w=="":
        return nullable(E)
    else:
        return matches(w[1:], dv(w[0], E))
if __name__ == "__main__":
    print(' matches("aa", re2ast("!((aaa)*)")) = ',\
          matches("aa", re2ast("!((aaa)*)")))
    re4 = '(a+b)*b(a+b)(a+b)(a+b)'
    nre4 = '!((a+b)*b(a+b)(a+b)(a+b))'
    print(' matches("aabaa", re2ast(re4)) = ', \
          matches("aabaa", re2ast(re4)))
    print(' matches("aabaa", re2ast(nre4)) = ', \
          matches("aabaa", re2ast(nre4)))
```

Nullability of $!E$*:* Regular expression $!E$ is nullable exactly when E is not.

Nullability of $E_1 + E_2$*:* Regular expression $E_1 + E_2$ is nullable if either expression is nullable.

Nullability of $E_1 E_2$*:* Regular expression $E_1 E_2$ is nullable if both expressions are nullable.

10.3 Implementation of Derivative-Based String Matching

Figure 10.3 takes regular expressions (presented as an abstract syntax tree or AST) and implements the pattern matcher. Function `nullable` implements the nullability predicate described in §10.2.2, and function `dv` implements the derivative rules defined in §10.2, Figure 10.2 for derivatives. Note that we use `@` to represent ε and `mty` to represent the empty RE, namely $\emptyset$. Given that the `dv` function is designed to return pairs, we return `("@","@")` in lieu of merely `"@"`, in case the derivative results in `"@"` (the same approach is taken, returning `("mty","mty")` instead of `"mty"`). This way, the `opr` function (which expects a pair to be given to it) will work uniformly for all derivatives. Finally, function `matches(w, E)` checks whether E is nullable when $w = \varepsilon$; else it recursively tries to match the rest of the input string `w` with the derivative of `E` with respect to `w[0]`, the first character of `w`.

The code in Figure 10.4 derives an abstract syntax tree from a string representing a regular expression (that is easier for a human to type-in). The tests at the end of this figure reveal the ASTs created for a few sample regular expressions.[3]

[3] See supplementary material for additional hands-on activity on derivative-based pattern matching at `https://bit.ly/Automata_Jove` under Derivative.

Exercise 10.3, RE derivatives

1. Extend your RegExp matcher and the `re2ast` parser to cover the negation (!) and conjunction (&) operators. These (respectively) negate and intersect regular expressions. Reading [35] can help you in this endeavor.

2. (Optional) Implement the RegExp to DFA generation algorithm described in [37]. □

> **Theorem 10.3:** The function `matches` in Figure 10.3 matches the regular expression E against the string w exactly when $w \in L(E)$.

Proof Sketch: By induction, and appealing to the derivative rules of Figure 10.2 and nullability definition given in §10.2.2. □

10.3.1 Derivatives: Closing Thoughts

The idea of using derivatives for regular expression matching goes through a fairly elegant syntax-directed process, and is pedagogically appealing,

Figure 10.4: A RegExp to AST Compiler. Extend suitably to cover negation implemented via ! and conjunction implemented via & as required by Exercise 1.

```python
from lex import lex
from yacc import yacc
tokens = ('EPS','STR','LPAREN','RPAREN','PLUS','STAR', 'NOT', 'AND')
#--- Tokens
t_PLUS    = r'\+'
t_AND     = r'\&'
t_STAR    = r'\*'
t_LPAREN  = r'\('
t_RPAREN  = r'\)'
t_EPS     = r'\@'
t_STR     = r'[a-zA-Z0-9]'
t_NOT     = r'\!'
t_ignore = " \t" # Ignored characters
def t_newline(t):
    r'\n+'
    t.lexer.lineno += t.value.count("\n")
def t_error(t):
    print("Illegal character '%s'" % t.value[0])
    t.lexer.skip(1)
#--- RegExp Parsing rules
precedence = (('left','PLUS'), ('left','STAR'), ('right','NOT'))
def p_expression_plus(t):
    '''expression : expression PLUS catexp'''
    t[0] = ('+', (t[1], t[3]))
def p_expression_plus_id(t):
    '''expression : catexp'''
    t[0] = t[1]
def p_expression_cat(t):
    '''catexp : catexp ordyexp'''
    t[0] = ('.', (t[1], t[2]))
def p_expression_cat_id(t):
    '''catexp : ordyexp'''
    t[0] = t[1]
def p_expression_ordy_star(t):
    'ordyexp : ordyexp STAR'
    t[0] = ('*', t[1])
def p_expression_ordy_paren(t):
    'ordyexp : LPAREN expression RPAREN'
    t[0] = t[2]
def p_expression_ordy_eps(t):
    'ordyexp : EPS'
    t[0] = ('@', '@')
def p_expression_ordy_str(t):
    'ordyexp : STR'
    t[0] = ('str', t[1])
def p_error(t):
    print("Syntax error at '%s'" % t.value)
def re2ast(s):
    """Convert a RegExp string to an abstract syntax-tree"""
    mylexer  = lex()
    myparser = yacc()
    myparseRETree = myparser.parse(s, lexer = mylexer)
    return myparseRETree
>>> re2ast("!((aaa)*)")
('!', ('*', ('.', (('.', (('str','a'), ('str','a'))), ('str','a')))))
>>> re4 = '(a+b)*b(a+b)(a+b)(a+b)'
>>> re2ast(re4)
('.', (('.', (('.', (('.', (('*', ('+', (('str', 'a'), ('str', 'b')))),
 ('str', 'b'))), ('+', (('str', 'a'), ('str', 'b'))))),
 ('+',(('str','a'), ('str','b'))))), ('+',(('str','a'), ('str','b')))))
```

especially given how straightforward it is to implement negation and intersection. In particular, negation does not rely on NFA determinization that can cause an exponential blowup of the number of states. This process also can be used to obtain DFA directly from regular expressions by taking the derived regular expressions as the names of states (Exercise 2). This process requires the use of "smart constructors" to do efficiently so that we are able to inexpensively determine when a previously generated regular expression is being re-generated through an alternate path (not detecting this duplication increases the number of DFA states). Last but not least, derivatives have also been shown to be a practical idea for context-free parsing [2].

11

Context-Free Languages and Grammars

Chapter Gist: *We present context-free languages through several examples including the familiar "nested parentheses language" and also examples from HTML and C++ (§11.1). We introduce context-free grammars and parse trees (§11.2) and then re-emphasize how to check that you have specified the correct grammar by ensuring consistency and completeness. We present one example grammar design, argue its consistency by induction, and employ the "hill-valley plot" to argue completeness (§11.4). We introduce the notion of ambiguity, show why it is bad, provide ideas to disambiguate (§11.5), and show when we cannot do so because the language itself is inherently ambiguous (§11.6). We show how to express DFA using CFGs, discuss linearity, and why purely linear grammars are DFAs in disguise (§11.7). Before CFG-based parsing was invented, compilers were inscrutably complex, and syntax errors had catastrophic results (§11.8). A Pumping Lemma for CFLs is presented (§11.9) and illustrated. While a language may not be context-free, its complement can be; we present a neat construction (§11.10).*

11.1 Context-Free Language Examples

Consider the set of strings s over parentheses, (and), such that it has an equal number of parentheses, but in every prefix of s,[1] the number of left parentheses is greater than or equal to the number of right parentheses. Examples include (()), (()()), ()(), and ε. This is our first example of a context-free language:

$$L_{Dyck} = \{\, s \in \{(,)\}^* \;:\; \#_((s) = \#_)(s) \text{ and in every prefix of } s, \#_(\geq \#_) \,\}$$

[1] "Every prefix of s" means "when you sweep along from left-to-right." As we sweep left-to-right, we are never at a point where the number of) so far exceeds the number of (. This is a canonical example of a context-free language, and is often called the **Dyck language**.

In addition to parentheses, one could also have other elements that serve to open and close contexts, for instance begin/end blocks, HTML tags that open/close HTML sections, mixtures of bracket types, etc. This pattern exists in arithmetic expressions and generalized bracketing with multiple bracket types:

```
(3 + -4 * (5 + 6 * 7 + 8 * 9 - 4))
{{ ( [[ { [ (( )) ] } ]] ) }}
```

Take a look at the nesting structures found in HTML (left) and C++ (right), that are similar to the generalized bracketing:

```
<!DOCTYPE html>              int main() {
<html>
  <body>                         try { std::vector<int> vec{3,4,3,1}; ...
  <h1>  Heading                  }
  </h1>                          catch ( ... ) {
  <p>   Paragraph                ...
  </p>                           }
</body>                      }
</html>
```

The study of context-free parsing helps us detect (and discard) syntactically malformed HTML/C++ programs; for those meeting the syntactic rules, we proceed to take semantic actions (e.g., draw a webpage or generate code to run the C++ program).

11.2 Context-Free Grammars and Parse Trees

We need a mechanism to specify these "nested bracketing" patterns. By now, we know that regular expressions will not do.[2] Taking a look at the generalized bracketing structure, it is clear that a recursive program will do[3] or a stack-based mechanism will do.[4] This chapter is about the recursive mechanism known as context-free grammars, while Chapter 12 is about the stack-based mechanism known as pushdown automata.

To see how natural a notation context-free grammars are, let us define the Dyck language in English. Here are the cases to consider:

- ε is in L_{Dyck}
- If S is in L_{Dyck}, so is (S). This covers cases such as $(())$
- If S is in L_{Dyck}, so is SS. This covers cases such as $()()$
- Nothing else is in L_{Dyck}

Translating these cases into a convenient syntactic notation, we have a **context-free grammar** for L_{Dyck}.[5]

[2] Why? Think of a very deep nest (...).

[3] Match an outer pair of parentheses, and recursively match inner nested matching structures.

[4] Push each "left" on the stack; when a "right" arrives, match with the stack-top.

[5] We employ " to denote ε in the ASCII syntax.

```
S -> '' | (S) | SS
```

11.2.1 Elements of Context-Free Grammars

A context-free grammar is a four-tuple (N, Σ, S, P), where
- N is a set of **nonterminals**. In L_{Dyck}, S is the only nonterminal.
- Σ is a set of **terminals**. In L_{Dyck}, the terminals are (and). The name "terminals" suggests places when the recursion of the context-free production rules *terminates*. ε itself can be viewed as a terminal, although strictly speaking, it is not. When we define P below, we will allow the right-hand sides of production rules to contain $\{(N \cup \Sigma)^*$. From that point of view, ε (ASCII ' ') is an *empty string of terminals*.
- S is the **start symbol** which is one of the nonterminals. In our example, the start symbol is S.
- P is a set of **production rules** which are of the form:
  - $N \rightarrow \{(N \cup \Sigma)^*$, and read "N derives a string of other N and Σ items." Such strings are called **sentential forms**. A terminal-only string is called a **sentence**.
    * Examples of sentential forms: (S), and (S)()(())SS
    * Examples of sentences: '', (), and ((()))()
  - As a shorthand, one can write multiple right-hand sides separated by vertical bars '|' as we already did. However in actuality, we have three production rules in our example (written without the vertical bar), namely:
    ```
    S -> ''
    S -> SS
    S -> (S)
    ```

11.2.2 Parse Trees, Language of a CFG

Parse trees are trees that depict how the start symbol of a context-free grammar (CFG) can be elaborated to yield any specific string in the language denoted by the CFG. As an example, consider the string s of balanced parentheses where $s = ((())())()$, and the following CFG:

```
S -> '' | (S) | SS
```

We can demonstrate that string s is in the language of L_{Dyck} by constructing a **derivation sequence** that ends with s. One can also achieve the same end by constructing the parse tree for this string (illustrated in Figure 11.1). Here is how we can write the derivation sequence[6] for $((())())()$:
- Start with the start symbol, which is S.

[6] A former professor of mine, Dr. Keshav Nori, used to call this "skyhook skyscraper construction," meaning you build a skyscraper by starting with the start symbol S held with a skyhook, and work toward the foundation.

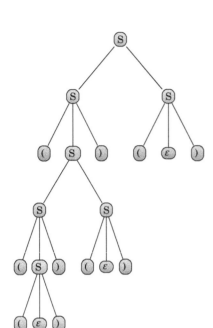

Figure 11.1: Parse Tree for string $((()())()($ that belongs to the Dyck Language. This *sentence* can be read off the parse tree by scanning the leaves left to right (the left-most leaf is of course left-most in the tree order—not its geometric position in the drawing).

- Apply one of the production rules, which rewrites S to a *sentential form*. A *sentential form* is a string over $(N \cup \Sigma)$. In our case, we choose to rewrite S to SS because we observe that
  - the string of interest, *i.e.* $((()())()$ is a concatenation of two instances of S, namely $((()()))$, and $()$, both of which are in the Dyck language. We notate this by `S => SS`
- Observe that the root node of the parse-tree in Figure 11.1 has two S-children. Thus, the first derivation in the construction of the parse tree is `S => SS`
- We now see that the best rule to apply is `S -> (S)`, as we must account for the leftmost (and its matching parenthesis. The sentential form we obtain is now shown by building the derivation: `SS => (S)S`, where the leftmost S yields `(S)`, as can also be seen from the parse tree.
- Building the derivation sequence in this manner, the whole derivation sequence now is as follows, ending in *sentence* $((()())()$:

> S => SS => (S)S => (SS)S => ((S)S)S =>
> (((S))S)S => ((("))S)S => (((())S)S => ((()(S))S =>
> ((()("))S => ((()())S => ((()())(S) => ((()())(") =>
> ((()())()

> **The language of a context-free grammar (CFG)** G is exactly all those sentences that can be obtained by constructing a *derivation sequence*. In mathematical notation, $L(G) = \{ w : S \Rightarrow^* w \}$. The language of a CFG is called a **context-free language.**

Regular languages are characterized by the "lasso shapes" (§4.6) and ultimate periodicity (§8.8). In contrast, context-free languages are characterized by the *tree shape*.

11.3 Avoiding Mistakes in Designing CFGs

It is important to keep in mind that context-free grammars are very often easy to write, but also quite error-prone. Consider these mistakes:

1. What happens if one provides no basis case for the inductive specification (no termination condition for the recursive specification)?

```
S -> (S) | SS  # This grammar is incorrect!
```

Answer: This grammar denotes the empty language $\emptyset$. This can be seen by studying how CFG rules help populate the associated CFL:

(a) Start by assuming that all right-hand side nonterminals denote the empty language ∅.

(b) Now, include the strings generated by the sentences present in the right-hand sides of productions.

- In our correct grammar, we would include '' (or ε) because that is the basis case (and a sentence). In our erroneous grammar, we would include *nothing* (*i.e.*, ∅) as there are no sentences on production right-hand sides.

(c) Now, since (S) is in our grammar's right-hand side, in the correct grammar, we can extend the language generated so far to (ε). This simplifies to ().

- In the erroneous grammar, we cannot so extend. That is, we might generate "(∅)", but because of the ∅ that is stuck in the middle, this language concatenation also yields ∅; thus, we do not make any progress.

(d) Thus, in general, only grammars that include at least one basis case can seed initial strings into the language, and then help build new strings around them. Here is how this approach proceeds with the correct grammar:

  i. Since () is inferred, we can apply the (S) rule and infer that (()) is in the language. Then we can infer ((())), (((()))), etc.

  ii. Also, since () is inferred, we can apply the SS rule and infer that ()() is in the language. Then we can infer ()()(), (())()(), ((()))()(), etc.

  iii. Keep applying the context-free grammar rules, and in the limit, **all and only** those strings you mean to include are in the language.

11.3.1 Completeness and Consistency

In general, context-free grammars must be shown to be **complete** (*i.e.*, they can infer **all** the strings intended to be in the language) and **consistent** (*i.e.*, they do not infer what is not in the intended language). Not having a sentence on the right-hand side of any production rule is an extreme form of *in*completeness. Other less severe forms of incompleteness are possible. For example, the following production rules omit the SS sentential form on the right-hand side of the production rules. These productions are consistent, but incomplete with respect to L_{Dyck} (Exercise: list two strings that are omitted, please):

```
S -> '' | (S)  # Incomplete (and hence incorrect)
```

Inconsistency results if the sentential form)S(is added to the right-hand side (Exercise: list two erroneous strings, please):

```
S -> '' | (S) | )S( | SS  # Inconsistent (and hence incorrect)
```

11.4   The Design of CFG, and the Hill/Valley Plot for Arguing Consistency and Completeness

[7] It should be clear that there are an infinite number of CFGs one can come up with even for one single CFL. For example, for the language ∅, one can come up with the production rule S -> S or *two* production rules S -> A and A -> S.

In Chapter 5, we spent quite some time elaborating *how to design* DFA. In the same manner, there are many steps one can follow to obtain the desired context-free *grammar* for a context-free *language*.[7]

The basic goal of designing a CFG is to come up with an elegant set of rules that are intuitively correct, and (upon being challenged) rigorously shown to be consistent and complete. Many such proofs can be elegantly structured around a rather nice picture which we call the "hill/valley" plot.[8]

Example: For our illustration, consider a different language, namely L_{eqab}, which is exactly the set of all strings that contain an equal number of a's and b's. We can derive a CFG for it by **growing the language inside-out**.[9] More specifically,

[8] In §11.5, we will discuss another *crucial* requirement that virtually all CFGs need to abide by in practice—namely *being unambiguous*— if at all possible. For most languages—except for those that are *inherently ambiguous languages* (§11.6), there always exists an unambiguous grammar.

- ε is in the language, and therefore we must include a production of the form S -> "
- Now, we set up the pattern aSbS, and think inductively in our minds; we also indicate the questions that naturally arise:
  - If S is consistent, then placing an equal amount of a's and b's keeps it consistent.
  - But do we settle for aSb or aSbS? We choose the latter, as it does not force a b-ending.
  - Do we include bSaS? We feel it is necessary, to avoid an a-only ending.
  - Do we throw in an SS case? While doing so won't cause inconsistency, it may be unnecessary for completeness (see below).

Let us assume we don't need the SS case, and present the CFG designed thus far:

[9] We will often observe that many CFGs can be quite naturally designed using this paradigm of growing a language "inside out." CFGs often treat a CFL as an onion, building nice layers around a wispy-thin core. It often is a double- or multi-cored onion.

```
S -> '' | aSbS | bSaS
```

We now prove that this CFG is consistent, and the SS production is not necessary (the CFG is complete with respect to L_{eqab} even without this production rule).

Consistency: By induction, make sure that the right-hand sides of productions are consistent.

- We have '' that is consistent (zero a's and b's)

Figure 11.2: Completeness via Induction Over Hill/Valley Plots

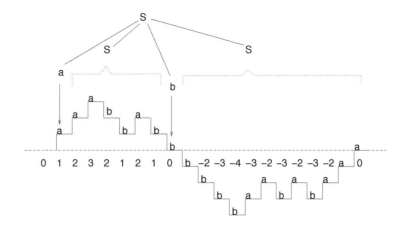

- If all instances of S on the right-hand side are consistent, then aSbS is consistent (it adds one a and one b), and so is bSaS
- Given all cases on the right-hand side are consistent, the grammar is consistent with respect to L_{eqab}.[10]

Completeness: Draw the "hill/valley" plot as in Figure 11.2. This plot presents a good way to visualize how the a-count grows and how the b-count helps offset it:

- The string may begin with a bunch of a's. We show positive growth when a's arrive.
- Then some b's arrive, reducing the excess a count.
- Now, b's may come as a barrage, sending the count negative.
- The diagram can oscillate about the X-axis till it comes to rest at the X-axis.

> Having a good visualization of the language in question is very helpful in arriving at a proof of completeness.

Proof of CFG Completeness

We assume by induction that all *consistent* strings of length $\leq n$ are "covered" by our rules—that is, such strings can all be derived by the given set of productions.[11] Our goal is to now show that longer strings can also be derived. Now consider a string s of length $n + 2$ (the next feasible length).

- We know that s may involve a "zero-crossing" (X-axis crossing) in the hill-valley plot; it may, however, not have such a crossing. We consider the case with zero-crossing now.
- Such a string can begin with an a, and then, at the point of the immediate next zero-crossing, it must have a b.
- After the zero-crossing, the string may ride below the X-axis and come back to rest at the X-axis in the end.

[10] Please keep in mind that a grammar that generates nothing at all is always consistent. This is like making no mistakes at all by choosing to do nothing at all. Therefore, while consistency is important, without completeness, we don't have something that covers all the desired cases.

[11] Thus, we show a little parse tree riding above the hills and valleys in Figure 11.2, and in this parse tree, assume that the two occurrences of S are able to derive the pieces under them.

See supplementary material at https://bit.ly/Automata_Jove under HillValley for a CFG whose design is actually made possible by a proof! That is, we will present a language, and ask you to design the CFG. It turns out that by trying to apply the hill/valley proof methodology, you can actually obtain the CFG! Any *ad hoc* attempt to obtain the CFG will prove difficult, and may contain errors.

- By induction, if you leave out the a and b just now mentioned, the remainder of the string is $\leq n$ in length and is consistent (equal a's and b's).
- This allows us to invoke the induction hypothesis and imagine a derivation from S in two places, as depicted in the figure.
- Now, the whole string can be derived by introducing *one more* S node, and writing the whole string as aSbS, as shown.
- Completeness is proven by considering the bSaS case, and also the case with no zero-crossings.

> Our proof now shows that the SS production is not necessary. Such grammar simplifications are important to carry out (of course, when we are absolutely sure).

Exercise 11.4, CFL completeness

1. Write a proof of completeness when no zero-crossings are involved in the hill-valley plot just mentioned.
2. Write down a context-free grammar for the language L_{a1b2} which is exactly all those strings over $\{a, b\}^*$ in which there are two b's for each occurrence of an a.
3. Argue that this context-free grammar is consistent.
4. Argue completeness as follows (only some of the cases are requested):
   - Draw a hill/valley plot in which you plot the string beginning with an a. For every a, imagine rising two steps, and for every b, falling one step.
   - Now consider the case of no zero-crossings. Argue completeness for this case. Hint: if we start with an a, we go up twice. How is it that we can land back on the X-axis without any zero-crossings (*i.e.* what must such a string end with)?
   - Now consider exactly one zero-crossing, and write the proof of completeness. □

11.5 Ambiguous Grammars, and Disambiguation

In everyday life, ambiguity leads to confusion;[12] with computers, it can result in unintended code being generated.

To illustrate this, we consider the task of parsing expressions. Let us arrive at a CFG for a simple expression language based on these details:

- Arithmetic expressions involving numbers 1, 2 and 3 are to be handled,
- The operators + and *, the unary minus operator ~, and parentheses for grouping expressions must be supported.

You may begin your design by saying these words:

- An expression is either the number 2 or the number 3
- or, it may be the unary minus of an expression

[12] Consider these ambiguous utterances: (1) "Time flies" (time goes fast, or a fly-race?); (2) "Fruit flies like a banana" (fruits can fly? bananas have wings? fruit-flies like bananas?); or (3) "Our mothers bore us" (I love my mom, and she did bear me, but she did not bore me when she bore me!)

- or, it may be two expressions added or multiplied
- or, it may be a parenthetic expression.

A CFG naturally follows (call this **CfgExp1**):

```
E -> 1 | 2 | 3 | ~E | E+E | E*E | (E)
```

We observe these facts:
- E is the only nonterminal.
- Σ consists of 1, 2, 3, ~, +, *, (and).[13]
- S is the start symbol.
- P is a set of seven elementary production rules.

Alas, given an expression 1+2*3, we can obtain two distinct parse trees as shown in Figure 11.3.

> A context-free grammar is **ambiguous** if it admits two distinct parse trees for some string $s \in \Sigma^*$.

Clearly, you know which tree to expect a compiler to be generating inside: the expected answer is 7 and not 9. However, the incorrect parse tree where + got higher precedence resulted in the wrong answer. Clearly, we must prevent such things from occurring within a compiler, the practical solution for which is discussed next.

11.5.1 Disambiguation

Disambiguation is the process by which one changes the grammar to *force* specific parse trees to be built—*without changing the language accepted*. Disambiguation proceeds by *layering* the grammar so that a nonterminal at the higher layer is forced to call a nonterminal at the lower layer. The nonterminal at the lower layer "packages up" some parses and it is only around these packages that higher level nonterminals can erect their parse-tree structures. Given all this, we know that

- '*' must be pushed to the lower layer, so that it gets processed before '+' (which lives in the higher layer).
- Similarly, the unary minus '~' and parenthesization must be situated within the lower layers.
- We denote *expressions* by E, and introduce the notion of a *term* (denoted by T) and *factor* (denoted by F) where E rides high in the hierarchy, T rides below it, and F rides at the lowest level.

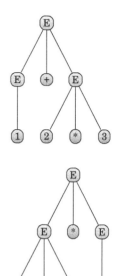

Figure 11.3: Two parses for the same expression (if more than one parse exists for *any* given string, then the **grammar is ambiguous**). The language is likely not inherently ambiguous.

[13] In practice, we will not have each number appearing as a separate terminal. Compilers employ a *scanner* that produces *tokens* which are treated as terminals; in this example, there will be a token called **number** that will be the single terminal representing all numbers. As you can imagine (and will see further in Chapter 12), scanners are realized using finite automata.

Following these principles, the grammar can be changed to the following (call this **CfgExp2**):

```
E -> E+T | T
T -> T*F | F
F -> 1 | 2 | 3 | ~F | (E)
```

As you can see, there are certain production rules that must be involved to process specific operators. For instance, processing a '+' involves the use of the E+T production rule. Now, the operator '+' has lower precedence than a '*'. Thus, when we have an expression involving a '+' and a '*', the processing of the '*' gets pushed into the lower layers.

Exercise 11.5.1, Parse trees

1. Argue that only the correct parse tree results by applying grammar **CfgExp2** to the sentences associated with the parse trees shown in Figure 11.3.
2. Develop all parse trees for the sentence $1+ \sim 2*3$ using **CfgExp1** and **CfgExp2**. Write down all possible answers produced by employing these grammars.
3. Argue that **CfgExp1** and **CfgExp2** denote the same context-free language. Your argument can be a narrative in English. □

Figure 11.4: Even this large expression has a unique parse under CfgExp2. Any guesses on how many parses would exist with CfgExp1? What answer is obtained under CfgExp2? List another answer that would have been returned under CfgExp1.

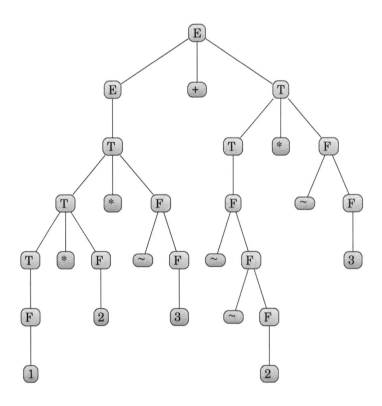

11.5.2 Disambiguation Is Crucial!

With disambiguation in place, even the expression

$$1 * 2* \sim 3+ \sim\sim 2* \sim 3$$

gets parsed as if the user parenthesized so:

$$((1 * 2) * (\sim 3)) + (((\sim (\sim 2)) * (\sim 3)))$$

It emerges to have exactly one parse under CfgExp2, as Figure 11.4 demonstrates by way of example.

11.5.3 Impossibility Results

In Part III of the book (*Concepts*), we will show two rather stunning *undecidability results* about CFGs:

- It is impossible to have an algorithm that, given a CFG, always halts (as all algorithms must do eventually) and prints "yes this CFG is ambiguous" or "no this CFG is unambiguous."
- It is impossible to have an algorithm that, given two CFGs, always halts and prints "yes these CFGs are language-equivalent" or "no these CFGs have a different language."

It is clear that we have left the comfortable territory of finite automata where similar questions could be algorithmically settled!

11.6 Inherently Ambiguous Languages

There are languages for which *every* CFG is ambiguous; they are called **inherently ambiguous** languages. Here is one such language:

$$L_{abORbc} = \{ a^i b^j c^k : (i = j) \text{ or } (j = k) \}$$

We can learn many valuable skills by developing a CFG for L_{abORbc}. Again we follow the "inside-out" approach.

1. Invent nonterminal names much like you would invent function names while programming:
   - Let Mab stand for "matches a and b", and likewise Mbc
   - Let Cs stand for "zero or more c", and likewise As
   - Then we can write this CFG fragment below

   ```
   S -> Mab Cs | As Mbc
   ```

2. Now invent an idiom to grow matched a's and b's inside-out:

   ```
   Mab -> a Mab b | ''
   ```

It is easy to see that if you pump on this production, you will get the "onion-esque" growth:

```
a Mab b
aa Mab bb
aaa Mab bbb
. . .
aaaaaaaaa Mab bbbbbbbbb
aaaaaaaaa " bbbbbbbbb ("poof", the Mab disappears)
aaaaaaaaabbbbbbbbb
```

3. Develop an idiom for As

> As -> a As | "

It is easy to see that if you pump on this production, you will get the "lasso-esque" growth:

```
a As
aa As
aaa As
. . .
aaaaaaaaa As
aaaaaaaaa " ("poof", the As disappears)
aaaaaaaa
```

(Observing the lasso-esque growth, we have actually discovered an idiom for describing DFA using CFG—something we wil pursue in §11.7.)

4. Thus, the final CFG (call it **CFGabORbc**) for L_{abORbc} is

> S -> Mab Cs | As Mbc
> Mab -> a Mab b | "
> Mbc -> b Mbc c | "
> As -> a As | "
> Cs -> c Cs | "

Exercise 11.6, Ambiguous parses

1. Parse aaabbbccc using **CFGabORbc** in all possible ways.
   (Comment: This question at least helps see why L_{abORbc} is inherently ambiguous. The *proof* that L_{abORbc} is inherently ambiguous is much more elaborate. We must not merely consider **CFGabORbc**, but *all possible* CFGs must be considered in such a proof, and all must be shown to result in ambiguity.)

2. Develop a CFG for the language that takes a string w from $\{0,1\}^*$, mirrors it to obtain w^R (reversal of w), and then concatenates w and w^R. Use the "onion approach."

$$L_{wwR} = \{\, ww^R : w \in \{0,1\}^* \,\}$$

Comment: In Chapter 12, we will show that any PDA implementing L_{wwR} will be nondeterministic, even though the CFG you obtain is not ambiguous. **Thus, ambiguity of a grammar implies nondeterminism on the part of the PDA**, as we shall see in the next chapter.

□

11.7 Expressing DFA via CFGs

Observing the lasso-esque pattern in §11.6, we have the beginnings of how to encode DFA using context-free grammars. For definiteness, consider the DFA for L_{eqc} from Chapter 5 shown in Figure 11.5. Clearly, states such as S01 can be encoded using the lasso pattern, as it spins on itself. Let us now describe a complete procedure that handles many more details not yet mentioned; in the process, we will also be converting this DFA into a CFG:

- We must begin in state IF, and that must be the grammar starting symbol also (name it the same, for convenience). Thus we have

  IF -> ...

- When in state IF, we must accept without any further ado. Thus, we include this production:

  IF -> "

- When in state IF, we must transition to F0 or F1 as shown:

  IF -> 0 F0 | 1 F1

- Now, F0 and F1 must also accept without further ado:

  F0 -> "
  F1 -> "

- F0 and F1 have loops, as well; they also can move to states S01 and S10 when (respectively) 1 (0) arrive:

  F0 -> 0 F0 | " | 1 S01
  F1 -> 1 F1 | " | 0 S10

- State S01 must not have the option to accept. It either stays at S01 (when 1 comes) without accepting, or moves to F0 when 0 comes (likewise for S10):

Figure 11.5: DFA for equal changes, or "begins and ends with the same" (we include ε in the language). Recall that state names starting with I are initial states, IF are initial and final, and other starting letters connote neither initial nor final. We will retain the same state names when we define CFG productions as well. You'll notice that F and IF nonterminals will have ε productions, as they are accepting states.

```
S01 -> 1 S01 | 0 F0
S10 -> 0 S10 | 1 F1
```

- This finishes the construction! The whole CFG (call it **BESameCFG**) is now shown:

```
IF -> " | 0 F0 | 1 F1
F0 -> " | 0 F0 | 1 S01
F1 -> " | 1 F1 | 0 S10
S01 -> 1 S01 | 0 F0
S10 -> 0 S10 | 1 F1
```

We now notice an interesting syntactic property of the grammar **BE-SameCFG**:

All its productions have exactly one nonterminal on the right-hand side, and this appears at the very end. Such grammars are **purely right-linear** (§11.7.1).

We now elaborate on such syntactic properties in the following section on *linearity*.

11.7.1 Purely *Right Linear Grammars*

If we have an equation $y = x$ for $x, y \in Real$, we say y is a *linear* function of x; when we have $y = x^2$, we say y is *nonlinear*. This is because two x's are multiplied. By way of analogy (at least for the nomenclature):

- When two nonterminals appear on the right-hand side of a *single production rule*, we call it a **nonlinear** production rule.
- If zero or one nonterminals appear, we call it a **linear** production rule. As a special case:
  - If the single nonterminal (if any) appears as the *last* item of a sentential form, we call it *right linear*.
  - If the single nonterminal (if any) appears as the *first* item of a sentential form, we call it *left linear*.
  - If no nonterminal appears at all, we can regard it either as right-linear or left-linear.

Now, if *all production rules* of a grammar are right-linear, we call it a **purely right-linear grammar**. It turns out that **BESameCFG** was purely right-linear. Here is a theorem.

Theorem 11.7.1: A CFG that is purely right-linear describes a regular language.

Proof Sketch: Observe that such grammars can be turned into a DFA. □

Please bear in mind that

- All regular languages are also context-free. This is because we can write a context-free grammar for any regular language.
- There are some context-free languages that are not regular:
  - The Dyck language is context-free but not regular—because we can apply the *Pumping Lemma* for regular languages and show that this language is not regular.

11.7.2   *Closure,* Purely *Left Linear, and Mixed Linearity*

> **Theorem 11.7.2(a):** Context-free languages are closed under union, Kleene-star, and reversal, but not intersection or complementation.

Proof Sketch: For union, Kleene-star and reversal, see the constructions discussed below. For "but not complementation," see §11.10. Intersection follows from DeMorgan's law.

Constructions for Union, Star, and Reversal of a CFG:

- *Union:* Given two CFGs with start symbols S1 and S2, a new CFG for the union can be created by simply introducing one new production:

  ```
  S -> S1 | S2
  ```

- *Kleene-star:* Given a CFG with start-symbol S, design a new CFG for the Kleene-star by introducing one new production:

  ```
  Star -> S Star | "
  ```

- *Reversal:*[14]

  The proof idea is to reverse each context-free production. We illustrate it on our example CFG **BESameCFG**. By reversing each production rule of that CFG, we obtain the following CFG whose language is the reverse of the language of **BESameCFG**:

  ```
  IFr -> " | F0r 0 | F1r 1
  F0r -> " | F0r 0 | S01r 1
  F1r -> " | F1r 1 | S10r 0
  S01r -> S01r 1 | F0r 0
  S10r -> S10r 0 | F1r 1
  ```

Let us call the above CFG **BESameCFGRev**.

We can see that **BESameCFGRev** is *purely left linear*, in the sense that the right-hand sides of each one of its production rules contains at most only one nonterminal, which occurs leftmost. We know that the language of **BESameCFG** is regular; thus, the language of **BESameCF-**

[14] In his book *Programming Pearls* [8], Jon Bentley demonstrates how to reverse a list AB: reverse A alone, then reverse B alone, then reverse the whole. That is, $(AB)^r = (B^r A^r)$. It is amply illustrated by our beloved canine friends below that perform a reversal as an ensemble by (1) first reversing individually, and (2) then reversing their places.

GRev must also be regular.[15] We have the following theorem:

> **Theorem 11.7.2(b):** A CFG that is purely left-linear describes a regular language.

Proof Sketch: By reversing this CFG, we obtain a right-linear grammar. □

Mixed Linearity Need Not be Regular

By employing mixed linearity, we can do things that obfuscate the structure of the CFG. More specifically, consider this grammar **MiniDyck** which denotes a *non-regular* language:[16]

```
S -> " | (S)
```

We can transform the above CFG without changing its language, obtaining three productions as shown below. The second production is right-linear, while the third production is left-linear (the first production can be considered to be either right-linear or left-linear).

```
S -> "
S -> (A
A -> S)
```

The main take-away observation is that *by mixing linearity, we can express a non-regular context-free language*. The intuitive idea is that production rules that have a mixed linear structure can situate nonterminals in the middle of non-empty strings of terminals. Thus, when these nonterminals "in the middle" recurse, they force us to count and tally the terminal strings surrounding them.

11.8   *Historical Importance of the Theory of Parsing*

All the things you studied in this chapter were not obvious to the computer scientists of the 1960s. In an early article on compiler construction, Donald Knuth mentions many crude rules that compiler writers had devised; for instance:

- They replaced every + by ")))+(((" and every * by "))*(("
- They introduced enough compensating parentheses to fix up the expression. The sequence of processing steps might have been:
  - 1 + 2 * 3
  - 1))) + (((2)) * ((3
  - (((1))) + (((2)) * ((3))) , and voilà!, the multiply got higher precedence assigned.

- To paraphrase Knuth a bit, "by sheer magic" the expressions emerged with the right precedences. But proving this (or being sure about this for all expressions) was a *nightmare*.

Clearly, we cannot leave important things such as operator precedence to chance; they must rest on firm theoretical grounds.

Avoid Brittle Syntax: Lost Space Craft to Venus

One must also develop robust CFGs where "innocuous" syntax errors such as a single forgotten comma must not turn the program into another syntactically legal program with a totally different meaning. Unfortunately, long ago, such brittle syntaxes were in vogue. There is this story of a spacecraft [24] called Mariner 1 launched to Venus in the 1960s which was supposed to have the following lines of code in Fortran:

```
DO 5 K=1,3
...body of DO...
5 CONTINUE
```

Unfortunately, the programmer typed a period "." in place of the comma, turning the statement into:

```
DO5K = 1.3
...body of DO...
5 CONTINUE
```

In the inadvertently altered program, the loop body will be executed exactly once, which changes the program's logic. Instead of performing three iterations of the loop, an accidentally introduced new variable "DO5K" gets assigned the fraction 1.3, with the CONTINUE statement (with or without a label) serving as a no-op. The spacecraft wasn't heard from ever since...[17]

[17] It may still be "in a coma," or "frantically looking for a missing comma."

11.8.1   Combating Inherent Ambiguity

There are many reasons why the existence of inherently ambiguous languages should/could upset you:
- That means *every* compiler can potentially produce two answers for the same input. Fortunately,
  - Most common computer languages (and also scripting languages) are not inherently ambiguous languages.
  - One can always throw in a sufficient number of keywords and "sugar" the language to make it unambiguous.[18]
- More fundamentally, we will show (in Chapter 12) that every CFG can

[18] The parser can "sink its teeth" into these keywords and parse unambiguously.

be converted to a pushdown automaton (PDA).

- This means that any PDA that is associated with a CFG that we come up with for an inherently ambiguous language is going to be *nondeterministic*.

- This means it is impossible to find a *deterministic* PDA that serves as the parser for an inherently ambiguous language.

- This means that the families of NPDA (nondeterministic PDA) and of DPDA (deterministic PDA) are distinct!

- Another reason to wistfully reflect on the territory of finite automata we have left behind—where, for every NFA, there is an equivalent DFA!

11.9   A Pumping Lemma for CFLs

Consider some of the languages studied in the exercises on Page 30 in §3.2 (reproduced below), and also some new languages. Let $\Sigma = \{0, 1\}$. Which of these are regular, which are context-free, and which are neither?

1. $L_{P0} = \{w : w \in \Sigma^*\}$
2. $L_{P1} = \{ww^R : w \in \Sigma^*\}$
3. $L_{P2} = \{waw^R : a \in (\{\varepsilon\} \cup \Sigma), w \in \Sigma^*\}$
4. $L_{eq01} = \{0^n 1^n : n \geq 0\}$
5. $L_{ww} = \{ww : w \in \Sigma^*\}$
6. $L_{w\#w} = \{w\#w : w \in \Sigma^*\}$, where # is a separator.
7. $L_{eq010} = \{0^n 1^n 0^n : n \geq 0\}$
8. $L_{eq012} = \{0^n 1^n 2^n : n \geq 0\}$

It should be clear that all but L_{P0} are non-regular.[19]

The new question now is *which of these languages are context-free?* It should be clear how to write context-free productions for all languages till L_{ww}. However, for L_{ww}, L_{eq010}, and L_{eq012}, attempts made are guaranteed to fail.[20] We prove this using a Pumping Lemma for context-free languages. To derive such a lemma, consider again a "very long string" $w \in L(G)$. It is easy to observe this: **very long strings in a CFL require very tall parse trees!** For example, consider the CFG:

```
S -> ( S ) | T | ''
T -> [ T ] | T T | ''
```

Here is an example derivation:

```
S => (S) => (( T )) => (( [ T ] )) => (( [  ]  ))
             ^                ^
             Occurrence-1  Occurrence-2
             Use T => [T]   Use T => ''
```

[19] As an exercise, you may wish to stop and prove, using the Pumping Lemma, that all but L_{P0} are non-regular.

[20] You might try for a little while and fail. You might also try to "grow these languages inside-out" and fail. There is however a better approach.

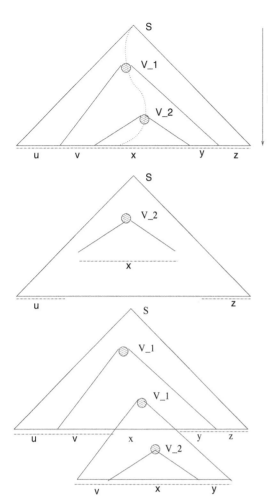

Height |V| + 1
max. branching factor = b

Figure 11.6: Depiction of a parse tree for the CFL Pumping Lemma. The upper drawing shows a very long path that repeats a nonterminal, with the lowest two repetitions occurring at V_2 and V_1 (similar to Occurrence-1 and Occurrence-2 as in the text). With respect to this drawing: (i) the middle drawing indicates what happens if the derivation for V_2 is applied in lieu of that of V_1, and (ii) the bottom drawing depicts what happens if the derivation for V_2 is replaced by that for V_1, which, in turn, contains a derivation for V_2. In our example, from the fact that $((([])))$ exists, we can infer that $(([^i]^i))$ exists for $i \geq 0$. Thus, $v = [$ and $y =]$, while $x = \varepsilon$.

In order to conveniently capture the conditions mentioned so far, it is good to resort to parse trees. Consider a CFG with $|V|$ nonterminals, and with the right-hand side of each rule containing at most b syntactic elements (terminals or nonterminals). Consider a b-ary tree built up to height $|V| + 1$, as shown in Figure 11.6. The string yielded on the frontier of the tree is $w = uvxyz$. The CFL Pumping Lemma can now be applied to deduce "pumped" strings as discussed under Theorem 11.9.

At Occurrence-1, we use the derivation T => [T] and at Occurrence-2, we use T => ".

There are two alternatives illustrated below:

1. Employ T => [T] even at Occurrence-2, and use T => " only in the last phase, or

2. Employ T => " at the beginning itself!

```
S => (S) => (( T )) => (( [ T ] )) => (( [[ T ]] )) => (( [[ ]] ))
                ^              ^              ^
           Occurrence-1  Occurrence-2   Here,
           Use T => [T]  Use T => [T]   use T => ''


S => (S) => (( T )) => (( ))
                ^
           Here, use T => ''
```

```
Given that this         We infer that this       OR, this
derivation exists:      derivation exists:       derivation exists:

==================      ==================       ==================

S => (S)                S => (S)                 S => (S)

   => (( T ))              => (( T ))               => (( T ))

   => (( [ T ] ))          => (( [ T ] ))           => ((   ))

   => (( [   ] ))          => (( [[ T ]] ))

                           => (( [[[ T ]]] ))

                           => ...

                           => (( [[[[[[[[[ T ]]]]]]]]] ))

                           => (( [[[[[[[[[   ]]]]]]]]] ))
```

Theorem 11.9: Given any CFG $G = (N, \Sigma, P, S)$, there exists a number N such that given a string w in $L(G)$ such that $|w| \geq N$, we can split w into $w = uvxyz$ such that $|vy| > 0$, $|vxy| \leq N$, and for every $i \geq 0$, $uv^i xy^i z \in L(G)$.

Proof Sketch: If there are two such parse trees for w, pick the one that has the fewest number of nodes. Now, if we grow the parse tree to height $|V| + 1$ (counting $|V|$ interior nodes and one leaf node), we are guaranteed to *force* a nonterminal to repeat along some path from the root to some leaf of the parse tree. The string $w = uvxyz$ is, in this case, of length $b^{|V|+1}$ (called N, the **pumping constant**, hereafter). Put another way, as soon as we select a string $w = uvxyz$ in $L(G)$, the parse tree for w would have a path from the root to some leaf that repeats a nonterminal.

Let V be the nonterminal that repeats. Call the higher occurrence V_1 and the lower occurrence V_2 (note that V_2 is contained within the parse tree rooted at V_1, as shown in Figure 11.6). If there are multiple instances of V that repeat, pick the lowest two instances, calling them V_1 and V_2, respectively. Now, the following facts hold true:

- $|vxy| \leq N$; if not, we would find two *other* nonterminals that exist lower in the parse tree than V_1 and V_2, thus violating the condition that V_1 and V_2 are the lowest two repeating instances of V.

- $|vy| \geq 1$; if not, we will have $w = uxz$, for which a shorter parse tree exists, namely the one where we directly employ V_2. That is, if $|vy| = 0$, then $vy = \varepsilon$, thus implying that $uvxyz = uxz$, and to generate that, we don't need to involve a parse tree containing both V_1 and V_2 (we can simply use V_2 at the higher position of V and be done).

- Now, by pumping v and y, we can obtain any desired degree of repetitions of v and y. By pumping down, this situation would imply $uxz \in$

$L(G)$, and by pumping up, this situation would imply $uv^i x y^i z \in L(G)$ for $i > 1$.[21]

11.9.1 Application of the CFL Pumping Lemma

We can apply this Pumping Lemma for CFGs in the same manner as we did for regular sets. For example, let us sketch that L_{ww} is not context-free:

- Suppose L_{ww} were a CFL.
- Then the CFL Pumping Lemma would apply.
- Let N be the pumping length associated with a CFG of this language.
- Consider the string $0^N 1^N 0^N 1^N$ which is in L_{ww}.
- The segments v and y of the Pumping Lemma are contained within the first $0^N 1^N$ block, in the middle $1^N 0^N$ block or in the last $0^N 1^N$ block, and in each of these cases, they could also have fallen entirely within a 0^N block or a 1^N block.
- In each case, by pumping up or down, we will then obtain a string that is not within L_{ww}. $\square$

> **Exercise 11.9.1, CFL Pumping Lemma**
> 1. Prove that L_{eq012} is not context-free.
> 2. Prove that L_{eq010} is not context-free.
> 3. Prove that **CFLs are not closed under intersection.** Specifically, show this:
>   (a) $L_{eq01Not2} = \{0^m 1^m 2^n : m, n \geq 0\}$ is a CFL
>   (b) $L_{eqNot012} = \{0^n 1^m 2^m : m, n \geq 0\}$ is a CFL
>   (c) Show that $L_{eq01Not2} \cap L_{eqNot012}$ is not context-free
> 4. Argue why this is true:
>
>> Given that CFLs are closed under union, if they are closed under complementation, then they would be closed under intersection. This immediately leads to a contradiction. Thus CFLs are not closed under complementation. $\square$

11.10 The Complement of a Non-CFL Can Be a CFL

Even though we showed (in §11.9.1) that L_{ww} is not a CFG, we can now show that its complement (call it L_{wwbar}) is context-free (we adapt this argument from [28]).

Any string in L_{wwbar} is one of these types:
- It is of an odd length
- Or it is of an even length, *i.e.* of the form $w_1 w_2$
  - where $|w_1| = |w_2|$, but
  - when we scan w_1 and w_2 using two "cursors," left to right
    * There will be a point where they differ (see illustration below)

[21] For more details, see [42] from which we adapt our proof.

See supplementary material at https://bit.ly/Automata_Jove under CFLPL which contains a detailed presentation of this proof, plus additional illustrations.

```
What w1, w2, and w1w2 look like:

   w1 : ~~~p~~~ 0 ~~~~~~~q~~~~~~

   w2 : ~~~p~~~ X ~~~~~~~q~~~~~~

w1 w2 : ~~~p~~~ 0 ~~~~~~~q~~~~~~ ~~~p~~~ X ~~~~~~~q~~~~~~
```

In other words, there would be some prefix called p that *may or may not* stay the same in both strings, but then w1 has a 0 while w2 has an X. Then there would be a suffix called q that, again, may or may not stay the same. Let us assume that both w_1 and w_2 are over $\{0,1\}^*$ and build a CFG to describe such w_1w_2 patterns, plus all odd-length patterns.

11.10.1   Growing "Inside-Out"

Unfortunately, the w1 w2 string does not show any signs of being growable inside-out. However, a slight transformation will allow us to render it into a shape which immediately suggests "inside-out" growth:

```
Instead of viewing w1 w2 as

  ~~~p~~~ 0 ~~~~~~q~~~~~~ ~~~p~~~ X ~~~~~~q~~~~~~


View w1 w2 as:

  ~~~p~~~ 0 ~~~p~~~ ~~~~~~q~~~~~~ X ~~~~~~q~~~~~~
```

This is a fine rearrangement because the substrings marked "p" and "q" are *arbitrary strings over* 0 *and* 1, and so $qp = pq$! We now hope that you see how you can write such strings via a CFG. We will give a start and let you finish in an exercise. All the CFG rules you have to finish are elided. Let 0 be realized as 0 and X as 1.

```
WWBar        -> Oddlen | EvenlenXO
EvenLenXO    -> GrowOMiddle GrowXMiddle
             | GrowXMiddle GrowOMiddle
GrowOMiddle -> B GrowOMiddle B | 0
B            -> 0 | 1
```

Exercise 11.10.1, CFG design

In all of the following questions, $\#_c(w)$ denotes the number of occurrences of character c in w.

1. Finish the specification of WWBar, a context-free grammar for L_{wwbar}.
   Hints:
   - Express a CFG for Oddlen
   - Write a CFG for GrowXMiddle

2. Test the grammar out by generating at least eight strings using it. Show the derivation sequences used.

3. Develop a CFG for the language of strings over $\Sigma = \{0, 1\}$ where the number of 1's is strictly greater than the number of 0's. *Hint:* Draw "hill/valley" plots covering various sub-cases that arise in such strings. Specifically, rise one step for every 1 and fall one step for every 0. All such plots then end up above the X axis. Cut-up such plots into chunks and develop a grammar covering the sequence of all such chunks.

4. Here is a proof that attempts to show that the language $L_{eq01} = \{0^n 1^n : n \geq 0\}$ is not context-free:

   (a) Assume there is a CFG with $|V|$ productions for this language.

   (b) Choose $N = 2^{|V|+1}$ as in the CFL Pumping Lemma

   (c) Consider the string $0^N 1^N \in L_{eq01}$

   (d) Consider the band of 0's in 0^N. It harbors a pump vxy which is surely less than or equal to N in length.

   (e) By pumping, we make the number of 0's differ from the number of 1's. Hence the language is not context-free.

   Find the flaw in the proof and propose a CFG for this language.

5. Argue that this language is a CFL by building a CFG for it. Answer for both cases of 'OP' listed below:

   $$L_{abcd} = \{a^i b^j c^k d^l : i, j, k, l \geq 0 \text{ and } ((i = j) \, OP \, (k = l))\}$$

   (a) (Case 1) OP is AND

   (b) (Case 2) OP is OR

6. Show that L_{acbd} is not context-free if OP is AND but is context-free if OP is OR:

   $$L_{acbd} = \{a^i c^k b^j d^l : i, j, k, l \geq 0 \text{ and } ((i = j) \, OP \, (k = l))\}$$

7. Someone proposes the following CFG for the language

   $$L_{abcd} = \{w : w \in \{a, b, c, d\}^* \text{ and } \#_a(w) = \#_b(w) \text{ and } \#_c(w) = \#_d(w)\}$$

   ```
   S -> a S b S | b S a S | '' | T
   T -> c T d T | d T c T | '' | S
   ```

   (a) Find one string in this CFG's language that is **not** in L_{abcd}.

   (b) Show that L_{abcd} is not context-free.   □

12

Pushdown Automata

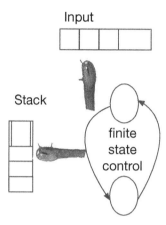

Input

Stack

finite
state
control

Figure 12.1: A Pushdown Automaton
shown with its input string and its stack.
The input of a PDA is read left-to-right
once. A PDA cannot rewind its read position back to an earlier point in the string
and re-read an input. This is the same as
with NFA. (Turing machines—discussed
in Chapter 13—on the other hand have
the ability to re-read their input tape.)
The stack can be pushed into and popped
from during each transition. We depict
the PDA as having two "eyes" that can be
moved independently (we have shown images of the actual eyes of a fiddler crab
that are mounted on long stems, and can
be moved independently).

Chapter Gist: *Pushdown Automata (PDA) receive an elaborate informal introduction (§12.1) followed by a formal one (§12.2). We explore a PDA for L_{Dyck} using Jove (§12.3), and follow this with in-depth descriptions of many curious executions of PDA on made-up examples (§12.4). These include stack-limited executions, and a discussion of preventing infinite-looping in Jove (§12.4.1). We then present more practical examples in the Automd markdown notation (§12.5). We discuss CFG to PDA conversion, including non-deterministic executions of PDA derived from ambiguous CFGs (§12.6). This book presents you with the opportunity to study three parsers, summarized in §12.7.*

12.1 Pushdown Automaton Basics

Pushdown automata (PDA) are machines that recognize the structure in
the input string with the aid of a finite-state control mechanism whose actions are governed by the string as well as the contents of an **unbounded
stack**. The stack contents are created by the PDA itself, and includes information gathered from the input seen so far and also symbols that the
PDA itself puts into the stack (to "remember" or "mark" important junc-
tures).[1] As a simple example, in order to recognize strings from L_{Dyck},
a PDA can use the stack to store left parentheses, '('. When right parentheses, ')', arrive in the input stream, the PDA can pop the stored left
parentheses and match against them.

A PDA (Figure 12.1) always starts from a single initial state with input
consisting of a string over Σ^* to be processed on the "input tape". PDA
transitions are governed by a somewhat complex **edge-label** type that
has the following structure:

oneInChrOrEps **,** oneStackSymOrEps **;** stringOfStackSyms

[1] We arrange it so that when the PDA is
"switched on," the stack already contains
#, the *bottom of the stack marker*. This
pre-arranged symbol is, strictly speaking,
not pushed in by the PDA; all others are
pushed in by the PDA. Things pushed into
the stack may come from the input alpha-
bet Σ or the stack alphabet Γ. Note that
$\Sigma \subset \Gamma$ because # is not allowed in Σ.

A PDA can, during any one of its transitions, read a single input symbol (as with an NFA or a DFA), if any are available to be read. A PDA can also **ignore the input** during a transition that specifies ε in the input position (as with an NFA). For these reasons, we call the part of the edge-label before the comma `oneInChrOrEps`.

Symbols from the input stream are read **left-to-right**. Each read advances the read position.[2] A PDA **cannot go back** and read an input it has already read (all these are similar to an NFA or a DFA).

In addition to the input stream, a PDA can also read a single *stack symbol* during any one of its transitions. Stack symbols come from the alphabet Γ which is a superset of Σ. In fact, it is a *proper* superset in that Γ always contains one special symbol, namely #, that is never allowed within Σ. Symbol # has the significance of being a **bottom-of-the-stack** marker. To reflect these cases, we call the part of the edge-label after the comma but before the semicolon `oneStackSymOrEps` or 'stack-read position' for short.

When a PDA starts, its stack contains just the symbol #, signifying that its stack is empty.[3] During any transition, the PDA's **stack can also be ignored** by a transition that specifies ε in the stack-read position. Only the symbol at the top of the stack can ever be read. Reading the stack top "pops" that symbol. *It is not possible to read the stack top without popping it.*[4] A PDA transition may also push a string of (*zero or more*) stack symbols during each transition.[5] This string is denoted by the component `stringOfStackSyms`.

[2] The "input eye" moves over to stare at the next input cell.

[3] With the bottom-of-the-stack marker saying "please do not look beyond me."

[4] The "stack eye" of a PDA always winds up staring at the top of the stack.

[5] We often say "stack this stack-string, which comes from Γ^*."

Example PDA for L_{Dyck}: A PDA for the Dyck language is given in Figure 12.2. The markdown description of this very PDA is also given below. We now walk through this example PDA.

```
pdaDyck = md2mc('''PDA
IF : (, # ; (# -> A  !! Push ( when stack top has #
A  : (, ( ; (( -> A  !! Push later-arriving ( if stack top is (
A  : ), ( ; '' -> A  !! Cancel ) and most recent (. Push nothing
A  : '',# ; #  -> IF !! When all ( ... ) match, accept. Head back
                    !! to state IF (ready to roll again!)''')
DOpdaDyck = dotObj_pda(pdaDyck, FuseEdges=True)
DOpdaDyck  # Draws the PDA
```

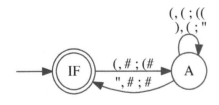

Figure 12.2: A PDA for the Dyck language

oneInChrOrEps: This position in a transition could be ε (shown as ' ', as in the A-to-IF transition), or non-ε. In our example, the first non-ε case is the '(' annotation labeling the IF-to-A transition. Another non-ε annotation is the '(' annotation labeling the A-to-A transition. The third non-ε annotation is the ')' annotation labeling the A-to-A transition.

In the A-to-IF transition, even with an actual input string present, a PDA taking such a transition must ignore the input, given that the oneInChrOrEps position is '' (ε). When non-ε labels are involved (*e.g.*, when a '(' or ')' is mentioned), taking those transitions means *actually reading those input symbols, and advancing the "input eye"*.

oneStackSymOrEps: If this position in a particular transition is '', the PDA ignores the stack string when this transition is taken. (No such transitions are present in our example PDA.) However, a single character might be mentioned in this position (in our example, both the A-to-A transitions mention a '(' while the A to IF transition mentions a '#'). In any of these cases, the PDA *must be able to read this character* from the top of the stack in order for the transition to be taken. Thus, in our example, the A-to-A transitions depend on the stack top being a '(', while the IF-to-A and A-to-IF transitions depend on this position being a '#'. During a transition governed by a non-ε entry sitting at the stack top, the stack top *must be* popped when the top-of-the-stack is read. The "stack eye" is then left staring at the *new* top-of-the-stack.

stringOfStackSyms: In our example PDA, there is one A-to-A transition with ε serving as the stringOfStackSyms entry (see the '' annotation). The other A-to-A transition carries the annotation '((' . In addition, there is an IF-to-A transition with annotation '(#', and finally an A-to-IF transition with annotation '#'. The important thing to note is that *the contents of this position do not influence whether the PDA takes its transition or not.*[6] It only specifies what gets pushed onto the stack during the transition.

Suppose the string stringOfStackSyms on a particular transition is of the form abcd. Then, when that transition is taken, string abcd is pushed onto the stack, where a ends up top-most on the stack, with b below a, c below b, and finally d below all of them. If stringOfStackSyms were to be ε (''), then nothing gets pushed.

Thus, in our example, the stack strings pushed are one of '((' (during an A-to-A transition), '(#' (during an IF-to-A transition), or a '#' (during an A-to-IF transition). Now, pushing the string '(#' during an IF-to-A transition means:

- Push '#' first;
- Then push '(', which now sits at the top of the stack.

Simulation of a PDA for L_{Dyck}

We now present a PDA for L_{Dyck} and describe its overall operation (see Figure 12.2). Jove's markdown allows us to easily input such a PDA. At the end of this markdown, we see that we are evaluating the dot object expression DOpdaDyck; this produces the PDA drawing of Figure 12.2 in the Jupyter console.[7] This PDA starts in state IF, and thus accepts ε

[6] We use the **;** as a separator to denote that what follows it is *after* the PDA has made a decision to transition. We will employ such a semicolon also in Chapter 13 on Turing machines.

[7] FuseEdges combines multiple transition labels into one, stacking them. To catch mistakes, it is good to see the full PDA drawn without FuseEdges first, making sure that all the intended edges are present, and *then* invoke FuseEdges.

(without reading any input). It can move from IF to A by reading '(' from the input and popping '#' from the top of the stack. However the stack is quickly restored to have '(#' in it. Thus, the PDA remembers the '(' that came in, after restoring the '#' below it. It stays in state A so long as '(' keep coming in the input stream (all these are stacked).[8] When a ')' comes, the PDA throws away one '(' from the top of the stack, never restoring it (essentially tallying the left and right parentheses). Now when '#' peeps from the top of the stack, the PDA restores this '#' and attains state IF, satisfied that the counts of the parentheses have matched.[9]

Before we detail PDAs, it is helpful to mention some high level facts and also remind the reader that context-free grammars can be converted to push-down automata, and vice versa. We will illustrate CFG to PDA conversion in this chapter. PDA to CFG conversion will be briefly mentioned in Part III of this book, at a high level.

Also, there are two types of PDA: **final-state acceptor** and **empty-stack acceptor**. Any final-state acceptor PDA can be converted to an empty-stack acceptor PDA, and vice versa. Most PDA we discuss are final-state acceptors. Empty-stack acceptor PDA matter because they are particularly amenable to being converted into CFGs.[10]

> Final-state acceptors and empty-stack acceptors differ only in one way: the manner in which the *acceptance of an input string* is defined. Otherwise, all the definitions in this chapter apply equally to both types of PDA.

12.2 Formal Description of PDA

Pushdown Automata are structures[11]

$$(Q, \Sigma, \Gamma, \Delta, q_0, z_0, F)$$

where

- Q is a finite non-empty set of states,
- Σ is a finite non-empty input alphabet,
- Γ is a finite non-empty stack alphabet (subsumes Σ),
- Δ is a transition function (see details below),
- q_0 is the starting state,
- z_0 is the initial stack's lone contents (for us, it is #), and
- F is the final set of states.

Δ's signature reflects how it can optionally consume an input symbol or pop a stack symbol, but **nondeterministically** select a pair consisting of a next state to attain and a string to be pushed onto the stack:

[8] We momentarily lose '(' on top of the stack; but in a femtosecond or less, the PDA pushes back '((', thus recording one new '('.

[9] It may find itself getting kicked back to state A for "another tour of duty" when another '(' comes in.

[10] Less painfully so...

[11] Most books (unnecessarily) differ on their definition of a PDA. We surveyed a wide array, finally settling on the one in Sipser's book, as it greatly reduces user burden. We also make things more convenient by preloading # on top of the stack.

$$(Q \times (\Sigma \cup \{\varepsilon\}) \times (\Gamma \cup \{\varepsilon\})) \rightarrow \mathscr{P}(Q \times \Gamma^*)$$

12.2.1 Acceptance, Deterministic PDA

Now we define PDA acceptance:

- An **instantaneous description** (ID) is a triple:
  (controlState, remainingInput, currentStack).
- A finite sequence of IDs where the first ID starts at *controlState* q_0 and each following ID is obtained by applying PDA's transition function Δ is a *computation*.
- A computation is *driven by string s* if it starts with an ID having *remainingInput s*, and ends with an ID with *remainingInput ε*.[12]
- A final-state acceptor PDA accepts an input string s when a computation driven by s ends in an ID (q_f, ε, g) for some $q_f \in F$ and for some $g \in \Gamma^*$. That is, the control reaches a final state and the input is "all gone." (The ending stack does not matter.)
- An empty-stack acceptor PDA accepts an input string s when a computation driven by s ends in an ID $(q_{any}, \varepsilon, \varepsilon)$ for $q_{any} \in Q$ (ending control state does not matter). The input and stack are "all gone."[13]

A Note on Deterministic PDA: A PDA is considered deterministic if it has exactly one enabled action whenever it enters any state. This high-level definition is sufficient for our purposes. The topic is a bit more elaborate than we have room to adequately discuss. Excellent descriptions of this topic may be found in many books (*e.g.*, Kozen's book "Automata and Computability.")

> In this book, we discuss only nondeterministic PDA. In fact, when we discuss Figures 12.12 and 12.13, you will understand why nondeterminism is essential for applications such as parsing based on arbitrary CFGs.

12.3 Exploring the PDA for L_{Dyck} Using Jove

We can simulate PDA on input strings through the `explore_pda` function as shown below. We then witness a computation consisting of IDs connected by transition arrows `->`. For better understanding, we add comments that begin with `#...` in the printouts that follow:

Run-1:

```
explore_pda("", pdaDyck)
```

[12] Input is "all gone."

[13] An optional parameter ACCEPT_S provided to Jove's `explore_pda` command selects the *accept by empty stack* policy. Otherwise, by default, Jove sticks with the *accept by final state* policy. In a sense, starting a Jove PDA with # on its stack is an "insurance policy" that it won't just accept by finding the starting stack to be empty. This helps keep your usage of Jove less error-prone.

```
String  accepted by your PDA in 1 ways :-)
Here are the ways:
Final state  ('IF', '', '#')      #... the starting state accepts
Reached as follows:               #... (notice # on initial stack)
-> ('IF', '', '#') .
```

Run-2:

```
explore_pda("()", pdaDyck)
```

```
String () accepted by your PDA in 1 ways :-)
Here are the ways:
Final state  ('IF', '', '#')
Reached as follows:
-> ('IF', '()', '#')          #... Starting ID
-> ('A', ')', '(#')           #... Consume (, push (, goto A
-> ('A', '', '#')             #... Tally with ), pop (
-> ('IF', '', '#') .          #... Accept
```

Run-3:

Finally, the input ()()(()) shows the sequence of push/pop actions and input actions. See Figure 12.3 (reproduced for your convenience).

```
explore_pda("()()(())", pdaDyck)
```

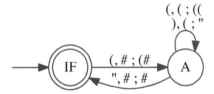

Figure 12.3: A PDA for the Dyck language (repeated for your convenience)

```
String ()()(()) accepted by your PDA in 1 ways :-)
Here are the ways:
Final state  ('IF', '', '#')
Reached as follows:
-> ('IF', '()()(())', '#')
-> ('A', ')()(())', '(#')
-> ('A', '()(())', '#')
-> ('IF', '()(())', '#')
-> ('A', ')(())', '(#')
-> ('A', '(())', '#')
-> ('IF', '(())', '#')
-> ('A', '())', '(#')
-> ('A', '))', '((#')
-> ('A', ')', '(#')
-> ('A', '', '#')
-> ('IF', '', '#') .
```

Run-4:

Finally, let us provide an input that must not be accepted:

    explore_pda("()()(()", pdaDyck)

```
String ()()(() rejected by your PDA :-(
Visited states are:
{('IF', '()()(()', '#'), ('A', ')()()', '(#'), ('A', '(()', '#'),
 ('A', '()', '(#'),        ('IF', '()(()', '#'), ('A', '', '(#'),
 ('A', ')()(()', '(#'), ('A', ')', '((#'),       ('IF', '(()', '#'),
 ('A', '()(()', '#')}
```

The printout shows that the PDA did go through many IDs, but found none to be accepting.

12.4   PDA Behavior Through Examples

Given how elaborate PDA behaviors are, we now present a series of examples that introduces PDA through Jove simulations.

Example: Consider pda1 (Figure 12.4) with its markdown description and its execution on the input shown in the explore_pda command:

```
pda1 = md2mc('''PDA
I : a, b ; c -> F    ''')
DOpda1 = dotObj_pda(pda1, FuseEdges=True)
DOpda1 # Draws the PDA


explore_pda("a", pda1)
String a rejected by your PDA :-(
Visited states are:
{('I', 'a', '#')}
```

The PDA rejects the string (is "stuck" at the ID shown in the visited states list above). The stack letter sought is 'b' whereas the stack-top has #.

Example: Consider pda2 (Figure 12.5) with its markdown description and its execution on the input shown in the explore_pda command:

```
pda2 = md2mc('''PDA
I : a , b  ; c  -> F
I : '', '' ; d  -> A
A : '', d  ; '' -> F    ''')
DOpda2 = dotObj_pda(pda2, FuseEdges=True)
```

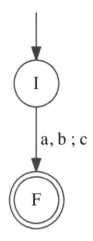

Figure 12.4: pda1, a simple PDA

```
DOpda2 # Draws the PDA


explore_pda("a", pda2)
String a rejected by your PDA :-(
Visited states are:
{('A', 'a', 'd#'), ('F', 'a', '#'), ('I', 'a', '#')}
```

Even though the ID in Visited states includes ('F', 'a', '#'), notice that the input isn't fully consumed (actually it was not consumed at all), and so merely reaching a final state isn't enough for acceptance.

Example: Consider pda3 (Figure 12.6) with its markdown description and its execution on the input shown in the explore_pda command:

```
pda3 = md2mc('''PDA
I : a , b ; c  -> F
I : '', '' ; d  -> A
A : a , d ; ''  -> F   ''')
DOpda3 = dotObj_pda(pda3, FuseEdges=True)
DOpda3 # Draws the PDA

explore_pda("a", pda3)
String a accepted by your PDA in 1 ways :-)
Here are the ways:
Final state  ('F', '', '#')
Reached as follows:
-> ('I', 'a', '#')
-> ('A', 'a', 'd#')
-> ('F', '', '#') .
```

Finally, we provide a path to acceptance. Notice how '#' is **not** popped during the I to A move, and in addition, d gets pushed. This d enables the move to F which consumes a, causing acceptance.

Example: Consider pda4 (Figure 12.7) with its markdown description and its execution on the input shown in the explore_pda command:

```
pda4 = md2mc('''PDA
I : a , # ; c  -> F
I : '', '' ; d  -> A
A : a , d ; ''  -> F   ''')
DOpda4 = dotObj_pda(pda4, FuseEdges=True)
DOpda4 # Draws the PDA
```

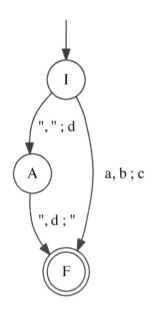

Figure 12.5: pda2, a bigger PDA

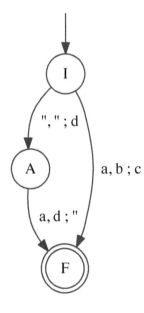

Figure 12.6: pda3, a variant of pda2

```
explore_pda("a", pda4)
String a accepted by your PDA in 2 ways :-)
Here are the ways:
Final state ('F', '', 'c')
Reached as follows:
-> ('I', 'a', '#')
-> ('F', '', 'c') .
Final state ('F', '', '#')
Reached as follows:
-> ('I', 'a', '#')
-> ('A', 'a', 'd#')
-> ('F', '', '#') .
```

Two paths to acceptance (two sequences of IDs) exist; essentially, state I proves to be *nondeterministic*.

Example: Consider pda5 (Figure 12.8) with its markdown description and its execution on the input shown in the explore_pda command:

```
pda5 = md2mc('''PDA
I : a , # ; c  -> F
I : '', '' ; d  -> A
A : a , d ; ''  -> F    ''')
DOpda5 = dotObj_pda(pda5, FuseEdges=True)
DOpda5 # Draws the PDA

explore_pda("a", pda5)
String a accepted by your PDA in 2 ways :-)
Here are the ways:
Final state ('F', '', 'c')
Reached as follows:
-> ('I', 'a', '#')
-> ('F', '', 'c') .
Final state ('F', '', '#')
Reached as follows:
-> ('I', 'a', '#')
-> ('A', 'a', 'd#')
-> ('F', '', '#') .
```

This PDA introduces a self-loop at A. Our PDA simulator in Jove is smart not to enter an infinite loop here. It still does not record additional accepting paths, as the self-loop is essentially ignored.

Example: Consider pda6 (Figure 12.9) with its markdown description and its execution on the input shown in the explore_pda command:

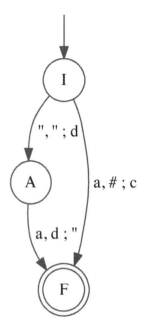

Figure 12.7: pda4, a variant of pda3

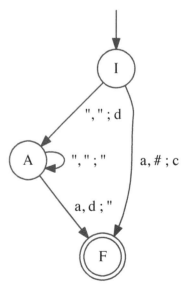

Figure 12.8: pda5, a variant of pda4

Figure 12.9: pda6, a non-trivial (made-up) PDA

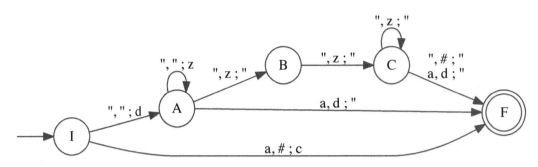

```
pda6 = md2mc('''PDA
I : a, #  ;  c  -> F
I : '', ''; d  -> A
A : '', ''; z  -> A
A : '', z ;  ''  -> B
B : '', z ;  ''  -> C
C : '', z ;  ''  -> C
C : '', # ;  ''  | a, d; ''  -> F    # Two paths from C to F
A : a, d ;  ''  -> F              ''')
DOpda6 = dotObj_pda(pda6, FuseEdges=True)
DOpda6 # Draws the PDA
```

The initial explore_pda("a", pda6) command only gave two accepting paths—*really baffling, right?*[14] Given that this does not "seem right," the user can invoke the "chatty" option, whereupon more information gets printed, as shown below:

```
terminal_id_path =
 [(('A', 'a', 'zd#'), [('I', 'a', '#'), ('A', 'a', 'd#')])]
final_id_path =
 [(('F', '', 'c'), [('I', 'a', '#')]),
  (('F', '', '#'), [('I', 'a', '#'), ('A', 'a', 'd#')])]
visited_ids =
{('A', 'a', 'd#'),('F', '', '#'),('I', 'a', '#'),('F', '', 'c')}
String a accepted by your PDA in 2 ways :-)
Here are the ways:
Final state ('F', '', 'c')
Reached as follows:
-> ('I', 'a', '#')
```

[14] See Figure 12.9 reproduced for your convenience. (1) There is a direct I to F path which can accept an a, and the stack condition matches. (2) Then, there is an A to F path seeking d on top of the stack, but the I to A transition puts d on top of the stack to enable it. (3) Finally, the I,A,B,C,F has the potential to push a d and then some number of zs, but this path can also remove two ds, thus exposing the d on top of the stack.

```
->  ('F', '', 'c') .
Final state  ('F', '', '#')
Reached as follows:
->  ('I', 'a', '#')
->  ('A', 'a', 'd#')
->  ('F', '', '#') .
```

- The `terminal_id_path` list is all the IDs from which the PDA has no transition enabled. Not only is the terminal ID (`'A'`, `'a'`, `'zd#'`) printed, but the path leading to this ID is also printed (namely, as a list [(`'I'`, `'a'`, `'#'`), (`'A'`, `'a'`, `'d#'`)]). This helps debug the situation.
- The `final_id_path` similarly prints the final ID together with the path leading to it.
- The `visited_ids` is a set of IDs that were visited (this is stored to help the PDA avoid looping).

We observe that the stack component of the ID did not grow very much.

12.4.1   Rerunning `pda6` with Larger Stack Allowed

We can invoke the `explore_pda` command with an additional option called `STKMAX` shown:

```
explore_pda("a", pda6, STKMAX = 3, chatty=True)
```

Here is what is going on. What we are doing in Jove is allowing the PDA to explore all the nondeterministic options in a breadth-first manner. Unfortunately, the PDA language is quite unrestricted. See state A which, without reading the input or stack, *keeps pushing* z onto the stack! In general, a PDA may do many "dangerous-looking" things with its stack:

- It may push a large number of symbols in a loop.
- It may decode delicate combinations of stack contents by sequentially looking for a collection of items to be on the stack.

Thus, **a naïvely written PDA simulator can easily go into an infinite loop!** However, you may complain saying that "the **acceptance problem for PDA is decidable**" (something that will be proven in Part III of our book). This basically says that[15]

> There is an algorithm that, given a PDA and an input, can print, in a finite amount of time "yes, this PDA will accept this input" or "no, this PDA won't accept this input."

Thus, it should be possible to tell whether an NPDA simulation will engage in an infinite non-accepting run (sequence of IDs that does not include the accept by final state or accept by final stack condition), or not. However, the key difficulty in writing a simulation tool that aims for

[15] The proof for this theorem will proceed as follows (see Part III for details). Obtain the string and represent it as a DFA. Intersect this DFA with the given PDA. See if the resulting PDA's language is empty or not.

simplicity is that *there isn't an easily implemented localize check at a state to detect whether a PDA is going to loop when a certain path is pursued.*

Thus, the following practical solution is adopted in Jove:[16]

> As a practical solution, Jove stores visited IDs. It also allows users to specify a constant called STKMAX. Jove then simulates till one of the newly generated IDs contains a stack string that is STKMAX longer than the stored ID's stack string. When that happens, simulation is cut off.

The printout below shows that with STKMAX = 3, Jove indeed explores many more IDs, and finds *three different ways* to accept an a input.

```
terminal_id_path =
[(('A', 'a', 'zzzd#'),
  [('I', 'a', '#'),('A', 'a', 'd#'),('A', 'a', 'zd#'),('A', 'a', 'zzd#')]),
  (('F', '', '#'),[('I', 'a', '#'),('A', 'a', 'd#'), ('A', 'a', 'zd#'),
    ('A', 'a', 'zzd#'), ('B', 'a', 'zd#'), ('C', 'a', 'd#')])]


final_id_path =
[(('F', '', 'c'),[('I', 'a', '#')]),
  (('F', '', '#'),[('I', 'a', '#'),('A', 'a', 'd#')]),
  (('F', '', '#'),[('I', 'a', '#'),('A', 'a', 'd#'), ('A', 'a', 'zd#'),
                   ('A', 'a', 'zzd#'),('B', 'a', 'zd#'),('C', 'a', 'd#')])]


visited_ids = {('B', 'a', 'd#'), ('A', 'a', 'zd#'), ('F', '', 'c'),
               ('B', 'a', 'zd#'), ('A', 'a', 'd#'), ('C', 'a', 'd#'),
               ('F', '', '#'), ('A', 'a', 'zzd#'), ('I', 'a', '#')}
String a accepted by your PDA in 3 ways :-)
Here are the ways:
Final state  ('F', '', 'c')
Reached as follows:
-> ('I', 'a', '#')
-> ('F', '', 'c') .
Final state  ('F', '', '#')
Reached as follows:
-> ('I', 'a', '#')
-> ('A', 'a', 'd#')
-> ('F', '', '#') .
Final state  ('F', '', '#')
Reached as follows:
-> ('I', 'a', '#')
-> ('A', 'a', 'd#')
-> ('A', 'a', 'zd#')
-> ('A', 'a', 'zzd#')
-> ('B', 'a', 'zd#')
-> ('C', 'a', 'd#')
-> ('F', '', '#') .
```

[16] An approach that maintains actual contexts (and avoids the STKMAX-based approach) is described in [7]. In this paper, Ball and Rajamani consider the problem of computing whether a statement label is reachable within (so called) *Boolean programs*. Boolean programs are C programs that employ a collection of mutually recursive functions, with the restriction that each function can employ only a finite number of finite variables in its local scope. Also, only a finite number of finite variables are permitted to be globals. It turns out that such programs are formally equivalent to nondeterministic PDA, and statement label reachability is equivalent to PDA-acceptance.

Ball and Rajamani's work was employed by Microsoft to check for (and actually find) deep-seated bugs in Windows device drivers by checking the driver code against a set of rules that define what it means for a device driver to properly interact with the Windows operating system kernel. This work won Ball and Rajamani the prestigious 2011 CAV (Computer-Aided Verification) Award in July 2011, and is detailed in [6].

Exercises

1. Briefly explain the first two accepting runs.
2. In detail, explain the last accepting run of this PDA, describing how each ID evolves into the next ID.

12.5 Toward More Practical PDA

Given a tool such as Jove, it is fun to solve design challenges and understand PDA behavior. Let us develop a PDA for the language (example from Sipser's book):

$$L_{abORac} = \{a^i b^j c^k : i, j, k \geq 0, \text{ and } (i = j) \text{ or } (i = k)\}$$

> PDA design is very low-level programming, and **highly error-prone**. Unless you adopt good practices laying out your PDA code and *commenting every line*, your code will be inscrutable, and nobody will be able to debug it for you.

Our Jove markdown notation encourages a convenient syntax that encourages comments. If you write this much in this notation, you immediately obtain the PDA diagram and can begin its simulation. Thus, you get the benefit of a text syntax and automatic (neat) layout generation:

```
f27sip = md2mc('''
PDA
!!-------------------------------------------------------------------------
!! This is a PDA adapted from Sipser's book. It either matches a's and b's,
!! ignoring c's; or matches a's and c's, ignoring b's in the middle.
!! Thus, the language is a^m b^m c^n union a^m b^n c^m.
!!-------------------------------------------------------------------------
iq2 : a  , ''  ; a      -> iq2    !! Stack a's.
iq2 : '' , ''  ; ''     -> q3,q5  !! Split non-det for a^m b^m c^n (q3)
                                  !! or a^m b^n c^m (q5).
q3  : b  , a   ; ''     -> q3     !! Match b's against a's.
q3  : '' , #   ; ''     -> fq4    !! Hope for acceptance when # surfaces.
fq4 : c  , ''  ; ''     -> fq4    !! Be happy so long as c's come.
                                  !! Will choke and reject if anything
                                  !! other than c's come.
q5  : b  , ''  ; ''     -> q5     !! Here, we are going to punt over b's, and
q5  : '' , ''  ; ''     -> q6     !! entertain c's matches against a's.
q6  : c  , a   ; ''     -> q6     !! OK to match so long as c's keep coming
q6  : '' , #   ; ''     -> fq7    !! when # surfaces, be ready to accept in
                                  !! state fq7. Anything else causes rejection.
!!-------------------------------------------------------------------------
''')
DOf27sip = dotObj_pda(f27sip, FuseEdges=True)
DOf27sip
```

```
explore_pda("aaabbbccc", f27sip)
String aaabbbccc accepted by your PDA in 2 ways :-)
Here are the ways:
Final state  ('fq4', '', '')
Reached as follows:
-> ('iq2', 'aaabbbccc', '#')
-> ('iq2', 'aabbbccc', 'a#')
-> ('iq2', 'abbbccc', 'aa#')
-> ('iq2', 'bbbccc', 'aaa#')
-> ('q3', 'bbbccc', 'aaa#')
-> ('q3', 'bbccc', 'aa#')
-> ('q3', 'bccc', 'a#')
-> ('q3', 'ccc', '#')
-> ('fq4', 'ccc', '')
-> ('fq4', 'cc', '')
-> ('fq4', 'c', '')
-> ('fq4', '', '') .
Final state  ('fq7', '', '')
Reached as follows:
-> ('iq2', 'aaabbbccc', '#')
-> ('iq2', 'aabbbccc', 'a#')
-> ('iq2', 'abbbccc', 'aa#')
-> ('iq2', 'bbbccc', 'aaa#')
-> ('q5', 'bbbccc', 'aaa#')
-> ('q5', 'bbccc', 'aaa#')
-> ('q5', 'bccc', 'aaa#')
-> ('q5', 'ccc', 'aaa#')
-> ('q6', 'ccc', 'aaa#')
-> ('q6', 'cc', 'aa#')
-> ('q6', 'c', 'a#')
-> ('q6', '', '#')
-> ('fq7', '', '') .
```

Figure 12.10: PDA for $a^i b^j c^k$ where $(i = j)$ or $(i = k)$. With equal counts of a, b, and c in the input, the PDA's nondeterminism results in two accepting paths, one that "explains" acceptance by tallying a's against b's, while the other "explains" acceptance by tallying a's against c's.

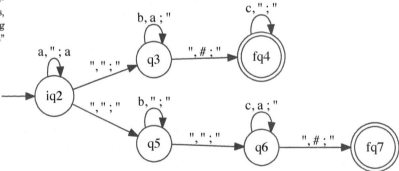

12.6   CFG to PDA Conversion

Figure 12.11: Markdown for the PDA of Figure 12.12

```
pdaEamb = md2mc('''PDA !! Encodes E -> E*E | E+E | ~E | (E) | 2 | 3
I : '', #  ; E#  -> M  !! Must parse E, so E goes on stack
M : '', E  ; ~E  -> M  !! Maybe this E will evolve to ~E
M : '', E  ; E+E -> M  !! .. or E+E (wish me luck!)
M : '', E  ; E*E -> M  !! .. or E*E
M : '', E  ; (E) -> M  !! .. or (E)
M : '', E  ; 2   -> M  !! .. or 2
M : '', E  ; 3   -> M  !! .. or 3
M : ~,  ~  ; ''  -> M  !! If stack top matches input, match!
M : 2,  2  ; ''  -> M  !! Another terminal match ..
M : 3,  3  ; ''  -> M  !! ..
M : (,  (  ; ''  -> M  !! ..
M : ),  )  ; ''  -> M  !! ..
M : +,  +  ; ''  -> M  !! ..
M : *,  *  ; ''  -> M  !! .. till all terminal matches done.
M : '', #  ; #   -> F  !! .. Input drained; parse is a success!
''')
DOpdaEamb = dotObj_pda(pdaEamb, FuseEdges=True)
DOpdaEamb # Draws the PDA
```

There is a very direct way to employ PDA to serve as parsing "engines." For illustration, see pdaEamb with its markdown description given in Figure 12.11 and transition graph in Figure 12.12. The CFG that is being converted is the following ambiguous grammar:

```
E -> E*E | E+E | ~E | (E) | 2 | 3
```

Here are the steps in the conversion algorithm.[17]

Input: A context-free grammar G with starting symbol S and production rules of the form

$$L \to R_1 R_2 \ldots R_n$$

where L is a nonterminal and R_i are either terminals or nonterminal.

Output: A PDA whose language is $L(G)$.

Method: Execute the following steps to build the desired PDA.

1. Create a 3-state PDA that starts at state I, ends in final state F, and has a middle state M (see Figure 12.12).

[17] See supplementary material at https://bit.ly/Automata_Jove under PDA2CFG for an algorithm to convert PDA to CFG, which is much more involved, but an elegant example of recursive functional programming. This material is included from my 2006 book [21].

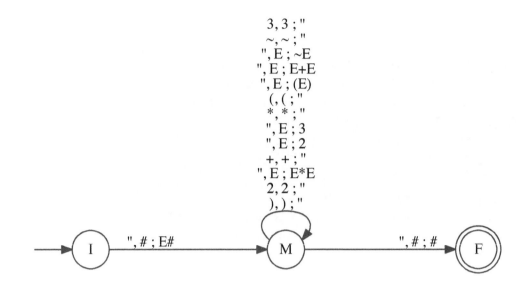

Figure 12.12: pdaEamb, a PDA embodying ambiguous parses.

2. Label the I to M transition with the edge label

$$'',\#;S\#$$

This is a transition taken without reading any input. It basically puts S on top of the PDA stack. Here, S is the *current parsing goal* or simply "goal."

Example: In Figure 12.12, we have a transition labeling the I to M move:

$$'',\#;E\#$$

This is because E is the starting symbol of this grammar.

3. The overall algorithm is geared toward removing a goal from the top of the stack and replacing it with its subgoals. To achieve this, we simply go by the CFG production rules. In general, each CFG production rule is of the form

$$L \to R_1 R_2 \ldots R_n$$

One can read this rule as follows:

(a) To parse L (the current goal), it is sufficient to parse $R_1 R_2 \ldots R_n$ in that order.

(b) Thus, whenever L is on top of the stack, one can replace the top of the stack with the string of subgoals

$$R_1R_2\ldots R_n$$

with R_1 being on top of the stack. Thus, we must introduce a PDA transition from M to M with edge label

$$'',L;R_1R_2\ldots R_n$$

This will help pop L and introduce the right-hand side of the CFG rule as the subgoals.[18]

Example: In our current example, from Figure 12.11, we know that we have a CFG rule of the form

$$\mathrm{E} \to \mathrm{E} * \mathrm{E}$$

Hence we add a transition going from M to M labeled by

$$'',\mathrm{E};\mathrm{E} * \mathrm{E}$$

4. For each terminal τ of the grammar, introduce a CFG transition from M to M of the form

$$\tau,\tau;''$$

This says that if the current parsing goal is τ, there is nothing one can do to decompose τ further (there are no rules associated with it). In fact, one must see τ in the input. This justifies the fact that we have such M to M transitions per terminal.

Example: In our example, given that 2 is a terminal, we must have a transition from M to M labeled by the edge label:

$$2,2;''$$

5. Finally, when the entire parsing is achieved, there will be no more parsing goals on top of the stack. Thus, state M must transition to state F when # is on top of the stack. Thus, we add a transition labeled by

$$'',\#;\#$$

from state M to state F.

This PDA starts out with E on top of the stack. It then behaves in a highly chaotic manner, trying to parse using every possible right-hand side of the grammar. This PDA will eventually produce a sequence of goal-to-subgoal replacements such that the relevant terminals appear on top of the stack. Whenever any terminal appears on top of the stack,

[18] A special case of this rule occurs when we have the rule $L \to \varepsilon$, in which case we pop L but introduce nothing on top of the stack. The corresponding PDA move will be $'',L;''$.

the PDA will consume an input token. In our current example, for the input expression 3+2*3, two non-deterministic evolutions are possible, as demonstrated by the following two ambiguous parses:

```
String 3+2*3 accepted by your PDA in 2 ways :-)
Here are the ways:
Final state  ('F', '', '#')
Reached as follows:
->  ('I', '3+2*3', '#')
->  ('M', '3+2*3', 'E#')
->  ('M', '3+2*3', 'E*E#')
->  ('M', '3+2*3', 'E+E*E#')
->  ('M', '3+2*3', '3+E*E#')
->  ('M', '+2*3', '+E*E#')
->  ('M', '2*3', 'E*E#')
->  ('M', '2*3', '2*E#')
->  ('M', '*3', '*E#')
->  ('M', '3', 'E#')
->  ('M', '3', '3#')
->  ('M', '', '#')
->  ('F', '', '#') .
Final state  ('F', '', '#')
Reached as follows:
->  ('I', '3+2*3', '#')
->  ('M', '3+2*3', 'E#')
->  ('M', '3+2*3', 'E+E#')
->  ('M', '3+2*3', '3+E#')
->  ('M', '+2*3', '+E#')
->  ('M', '2*3', 'E#')
->  ('M', '2*3', 'E*E#')
->  ('M', '2*3', '2*E#')
->  ('M', '*3', '*E#')
->  ('M', '3', 'E#')
->  ('M', '3', '3#')
->  ('M', '', '#')
->  ('F', '', '#') .


~~~


explore_pda("3+2*3+2*3", pdaEamb, STKMAX=7)
String 3+2*3+2*3 accepted by your PDA in 13 ways :-)
...
```

We see that even some short strings have dozens of parses! Clearly, disambiguated grammars will only have single parses, as now discussed.

12.6.1 Disambiguation

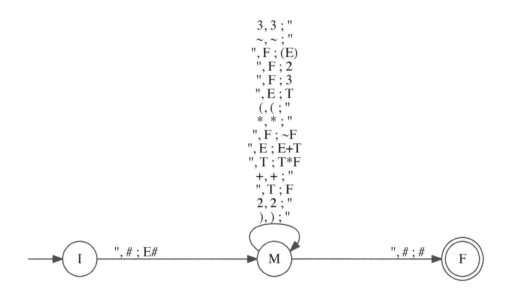

The PDA of Figure 12.13 uses the disambiguated *layered* CFG of Chapter 11. It produces a *single* parse even for a long string. It also runs fast, even with a high STKMAX such as 10 (for example), as the layered grammar forces the required precedence, thus avoiding ambiguity. Employing the earlier ambiguous grammar with a STKMAX of seven (7) gives rise to 36 parses! There is even a perceptible delay – beware that a STKMAX of 8 will take extraordinarily long.[19]

Figure 12.13: pdaE, a PDA that uses the disambiguated grammar. Notice that even here, we pretty much turn each production rule into an edge-label of the M to M transition of the PDA.

[19] We went and had coffee, and when we came back, we got 36 parses. So that is the total number of parses, indeed! We did not dare try a longer string.

```
explore_pda("3*2*~3+~~3*~3", pdaE, STKMAX=10)
String 3*2*~3+~~3*~3 accepted by your PDA in 1 ways :-)
...can push to longer inputs or STKMAX w/o worries...
explore_pda("3*2*~3+~~3*~3", pdaEamb, STKMAX=7)
String 3*2*~3+~~3*~3 accepted by your PDA in 36 ways :-)
...this took about a dozen seconds...
explore_pda("3*2*~3+~~3*~3", pdaEamb, STKMAX=8)
String 3*2*~3+~~3*~3 accepted by your PDA in 36 ways :-)
...this took time for a coffee; do not go higher STKMAX...
```

Exercise 12.6.1, PDA Design

1. Develop a final-state acceptor PDA from scratch (*i.e.*, not using the CFG to PDA conversion) for the language L_{a1b2} introduced in Exercise 2 on Page 144. (This is the set of strings with two b's for every a.)

2. Develop an empty-stack acceptor PDA from scratch (*i.e.*, not using the CFG to PDA conversion) for the language L_{a1b2}.

3. *Directly* translate the CFG for L_{eqab} presented in §11.3 into a PDA. Parse the first 10 strings in numeric order over a and b, thus checking that the parses are correct.

4. Now add the SS production to the CFGs, and reflect that in an extra transition of the PDA. Observe that no additional strings are being accepted. Observe the computations generated and argue why SS does not seem to be playing a role. (If it played a role in the PDA's actions, demonstrate that for the same string, the PDA without the SS production also ended up discovering an equivalent parse – with respect to keeping the counts.)

5. Develop a PDA for the language of odd length strings over $\{0, 1\}$ with a zero in the middle. Hint: argue whether nondeterminism is essential to solve this problem. Then design your PDA accordingly.

6. Consider the language

$$L_{abc} = \{w \mid w \in \{a, b, c\}^* \text{ and } \#_a(w) = \#_b(w) = \#_c(w)\}$$

   (a) Argue that the intersection of a PDA and a DFA is a PDA. *Hint:* Provide a product construction algorithm in pseudo-code form, very similar to the DFA product construction presented in §6.2. Specifically, move the PDA and the DFA from their current pair of control states to their next pair of control states if the PDA and DFA move on the same input symbol. When the PDA takes ε moves, do not move the DFA; keep it rooted at its current control state – until the PDA comes to a control state where it moves on an input symbol. The stack is updated as per the PDA actions.

   (b) Let us (for the moment) pretend that there is a PDA for L_{abc}. Now, write down the language you will get as a result of the operation

$$L_{abc} \cap a^* b^* c^*$$

   Based on our pretend position, there must be a PDA for the language resulting from this intersection.

   (c) Using the CFL Pumping Lemma, argue that the language resulting from this intersection is not context-free.

   (d) Hence argue that L_{abc} cannot have a PDA (our pretension is incorrect).

7. Argue that this language is a CFL by building a PDA for it. Answer for both cases of 'OP' listed below:

$$L_{abcd} = \{a^i b^j c^k d^l \; : \; i,j,k,l \geq 0 \text{ and } ((i = j) \, \text{OP} \, (k = l))\}$$

   (a) (Case 1) OP is AND
   (b) (Case 2) OP is OR

8. (a) Design a context-free grammar for the language of strings over $\{a,b,c\}$ where the number of a's equals twice the number of b's plus the number of c's.

$$\{w \in \{a,b,c\}^* \; : \; \#_a = 2\#_b + \#_c\}$$

   (b) Then convert this CFG into a PDA using the direct CFG to PDA conversion method (done by hand). Administer a sufficient number of tests using Jove to test this PDA out.

   (c) Now design a PDA for this language directly (without resorting to this conversion). Compare the PDAs in terms of their ease, as well as recognition times (try and feed longer strings and estimate which PDA runs faster). Explain the reasons for any noticed differences between the speeds of these PDA.   □

12.7   Practical Knowledge Imparted by Jove: Three Parsers

By experimenting with Jove and reading its code, you will emerge well prepared to take many central CS courses such as Compilers. In particular, we offer you the opportunity to study *three* parsers.

The **first parser** is in §8.5 where we detailed the regular expression compiler that includes a scanner (lexer) for tokens, and a parser that recognizes regular expressions. The "code" emitted by this parser consists of NFA representations. The start symbol is expression (the first p_ . . function), and one can see an uncanny resemblance to the disambiguated expression grammar introduced in §11.5. Disambiguation is crucial for regular expressions, as RE concatenation has higher precedence than RE union (+). The parser generator PLY assigns STAR higher precedence (listed later) and both PLUS and STAR are left associative as well.

The **second parser** is within a mini-compiler and converts a simple regular-expression syntax (§10.3). The "code" emitted by this compiler consists of NFA representations. Perhaps this is the simpler parser to first study, as the whole mini-compiler fits within a page.

The **third parser** is within module Def_md2mc where we parse our entire markdown language and generate DFA, NFA, PDA or Turing machine representations. This parser is characterized by many more features than present in the other two parsers just mentioned:

• It filters out comments.

- It keeps track of line numbers, and upon a syntax error, it at least prints the offending line number. It then does a best-effort resetting of line numbers for your next round of experimentation.
- The grammar for DFA, NFA, PDA and Turing machines differ slightly. This is a practical reality, and requires thoughtful handling:
  - One option would have been to have four distinct families of production rules, and to generate code under all those productions. The advantage would be that the production rules only generate code pertinent to each machine type. A huge disadvantage would be having four times the number of rules.
  - We opted for fusing the rules into very similar subgroups and only made localized differences. While the semantic attribute handling becomes a little more involved (some attributes such as stack symbol are not relevant except for PDA), the vast reduction in the number of parser rules, and the associated parser-generator headaches makes this version more maintainable.

Textual Syntax: Importance in CS

In conclusion, despite the explosion of input and output devices, computer science continues to rely on text scanning and parsing to support the vast number of programming languages, scripting languages and data handling languages. Text still rules supreme, and correct plus efficient text handling for very large alphabets continues to be cutting-edge CS research that also immediately ties into practice.

13

Turing Machines

> **Chapter Gist:** *We begin with a very brief historical account of Turing machines §13.1. A few universal computing devices (equivalent in power to Turing machines) are discussed (§13.2), including how one simulates a TM using two stacks and finite-state control. We then present a formal definition of TMs (§13.3), a few simple TM examples (§13.4), examples of medium complexity (§13.5), and finally a nontrivial TM that implements the famous "$3x+1$ function" (§13.6). We then present the Chomsky hierarchy (§13.7) that ties together machines, languages, and grammars. A formal notion of a TM's ID finishes this chapter (§13.8).*

13.1 Brief History of Turing Machines

In the early part of the 20th century, scores of scientists, notably Kurt Gödel, Alonzo Church and Alan Turing, attempted to define the limits of *effective computability*. In the words of Turing himself [45] (and also summarized in Andrew Hodges' biography on Turing [25]), the notion of *computation* was described as follows:

Computing is normally done by writing certain symbols on paper. We may suppose this paper is divided into squares like a child's arithmetic book. In elementary arithmetic the two-dimensional character of the paper is sometimes used. But such a use is always avoidable, and I think that it will be agreed that the two-dimensional character of paper is no essential of computation. I assume then that the computation is carried out on one-dimensional paper, i.e., on a tape divided into squares. I shall also suppose that the number of symbols which may be printed is finite ... The behavior of the [human] computer at any moment is determined by the symbols which he is observing, and his state of mind at that moment.

The historic term for the underlying quest that scientists of Turing's era were engaged in was known as *Entscheidungsproblem* (meaning "the decision problem" in German). They were trying to settle one of Hilbert's challenges (Chapter 1), namely:

> Prove that there is an algorithm—a systematic and mechanical *procedure* (see Page 4 for a discussion) that terminates on any input—to decide the truth of *any* logical statement in mathematics.

Gödel settled this challenge in the negative by showing that sufficiently expressive mathematical logical systems that are powerful enough to encode statements in mathematics contain true sentences whose proof cannot be demonstrated within the same logical system.[1]

In related developments, Turing proved that the formalism of his machines is sufficiently general to encode *any* mechanizable computational procedure (hereafter "procedure"). Turing also proposed the notion of *universal Turing machines* capable of simulating the workings of other Turing machines.[2]

Now, given that Turing machines are powerful enough to encode *any* procedure, it is tempting to imagine a single Turing machine **algorithm** (call it H, standing for *Halting decider*) that, given any Turing machine M with its input w, can decide whether M when run on w will halt.[3] In Chapter 15, we will prove that a Turing machine such as H cannot exist, or in other words, *the halting problem of Turing machines is undecidable*.

Alonzo Church and Alan Turing are both credited with stating emphatically that all prevalent notions of universal computability in existence at that time and shown equivalent together defined the fundamental limit of effective computability. Their hypothesis nowadays goes by the name *Church-Turing thesis*. Church accepted that Turing's definition gave a compelling, intuitive reason for why his thesis was true. This situation has not changed since the late 1930s. Meanwhile hundreds of universal computing machines have been proposed and proven to be universal.[4]

13.2   Universal Computing Devices

Universal computing formalisms such as Turing machines mainly serve as vehicles for writing proofs. For example, in Chapter 11, we stated that the problem of checking whether two context-free grammars are equivalent is *not algorithmically solvable*. Such proofs are formulated in terms of decision problems with respect to Turing machines. Nobody would attempt to write such proofs in terms of practical computers.[5] On the other hand, if something is algorithmically solvable and represents a useful algorithm, one would anyhow use a real computer to run the algorithm fast.

---

[1] A related fact is that because of the expressiveness of these logics, one can write a formula ϕ in these logics that asserts that ϕ (*i.e.*, *itself*) is false.

[2] Universal Turing machines are nothing but computers that can simulate other computers as well as their program executions. One would nowadays call them *interpreters* or *virtual machines*.

[3] This is the *grader's dilemma*: will a student-submitted program P run on an input i halt, or go into an infinite loop? There is no way for the TA to find out other than to take a chance and run P on input i!

[4] Turing, as a scientist and a person, continues to evoke a deep sense of mystique, often overshadowing other luminaries in the pantheon of computability theory. His central role in cracking the Enigma code that ultimately led to the Allied victory against the Nazis is widely acknowledged [44]. Turing is obviously more of a "celluloid celebrity," and was eminently portrayed in the highly acclaimed 2014 movie '*The Imitation Game*' by actor Benedict Cumberbatch. This movie provides another glimpse into Turing's life and his sufferings at the hands of an intolerant society.

[5] Computers such as the MacBook Pro on which I'm typing this book. A MacBook may be a quadrillion times faster than the best Turing machine one can build. Yet, what is inherently undecidable remains undecidable despite this speed advantage. Put another way, a MacBook Pro can't "compute itself out of an infinite loop" just because it is fast!

The Long List of Universal Computers

A long list of computational devices have been shown to be (or can be argued to be) universal; here are a few examples:[6]

- Deterministic or nondeterministic finite-state control equipped with a doubly infinite tape, a singly infinite tape, or multiple infinite tapes.[7]
- Finite-state control equipped with **two** unbounded stacks (one can simulate an infinite tape using two stacks).
- Finite-state control equipped with **one** unbounded FIFO queue (one can simulate a tape by "rotating the queue contents around").
- Finite-state control equipped with two infinite unbounded counters that can be incremented, decremented, cleared, copied one into the other, and exchanged.
- Mike Davey's mechanical Turing machine from Chapter 1.[8]
- Any one of these actual computers (all assumed to have infinite memory): one-instruction computers (search for "One instruction set computer" on Wikipedia); the computer inside a Furbee doll; world's fastest computer that can perform 10^{18} operations a second.

The above list includes actual computers as well as conceptual devices. We will not have the occasion to prove in detail why these mechanisms are equivalent to a Turing machine. The basic idea is to argue that given any of the above mechanisms, one can simulate the actions of a Turing machine using them. Let us illustrate this idea by example: given two unbounded stacks and a finite-state controller, how to mimic the actions of a real Turing machine.

Simulating an Infinite TM Tape using Two Unbounded Stacks

Let the stack pair be denoted by the pair of strings L and R where L is a stack facing right, and R is a stack facing left. Pictorially, let $L = [ab)$ with b being on top and $R = (qp]$ with q being on top, and the whole tape then looks like $[ab)(qp]$. We maintained the invariant that the Turing machine is always looking at the top of the righthand-side stack, *i.e.* q in this example. Then, the following simulations can be carried out:

- *Writing x on the current tape cell and moving right:* This results in a transition of the following form:

$$[ab)(qp] \longrightarrow [abx)(p]$$

This is achieved by pushing x on L and popping R.

- *Writing x on the current tape cell and moving left:* This results in a transition of the following form:

$$[ab)(qp] \longrightarrow [a)(bxp]$$

This is achieved by capturing L's top (which is b), popping L, then

[6] These are often called "Turing-complete" devices.

[7] Both deterministic and nondeterministic Turing machines have the same computational power; however, their time complexities are in different classes; see Chapter 16.

[8] Mr. Davey estimates taking "only" 870 of the 1000-foot tape rolls shown in Figure 1.2 to store a megabyte of data.

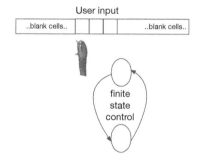

Figure 13.1: A Turing machine shown with its input string on a doubly-infinite tape. This is the type of TMs we shall be studying.

pushing x and b onto R.

- *Writing x and staying in the same position:* This results in the transition

$$[ab)(qp] \longrightarrow [ab)(xp]$$

This is achieved by popping R and pushing x onto R.

Thus, we can see that two unbounded stacks can simulate an infinite tape. The TMs we employ are the ones with exactly one doubly-infinite tape (Figure 13.1).

Exercise 13.2, Turing Machine Simulation

1. Write a simulation of a Turing machine's infinite tape if you only have an infinite FIFO queue. Imagine that we have access to the head and tail of this queue, and can perform the operations *enqueue(Q,x)* which adds x to the tail of Q, *dequeue(Q)* that removes the head element of Q, and *front(Q)* that does not modify Q but returns the element at its front.

2. Simulate the actions of a multi-tape Turing machine on a Turing machine with a singly infinite tape. *Hint:* For k tapes in the multi-tape TM, arrange the tape of the singly infinite tape TM by interleaving the tapes of the multi-tape TM. For example, if the multi-tape TM has tape contents $a_1b_1c_1d_1$ on Tape-1 and $a_2b_2c_2d_2$ on Tape-2, arrange the tape of the singly infinite tape TM to contain $a_1a_2b_1b_2c_1c_2d_1d_2$. Now, each action of the multi-tape TM could have affected its tapes differently (say write x_1 and move right; write x_2 and move left). Simulate these actions suitably on a single tape. $\qquad\qquad\square$

We now turn our attention to the study of Turing machines that have exactly one doubly-infinite tape, as realized in Jove.

13.3 *Formal Definition of Turing Machines*

Turing Machines are structures $(Q, \Sigma, \Gamma, \Delta, q_0, B, F)$ where

- Q is a finite non-empty set of states ("program locations or labels").
- Σ is a finite non-empty **input alphabet**.
- Γ is a finite non-empty **tape alphabet**. Γ is a proper superset of Σ, as we allow the blank tape cell in $\Gamma - \Sigma$, helping to model unwritten parts of the tape.
- $q_0 \in Q$ is the (unique) start state of the Turing machine.
- Generically, "B" represents the 'blank' tape cell. **In Jove**, we use '.' (period) for blanks, so that they stand out in Jove simulations.
- $F \subseteq Q$ is the set of final (accepting) states.
- Computations are set up by writing the user-given input on the tape. **To feed ε as input to a TM, leave the tape entirely blank.**

- Given that we are using a doubly infinite tape, we situate the user-given input under the TM head, and spreading right (details to be provided momentarily, when we describe *instantaneous descriptions*, or ID). To the left of the head, we arrange for an infinite sequence of blanks.[9]

- Nondeterministic TMs in Jove convey the fuel equally to all the nondeterministic threads being spawned. Thus, a simulation proceeds till (at least) the point at which all threads run out of fuel. Using this notion, we can naturally as well as rigorously define notions such as *non-deterministic runtime*.[10]

- For a deterministic TM, Δ is a transition function that takes a control state, and the current tape symbol being scanned, and generates a single replacement triple of a tape symbol, a next control state, and a head direction. More specifically, the signature of Δ is:

$$\Delta : Q \times \Gamma \to Q \times \Gamma \times \{L, R, S\}$$

For a nondeterministic TM, it generates a set of such triples, and has the following signature (here, $\mathscr{P}$ is the powerset operator):

$$\Delta : Q \times \Gamma \to \mathscr{P}(Q \times \Gamma \times \{L, R, S\})$$

- We will be dealing with nondeterministic Turing machines, in general. However, Jove's language is versatile enough to model both types of transitions for the same machine. A concrete example coming from Figure 13.13 is as follows:
  - A line of the form

    ```
    q5 : Y ; 3, R -> q11
    ```

    means that this TM, in state q5, can see if Y is under the tape head. If so, it replaces Y by 3, and moves its head right, and transitions to state q11.
  - A line of the form

    ```
    q10 : 1;Y,L | .;.,L | 0;X,L -> q8
    ```

    means that this TM, in state q10, has multiple options.
    * If it sees a 1 under the tape head, it replaces it with a Y and moves left.
    * If it sees a blank (.), it replaces it with a blank itself, and moves left.
    * If it sees a 0, it replaces it with an X and moves left.

- A TM is "stuck" if it cannot fire any transition from a given (q, i) pair where $q \in Q$ and $i \in \Gamma$. Such a TM is said to have **halted**. Furthermore,
  - If $q \in F$, it is said to have **accepted** its input.
  - Otherwise ($q \in (Q \setminus F)$), it has **rejected** its input.

[9] In Jove, we of course don't allocate an infinite tape to begin with. We employ a finitary tape, and allocate blanks on demand, as follows: when a TM is about to "step off" its finitary tape, we allocate 8 more cells! This is like demand-paging in Operating Systems (imagine a marathon runner being given a meter of turf initially, with additional turf added as and when necessary).

Another feature of Jove's TMs is that they come with a fuel tank with finite capacity. We initially top-up the tank, and let the TM run till it runs out of fuel. This is to prevent truly "runaway" simulations that can easily infinitely loop.

[10] Essentially, non-deterministic time (Chapter 16) gives a bound on the number of steps taken by any computational path.

- All that is required for an input i to be accepted or rejected is:
  - The TM *starts* in configuration (q_0, i).
  - Later, the TM is **found to have halted.**

  Typically, in the interim, the TM will read its input, or at least "take a nibble at it." It is also allowed to re-read its input any number of times. However, **it is not necessary for it to have read even one single character of the input string.**[11]

  Clearly, this is a major difference with respect to NFA, DFA, and PDA where: (1) the input must be fully read before the **accept** or **reject** status can be declared; (2) the input is always scanned left-to-right and that too exactly once.
- The instantaneous description (ID)[12] of a TM is the (q, h, i) triple, with h being the index into the input i (head position).[13]
- **Accepting computations** start from an ID $(q_0, 0, w)$ where $q_0 \in Q$, w is the input, and the head initially is staring at the leftmost cell of the input, namely $w[0]$. TMs begin with an infinite number of blank (.) cells to the left and right of w.[14] Accepting computations end in an ID (q_f, h, g) where $q_f \in F$, $g \in \Gamma^*$, and $h \in Nat$ is some head position.

> The language of a Turing machine—whether it be deterministic or not—is the set of strings accepted by it. More specifically for a non-deterministic Turing machine, its language is the set of all strings that result in an accepting computation along some nondeterministic computational path.

Each transition in a TM-diagram (e.g., Figure 13.10) has edge-labels:

> oneInChrOrBlnk $\overset{\bullet}{}$ oneOutChrOrBlnk headMoveSpec

where oneInChrOrBlnk and oneOutChrOrBlnk are a single character from Γ (or the blank symbol '.'), and the headMoveSpec is one of S (same), L (go left), or R (go right). We now present several example TMs.

---

[11] This definition makes sense for a Turing machine simply because *it is impossible to definitively establish* whether a TM engaged in a big loop or random zig-zag motion on its tape (also termed "looping") will ever return to read any of (or any *more of*) its input. Thus, *insisting* that a TM read its input exactly once (and that too fully) seems pointless. Also, "read" could be a vacuous term: any TM can be, when "switched on," made to read its input exactly once, and copy it onto a vacant spot of its tape. After that, the TM may behave like a "normal" TM—taking occasional nibbles at the copied-over input or ignoring it totally.

[12] A more formal version of IDs is discussed in §13.8.

[13] In Jove, we also keep the remaining fuel in the ID, making it (q, h, i, f).

[14] In Jove, users can simply enclose w inside a string and submit. It is not necessary to pad blanks! As said earlier, Jove will allocate blanks as/when necessary.

13.4 Examples of Simple TMs

We begin with two example TMs, one that simply flips the bits on the input tape (§13.4.1) and another that checks whether the tape contains the pattern "101" (§13.4.2).

13.4.1 A Simple DTM that Flips Bits

Our first example Turing machine starts in state I and stays in this state so long as it encounters a 0 or a 1 (see the state transition diagram of Figure 13.2). After inverting this bit, the TM's head goes one step right on the tape; the TM then resumes its work. When in state I if it encounters a blank ('.'), the TM enters state F where it gets stuck. A run of this TM (call it *Flipper*) is also shown. Please answer the questions in the exercises below to learn more about machine *Flipper*.

Exercise 13.4.1, Flipper TM

1. How do we initialize the tape of *Flipper* to contain ε?
2. What is the language of *Flipper*? □

13.4.2 TMs that check if a string contains 101

While *Flipper* illustrated some of the mechanics of specifying Turing machines, it did not emphasize why algorithms stated in terms of Turing machines matter. We shall now present such details – specifically the notion of a *deterministic* algorithm versus a *nondeterministic* algorithm – with the help of a very simple example. A few salient observations will be pointed out: (1) Nondeterministic Turing machines will often be far more succinct and easy to specify; (2) String *s* is in the language of an NDTM if it results in at least one nondeterministic accepting computation.

```
Allocating  8  tape cells to
the RIGHT!

Detailing the halted configs
now.

Accepted at
('F', 6, '101100........', 93)
 via ..
 ->('I', 0, '010011', 100)
 ->('I', 1, '110011', 99)
 ->('I', 2, '100011', 98)
 ->('I', 3, '101011', 97)
 ->('I', 4, '101111', 96)
 ->('I', 5, '101101', 95)
 ->('I', 6, '101100', 94)
 ->('F', 6, '101100...', 93)
```

Figure 13.2: A TM called *Flipper* that flips a given bit sequence. It starts at state I with the head position being 0. It is staring at the left end of the string 010011. It has initially a fuel tank with 100 thimbles of fuel.

Governed by the I to I move that asks it to flip a 0 to a 1, this TM then takes a step to the right, and the string is now 110011. Notice that the head position is now 1 and the remaining fuel is 99 thimbles. So it is now staring at 110011[1], *i.e.* 1.

Governed by the I to I move that asks it to flip a 1 to a 0, this TM then takes a step to the right, and the string is now 100011. Notice that the head position is now 2, and the remaining fuel is 98 thimbles. So it is now staring at 100011[2], *i.e.* 0.

This process continues till the TM's head position becomes 6. At this point, the TM sees a blank, and executes the I to F move. The blank is replaced by another blank, and the head position is S (same). The machine gets *stuck* at state F, which causes it to accept the input.

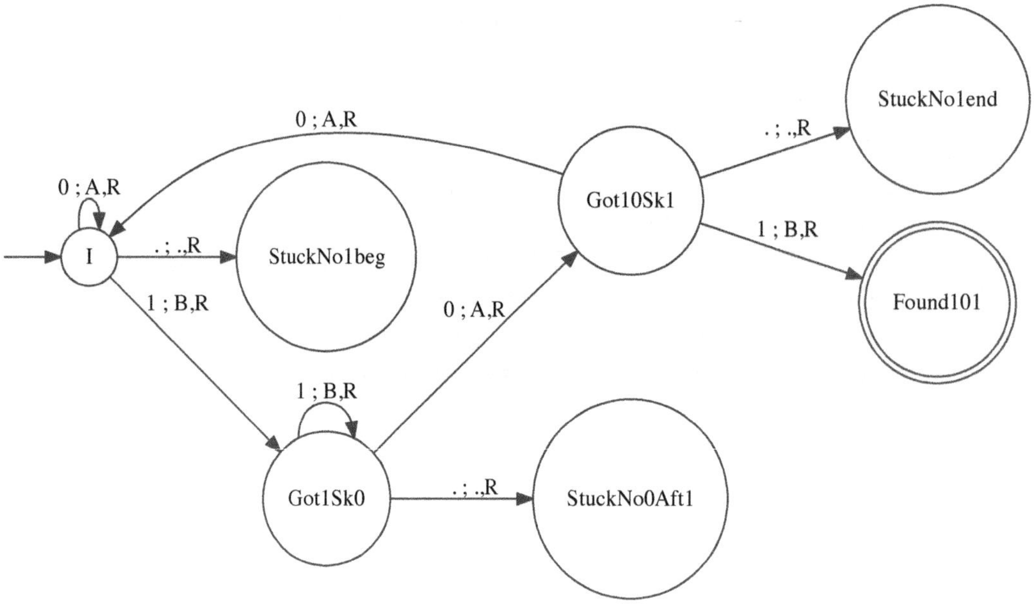

Figure 13.3: Transition diagram for a DTM that looks for 101 within given w.

```
!! Sweep, starting at the beginning, looking for 101
whas101DTM = md2mc('''TM


I : 1; B, R -> Got1Seek0    !! Partial success; climb to next stage
I : 0; A, R -> I            !! Continue hunting for 1
I : .; ., R -> StuckNo1beg  !! Ran off end; REJECT


Got1Seek0 : 0; A, R -> Got10Seek1 !! More success; climb onto next state
Got1Seek0 : 1; B, R -> Got1Seek0  !! Didn't find 0, but starts with 1; so Seek0
Got1Seek0 : .; ., R -> StuckNo0Aft1 !! Ran off end w/o finding 0; so REJECT


Got10Seek1 : 1; B, R -> Found101    !! Successfully found 101. ACCEPT!
Got10Seek1 : 0; A, R -> I           !! Failure finding 1; start over
Got10Seek1 : .; ., R -> StuckNo1end !! Ran off end; so REJECT
''')
```

Figure 13.4: Markdown for a DTM that looks for 101 within given w.

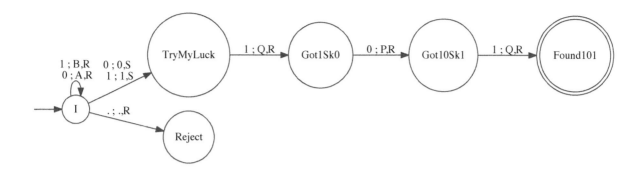

Figure 13.5: Transition diagram for an NDTM that looks for 101 within given w.

Figure 13.3 presents the DTM for this problem. The algorithm is explained in Figure 13.4, and consists of starting from state I, seeking 1 first, then the 0 coming after the 1 to get to 10, and finally the last 1 bit. Bit 0 is converted to an A and 1 is converted to an B so that the user may see the extent of the tape moved over before the first 101 is located. Any failure to finish this pattern results in the machine running off the end, and rejecting. A typical accepting run is shown in Figure 13.7.

The nondeterministic algorithm in Figure 13.5 is explained in Figure 13.6, and banks on taking a chance: it keeps converting 0 to an A and 1 to a B, moving right (these conversions are to leave a trail of operation behind on the tape).

Boom! The NDTM suddenly decides to take a transition to the TryMyLuck state! There, it expects to see a 101 pattern. If this sequence is seen, the machine accepts. All wrong guesses "get killed," making the machine reject. The full computational history of running this machine on input 0010101 is shown in Figure 13.8. This example underscores the inherent ease of specifying a machine that adopts a nondeterministic algorithm.

```
!! Choose a spot at random
!! Check for 101 to be there
whas101NDTM = md2mc('''TM
I : 0; A, R | 1; B, R -> I
I : 0; 0, S -> TryMyLuck
I : 1; 1, S -> TryMyLuck
I : .; ., R -> Reject
TryMyLuck: 1; Q, R -> Got1Sk0
Got1Sk0  : 0; P, R -> Got10Sk1
Got10Sk1 : 1; Q, R -> Found101
''')
```

Figure 13.6: Markdown for an NDTM that looks for 101 within given w.

```
Detailing the halted configs.
Accepted at
('Found101', 5, 'AABAB01', 5)
 via ..
->('I', 0, '0010101', 10)
->('I', 1, 'A010101',  9)
->('I', 2, 'AA10101',  8)
->('Got1Sk0', 3,'AAB0101',7)
->('Got10Sk1',4,'AABA101',6)
->('Found101',5,'AABAB01',5)
```

Figure 13.7: Accepting run of the DTM for the problem of checking whether 101 occurs in input w.

```
explore_tm(whas101NDTM, "0010101", 10)
...tape allocation suppressed...
Detailing the halted configs now.
Rejected at  ('TryMyLuck', 0, '0010101', 9)
 via ..
 ->('I', 0, '0010101', 10)
 ->('TryMyLuck', 0, '0010101', 9)
Rejected at  ('TryMyLuck', 1, 'A010101', 8)          Rejected at  ('TryMyLuck', 5, 'AABAB01', 4)
 via ..                                               via ..
 ->('I', 0, '0010101', 10)                            ->('I', 0, '0010101', 10)
 ->('I', 1, 'A010101', 9)                             ->('I', 1, 'A010101', 9)
 ->('TryMyLuck', 1, 'A010101', 8)                     ->('I', 2, 'AA10101', 8)
Accepted at  ('Found101', 4, 'AAQPQ01', 4)            ->('I', 3, 'AAB0101', 7)
 via ..                                               ->('I', 4, 'AABA101', 6)
 ->('I', 0, '0010101', 10)                            ->('I', 5, 'AABAB01', 5)
 ->('I', 1, 'A010101', 9)                             ->('TryMyLuck', 5, 'AABAB01', 4)
 ->('I', 2, 'AA10101', 8)                            Rejected at  ('Got1Sk0', 7, 'AABABAQ', 2)
 ->('TryMyLuck', 2, 'AA10101', 7)                     via ..
 ->('Got1Sk0', 3, 'AAQ0101', 6)                       ->('I', 0, '0010101', 10)
 ->('Got10Sk1', 4, 'AAQP101', 5)                      ->('I', 1, 'A010101', 9)
 ->('Found101', 4, 'AAQPQ01', 4)                      ->('I', 2, 'AA10101', 8)
Rejected at  ('TryMyLuck', 3, 'AAB0101', 6)           ->('I', 3, 'AAB0101', 7)
 via ..                                               ->('I', 4, 'AABA101', 6)
 ->('I', 0, '0010101', 10)                            ->('I', 5, 'AABAB01', 5)
 ->('I', 1, 'A010101', 9)                             ->('I', 6, 'AABABA1', 4)
 ->('I', 2, 'AA10101', 8)                             ->('TryMyLuck', 6, 'AABABA1', 3)
 ->('I', 3, 'AAB0101', 7)                             ->('Got1Sk0', 7, 'AABABAQ', 2)
 ->('TryMyLuck', 3, 'AAB0101', 6)                    Rejected at  ('Reject', 7, 'AABABAB........', 2)
Accepted at  ('Found101', 6, 'AABAQPQ', 2)            via ..
 via ..                                               ->('I', 0, '0010101', 10)
 ->('I', 0, '0010101', 10)                            ->('I', 1, 'A010101', 9)
 ->('I', 1, 'A010101', 9)                             ->('I', 2, 'AA10101', 8)
 ->('I', 2, 'AA10101', 8)                             ->('I', 3, 'AAB0101', 7)
 ->('I', 3, 'AAB0101', 7)                             ->('I', 4, 'AABA101', 6)
 ->('I', 4, 'AABA101', 6)                             ->('I', 5, 'AABAB01', 5)
 ->('TryMyLuck', 4, 'AABA101', 5)                     ->('I', 6, 'AABABA1', 4)
 ->('Got1Sk0', 5, 'AABAQ01', 4)                       ->('I', 7, 'AABABAB', 3)
 ->('Got10Sk1', 6, 'AABAQP1', 3)                      ->('Reject', 7, 'AABABAB........', 2)
 ->('Found101', 6, 'AABAQPQ', 2)
```

Figure 13.8: Run of the NDTM for 101 ∈ w with 10 units of fuel. Notice the full nondeterministic computational tree generated. Two accepting runs are obtained.

13.5   A DTM for w#w and an NDTM for ww

We will now build Turing machines that recognize patterns in which a string $w \in \{0,1\}^*$ is followed by an identical w. To simplify our DTM, we will design it to recognize strings of the form $w\#w$.[15] It must be clear to the reader that the language of strings of the form $w\#w$ or ww cannot be recognized by DFA or PDA.[16]

13.5.1   A DTM Recognizing w#w

The TM presented via the markdown in Figure 13.9 can be converted to the drawing in Figure 13.10 and run as shown in the same figure. This TM recognizes the language $w\#w$ for $w \in \{0,1\}^*$. In §11.9, an exercise asks you to show that this language is not context-free, via the Context-Free Language Pumping Lemma. However, this language can be easily recognized by a Turing machine.[17] The commands to create and run this TM are now presented:

```
wpw_tm = md2mc(src='File', fname="tmfiles/wpw.tm")
DOwpw = dotObj_tm(wpw_tm, FuseEdges = True)
DOwpw  # This draws the TM transition diagram


# Run the TM on input shown (fuel given = 120 thimbles)
explore_tm(wpw_tm, "001#001", 120)
```

13.5.2   A Nondeterministic TM Recognizing ww

If the input has no middle marker, *i.e.* the entire input given is ww, can we process it using a deterministic TM? The answer is of course yes: *given that the TM is a universal device, it is possible to find the halfway point in string ww through search,* and then match around it.

We will however employ the power of nondeterminism to guess a plausible midpoint and then match around that midpoint; doing so helps illustrate NDTMs, and also makes the coding much easier. There are four incorrect markings of the "winged" midpoint that fail, as can be seen in detail through a Jove simulation (excerpts provided in Figure 13.11, with markdown provided with comments in Figure 13.13).

The accepting computational history is 25 steps long, while the rejecting histories consume far less fuel before biting the dust. It can be seen that when a nondeterministic path split is about to happen, the currently remaining fuel is equally split.

```
explore_tm(wwndtm_md, "001001", 170)
```

[15] The reader is invited to design a DTM that does not employ such a separator. DTMs and NDTMs are equivalent in power, and this proof involves simulating nondeterminism somehow, say using multiple threads, as in §13.5.3. However, for strings of the form ww, a direct deterministic algorithm can be employed (without simulating nondeterminism). One can estimate its complexity and compare it with the complexity of the NDTM simulation. See Exercises 13.5.3.

[16] Why in each case? Also if $w \in \{0\}^*$, how does this result change?

[17] It is context-sensitive, and so, it can be recognized by a linear bounded automaton as well.

13.5.3 Nondeterminism does not increase a TM's Expressive Power

It can be shown that the presence of nondeterminism does not add any expressive power to a TM. One can design a DTM with a few extra tapes that store the nondeterministic options, and explores them one by one, systematically:

- Imagine writing a multi-threaded version of an NDTM where each separate thread is one nondeterministic choice, and the NDTM accepts if/when one of the threads accepts.
- The DTM simulation simulates the threads in a breadth-first manner. It steps each thread forward one step, taking a round-robin scheduling approach. That is, the simulation must not be "fixated" on "finishing one thread" before moving on to another thread.[18]

Jove's step_tm function is given in Figure 13.12. Simulating a Turing machine is basically quite simple: take the current ID and calculate the next set of IDs using the TM's Delta function. The additional details are to maintain a computational path, allocate tape when we "step beyond" the current tape, and also decrement the fuel per nondeterministic path.

---

[18] If we are fixated on one thread and that thread goes into an infinite loop, we would never discover the fact that another (nondeterministic) thread might have accepted. An NDTM simulation on a DTM can be abandoned as soon as one of the threads accepts (this can help avoid infinite loops hidden in some threads). For this benefit to be accrued, a fair scheduling approach (*e.g.*, round-robin) must be taken.

Exercise 13.5.3, DTM design

1. Design a DTM to recognize $\{ww \mid w \in \{0,1\}\}$ using an $O(N^2)$ deterministic algorithm. Proceed by writing pseudo-code and then convert it into a TM program. Enter it in Jove, simulate, and demonstrate that your algorithm works on a sufficient number of interesting test cases.

```
TM !! This is a DTM for recognizing strings of the form w#w where w is in {0,1}*.
  Iq0: 0  ; X  , R  -> q1            !! Convert 0s to Xs. Matching 0s are then
                                     !! sought to the right of the #
  Iq0: 1  ; Y  , R  -> q7            !! Convert 1s to Ys. Matching 1s then
                                     !! sought to the right of the #
  Iq : #  ; #  , R  -> q5            !! OR we see # right away, then accept
                                     !! right away, as we then have eps#eps
  q5 : X ; X,R | Y ; Y,R -> q5       !! In q5, we skip over X and Y
                                     !! (eq. num of X,Y lie to the left of #)
  q5 : .  ; .  , R  -> Fq6           !! Accept when we see a blank (.)

  q1 : 0 ; 0,R | 1 ; 1,R -> q1       !! In q1, skip over remaining 0/1
  q1 : #  ; #  , R  -> q2            !! But upon seeing #, look for matching 0

  q2 : X ; X,R | Y ; Y,R -> q2       !! All X,Y are "spent stuff" to skip over
  q2 : 0  ; X  , L  -> q3            !! When we find a matching 0, turn that to
                                     !! an X, and sweep left to do the next pass
  q3 : X ; X,L | Y ; Y,L -> q3       !! In q3, we move over all past X, Y
  q3 : #  ; #  , L  -> q4            !! but when we reach the middle marker,
                                     !! we know that the next action is to seek
                                     !! the next unprocessed 0 or 1
  q4 : 0 ; 0,L | 1 ; 1,L -> q4       !! In q4, wait till we hit the leftmost 0/1
  q4 : X ; X,R | Y ; Y,R -> Iq0      !! When we hit X,Y, we know that we've
                                     !! found the leftmost 0/1. Begin another pass.
  q7 : 0 ; 0,R | 1 ; 1,R -> q7       !! q7 is similar to q1
  q7 : #  ; #  , R  -> q8            !! and q8 is similar to q2


  q8 : X ; X,R | Y ; Y,R -> q8
  q8 : 1  ; Y  , L  -> q3
```

Figure 13.9: The DTM for *w#w* in Jove's markdown notation. Detailed comments help others understand TM code. In this sense, textual input can prove superior to TM diagrams beyond a certain size. TM and PDA programming is lower level than assembly programming, and so extreme clarity of expression plus clear comments are essential.

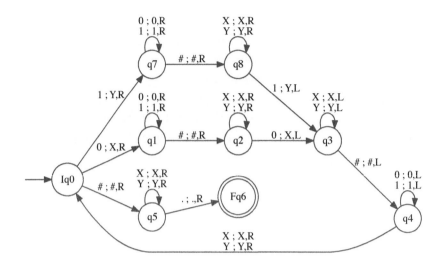

```
Allocating  8  tape cells to the RIGHT!
Detailing the halted configs now.
Accepted at  ('Fq6', 8, 'XXY#XXY........', 88) ... via ..
 ->('Iq0', 0, '001#001', 120)    ->('q1', 3, 'XX1#X01', 109)    ->('q8', 5, 'XXY#XX1', 99)
 ->('q1', 1, 'X01#001', 119)     ->('q2', 4, 'XX1#X01', 108)    ->('q8', 6, 'XXY#XX1', 98)
 ->('q1', 2, 'X01#001', 118)     ->('q2', 5, 'XX1#X01', 107)    ->('q3', 5, 'XXY#XXY', 97)
 ->('q1', 3, 'X01#001', 117)     ->('q3', 4, 'XX1#XX1', 106)    ->('q3', 4, 'XXY#XXY', 96)
 ->('q2', 4, 'X01#001', 116)     ->('q3', 3, 'XX1#XX1', 105)    ->('q3', 3, 'XXY#XXY', 95)
 ->('q3', 3, 'X01#X01', 115)     ->('q4', 2, 'XX1#XX1', 104)    ->('q4', 2, 'XXY#XXY', 94)
 ->('q4', 2, 'X01#X01', 114)     ->('q4', 1, 'XX1#XX1', 103)    ->('Iq0', 3, 'XXY#XXY', 93)
 ->('q4', 1, 'X01#X01', 113)     ->('Iq0', 2, 'XX1#XX1', 102)   ->('q5', 4, 'XXY#XXY', 92)
 ->('q4', 0, 'X01#X01', 112)     ->('q7', 3, 'XXY#XX1', 101)    ->('q5', 5, 'XXY#XXY', 91)
 ->('Iq0', 1, 'X01#X01', 111)    ->('q8', 4, 'XXY#XX1', 100)    ->('q5', 6, 'XXY#XXY', 90)
 ->('q1', 2, 'XX1#X01', 110)                                    ->('q5', 7, 'XXY#XXY', 89)
                                                      -> ('Fq6', 8, 'XXY#XXY........', 88)
```

Figure 13.10: Run of the *w#w* string on the DTM shown results in acceptance at state Fq6, with 88 thimbles of fuel left. While in state q5, the machine allocates tape cells (the "..."), finds the blank being sought, and accepts in state Fq6.

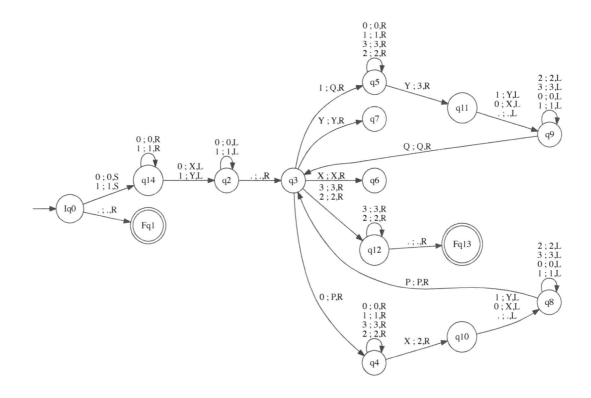

```
Accepted at  ('Fq13', 13, '........PQ23........', 76)  via ..

 ->('Iq0', 0, '0101', 100)        ->('q8', 8, '........P12Y', 87)
 ->('q14', 0, '0101', 99)         ->('q3', 9, '........P12Y', 86)
 ->('q14', 1, '0101', 98)         ->('q5', 10, '........PQ2Y', 85)
 ->('q14', 2, '0101', 97)         ->('q5', 11, '........PQ2Y', 84)
 ->('q2', 1, '01X1', 96)          ->('q11', 12, '........PQ23', 83)
 ->('q2', 0, '01X1', 95)          ->('q9', 11, '........PQ23........', 82)
 ->('q2', 7, '........01X1', 94)  ->('q9', 10, '........PQ23........', 81)
 ->('q3', 8, '........01X1', 93)  ->('q9', 9, '........PQ23........', 80)
 ->('q4', 9, '........P1X1', 92)  ->('q3', 10, '........PQ23........', 79)
 ->('q4', 10, '........P1X1', 91) ->('q12', 11, '........PQ23........', 78)
 ->('q10', 11, '........P121', 90)->('q12', 12, '........PQ23........', 77)
 ->('q8', 10, '........P12Y', 89) ->('Fq13', 13, '........PQ23........', 76)
 ->('q8', 9, '........P12Y', 88)


Rejected at  ('q14', 4, '0101', 95)      Rejected at  ('q4', 9, '........PY01', 94)
Rejected at  ('q4', 11, '........P10Y', 88)  Rejected at  ('q6', 9, '........X101', 96)
```

Figure 13.11: Simulation of *ww* on the NDTM shown.

```
TAPE_ALLOC_SIZE = 8
def step_tm(T, q_hi_tape_fuel, path, haltList):
    """Given a TM (T), an ID (q,hi,tape,fuel) where q is the current state, hi is the head index,
       tape is the current tape (string) and fuel is the number of steps that the TM is eligible
       to execute, and given a computational path executed so far, return the new set of IDs and
       extend the computations along each nondeterministic path.
    """

    (q, hi, tape, fuel) = q_hi_tape_fuel
    extpath             = path + [q_hi_tape_fuel]

    nl_id_path          = []
    if (hi == len(tape)): # Going beyond end of allocated tape; allocate more!
        print("Allocating ", TAPE_ALLOC_SIZE, " tape cells to the RIGHT!")
        tape = tape + T["B"]*TAPE_ALLOC_SIZE
    if (q, tape[hi]) not in T["Delta"]: # No move on (q, tape[hi]), so record ID+path in haltList; return
        return (nl_id_path, haltList + [(q_hi_tape_fuel, path)])
    l_nq_ng_dirn        = T["Delta"][(q, tape[hi])]
    for nq_ng_dirn in l_nq_ng_dirn:
        (nq, ng, dirn) = nq_ng_dirn # Head attempts to move to the left of the left-end
        if (hi==0) and (dirn=="L"):
            print("Allocating ", TAPE_ALLOC_SIZE, " tape cells to the LEFT!")
            ntape = T["B"]*TAPE_ALLOC_SIZE + ng + tape[1:]
            nhi   = TAPE_ALLOC_SIZE - 1 # Do the left move too!
        else:
            ntape = tape[0:hi] +  ng  + tape[hi+1:len(tape)]
            nhi = (hi+1 if dirn=="R"
                    else ((hi-1) if dirn=="L"
                          else (hi if dirn=="S"
                                else print("Illegal direction!"))))
        if (fuel > 0):
            nl_id_path += [((nq, nhi, ntape, fuel-1), extpath)]
    return (nl_id_path, haltList)
```

Figure 13.12: Function step_tm

```
TM  !! This is a TM for ww processing. Guesses midpoint using nondet.
Iq0    : 0 ; 0 , S -> q14 !! This simulates the TM taking a guess
Iq0    : 1 ; 1 , S -> q14 !! that it hasn't seen the midpoint. It moves to q14
Iq0    : . ; . , R -> Fq1 !! Yay! shortest acceptance is for eps eps, i.e., facing a sea of
                          !! blanks that encodes an epsilon followed by another epsilon.
q14    : 0 ; 0 , R  -> q14 !! The TM skips over 0s or 1s for a while, and then chooses a cell,
q14    : 0 ; X , L  -> q2  !! declaring it the midpoint, or more specifically FIRST CHARACTER PAST MIDPOINT
                           !! by marking it 'X' and then moves to q2 (to march around the chosen midpoint).
q14    : 1 ; 1 , R  -> q14 !! Similar actions as with 0 in state q14, except that it "dings" the
q14    : 1 ; Y , L  -> q2  !!  "1" with a "Y" to mark it the FIRST CHARACTER PAST MIDPOINT.
!!-- Then we march around it. While the separate use of "X" and "Y" may not be necessary, it
!!-- improves understandability when you finally see the result of TM executions.
q2     : 0 ; 0 , L  -> q2  !! The TM is now winding back, seeking the
q2     : 1 ; 1 , L  -> q2  !! left-end of the tape till it hits a '.' (blank).
q2     : . ; . , R  -> q3  !! When that happens, the TM goes to state q3 to begin its work of "matching around."
!! Below, we describe the q3,q5,q11,q9,q3 loop well (The loop q3,q4,q10,q8,q3 is similar).
q3     : X ; X , R    -> q6  !! This state is a stuck state (no progress)
!! We came to q3 because we dinged a 0->X or a 1->Y while in q14; so its matching
!! "partner" 0 or 1 must be found to the left. Unfortunately, we are finding an
!! X or a Y.  Thus, no "match around the middle" is likely to happen.
q3     : Y ; Y , R    -> q7  !! This state is ALSO a stuck state for similar reasons as described at q3 : X ; X ...
!! Description of the q3,q5,q11,q9,q3 loop :
!! Upon seeing a 1, change to Q. Then MUST see a  matching Y, then change to 3, and go right,
!! and to state q5. We do this because 'Y' represents what was '1' and got marked as "midpoint"
q3     : 1 ; Q , R    -> q5
!!-- What will happen in q5,q11,q9,q3 --
!! So we have to get past this assumed midpoint and choose the next "one past midpoint that
!! has not been seen so far". We enter q11 to then ding a matching 0 to X or 1 to Y, moving left.
!! A blank sends us leftwards, as well.
!! We sweep left till we hit a Q. We MUST see a Q because we entered "this lobe" by dinging a 1->Q.
!! The process repeats from state q3.
q5     : 0;0,R | 1;1,R | 2;2,R | 3;3,R -> q5  !! punt the 0/1/2/3; we need a "Y".
q5     : Y ; 3, R                      -> q11 !! ah-ha, got a Y. Ding to 3, seek 0/1/.
q11    : 1;Y,L | .;.,L | 0;X,L         -> q9  !! phew! got to sweep left now!
q9     : 0;0,L | 1;1,L | 2;2,L | 3;3,L -> q9  !! whee! going left!
q9     : Q ; Q , R                     -> q3  !! Boiinggg - now gonna go right!
!! Description of the q3,q4,q10,q8,q3 loop :
q3     : 0 ; P , R    -> q4   !! This is similar to q3 : 1 ; Q , R -> q5 above
q4     : 0;0,R | 1;1,R | 2;2,R | 3;3,R -> q4  !! punt the 0/1/2/3; we need an "X".
q4     : X ; 2, R                      -> q10 !! ah-ha, got an X. Ding to 2, seek 0/1/.
q10    : 1;Y,L | .;.,L | 0;X,L         -> q8  !! phew! got to sweep left now!
q8     : 0;0,L | 1;1,L | 2;2,L | 3;3,L -> q8  !! whee! going left!
q8     : P ; P , R                     -> q3  !! Boiinggg - now gonna go right!
!! Seeing every sign of acceptance. We are seeing piles of 2 and 3. ALSO did not get stuck in
!! q6 or q7. That means all the matches went fine.
q3     : 2;2,R | 3;3,R -> q12
q12    : 2 ; 2 , R | 3 ; 3 , R        -> q12 !! Skip over piles of past 2s and 3s
q12    : . ; . , R                    -> Fq13!! ** Yay, acceptance when we hit a blank! **
```

Figure 13.13: Markdown for our NDTM
that recognizes ww, $w \in \{0,1\}^*$.

13.6   Example: A TM that Works on the Collatz Problem

The following function `tep1` was proposed by German mathematician Lothar Collatz in 1937:

```
def tep1(x):
    if (x==1):
        return 1
    elif (x%2 == 0):
        return tep1(x/2)
    else:
        return tep1(3*x+1)
>>> tep1(3)
1
>>> tep1(191)
1
>>> tep1(19192949297080)
1
>>> tep1(191929492970809272397923472398492839482)
1
>>> tep1(19192949297080927239792347239849283948293 8492)
1
```

[19] A crowdsourced attempt to find a counterexample is underway. You may contribute to this effort by donating idle cycles of your own machine, by visiting http://www.ericr.nl/wondrous/search.html.

[20] One website has claimed that this is related to the undecidability of the Halting problem studied in Part III. This is not true. `tep1` just happens to be a curious function. One can couch many conjectures—even Goldbach's conjecture that every integer above 2 can be expressed as the sum of two primes—as such a program, and make the program's halting conditional on a violation of the conjecture.

No matter what we feed it, `tep1` seems to halt (returning 1).[19] *Unfortunately*, it is open whether this holds true for all natural numbers.[20] We present a TM for the Collatz problem and its simulation on input 6 in binary (*i.e.*, 0110) in Figure 13.14.

See additional examples in our Jove distribution including a TM that doubles a number given in decimal and a TM that carries out binary addition.

13.6.1   Markdown for the Collatz Problem TM with Comments

[21] We convert such markdowns to internal formats ("machines") through the **md2mc** command (§B.1.5).

Our Automd-style markdown description[21] of the coding of the Collatz problem appears in Figure 13.15. This description embodies very good TM coding practices (meaningful state names and well-placed comments).

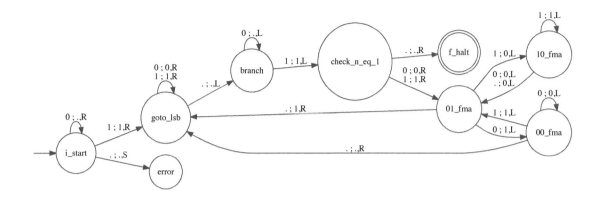

```
# Will loop if the Collatz ("3x+1") program will ever loop!
explore_tm(collatz_tm, "0110", 100)
Allocating  8  tape cells to the RIGHT!
Allocating  8  tape cells to the LEFT!
Detailing the halted configs now.
Accepted at  ('f_halt', 5, '.....1...............', 65)
 via ..
 ->('i_start', 0, '0110', 100)          ->('01_fma', 9, '.......101..........', 81)
 ->('i_start', 1, '.110', 99)           ->('10_fma', 8, '.......100..........', 80)
 ->('goto_lsb', 2, '.110', 98)          ->('01_fma', 7, '.......100..........', 79)
 ->('goto_lsb', 3, '.110', 97)          ->('10_fma', 6, '.......000..........', 78)
 ->('goto_lsb', 4, '.110', 96)          ->('01_fma', 5, '......0000..........', 77)
 ->('branch', 3, '.110........', 95)     ->('goto_lsb', 6, '.....10000..........', 76)
 ->('branch', 2, '.11.........', 94)     ->('goto_lsb', 7, '.....10000..........', 75)
 ->('check_n_eq_1', 1, '.11.........', 93)  ->('goto_lsb', 8, '.....10000..........', 74)
 ->('01_fma', 2, '.11.........', 92)     ->('goto_lsb', 9, '.....10000..........', 73)
 ->('10_fma', 1, '.10.........', 91)     ->('goto_lsb', 10, '.....10000..........', 72)
 ->('10_fma', 0, '.10.........', 90)     ->('branch', 9, '.....10000..........', 71)
 ->('01_fma', 7, '........010.........', 89)  ->('branch', 8, '.....1000...........', 70)
 ->('goto_lsb', 8, '.......1010.........', 88)  ->('branch', 7, '.....100............', 69)
 ->('goto_lsb', 9, '.......1010.........', 87)  ->('branch', 6, '.....10.............', 68)
 ->('goto_lsb', 10, '......1010.........', 86)  ->('branch', 5, '.....1..............', 67)
 ->('goto_lsb', 11, '......1010.........', 85)  ->('check_n_eq_1', 4, '.....1...............', 66)
 ->('branch', 10, '......1010.........', 84)   ->('f_halt', 5, '.....1...............', 65)
 ->('branch', 9, '.......101..........', 83)
 ->('check_n_eq_1', 8, '.......101..........', 82)
```

Figure 13.14: A simulation of the Collatz
conjecture using a DTM

```
TM !!
i_start      : 0; ., R -> i_start         !! erase this zero and try to find more
i_start      : 1; 1, R -> goto_lsb        !! we have a proper number, go to the lsb
i_start      : .; ., S -> error           !! error on no input or input == 0
goto_lsb     : 0; 0,R | 1; 1,R -> goto_lsb !! scan off the right edge of the number
goto_lsb     : .; .,L -> branch           !! take a step back to be on the lsb and start branch
branch       : 0; .,L -> branch           !! number is even, divide by two and re-branch
branch       : 1; 1,L -> check_n_eq_1     !! number is odd, check if it is 1
check_n_eq_1 : 0; 0,R | 1; 1,R -> 01_fma  !! number wasn't 1, goto 3n+1
check_n_eq_1 : .; .,R -> f_halt           !! number was 1, halt
!! carrying 0 we see a 0 so write 0 and carry 0 forward
00_fma       : 0; 0,L -> 00_fma
!! carrying 0 we see a 1 (times 3 is 11) so write 1 and carry 1 forward
00_fma       : 1; 1,L -> 01_fma
!! reached the end of the number, go back to the start
00_fma       : .; .,R -> goto_lsb
!! carrying 1 we see a 0 so write 1 and carry 0 forward
01_fma       : 0; 1,L -> 00_fma
!! carrying 1 we see a 1 (times 3 is 11, plus our carry is 100) so write 0 and carry 10 forward
01_fma       : 1; 0,L -> 10_fma
!! reached the end of the number, write our 1 and go back to the start
01_fma       : .; 1,R -> goto_lsb
!! carrying 10 we see a 0, so write 0 and carry 1 forward
10_fma       : 0; 0,L -> 01_fma
!! carrying 10 we see a 1 (times 3 is 11, plus our carry is 101), so write 1 and carry 10 forward
10_fma       : 1; 1,L -> 10_fma
!! reached the end of the number, write a 0 from our 10 and carry 1
10_fma       : .; 0,L -> 01_fma
```

Figure 13.15: Markdown for TM that encodes the Collatz problem

13.7 The Chomsky Hierarchy

Machines	Languages	Nature of Grammar
DFA/NFA	Regular	Purely left-/right- linear productions
DPDA	Deterministic CFL	Each LHS has one nonterminal. The productions are deterministic.
NPDA (or "PDA")	CFL	Each LHS has only one nonterminal.
LBA	Context Sensitive Languages	LHS may have length > 1, but $|LHS| \le |RHS|$, ignoring ε productions.
DTM/NDTM	Recursively Enumerable	General grammars ($|LHS| \ge |RHS|$ allowed).

Figure 13.16: Situation of TMs in the Chomsky Hierarchy.

We will now define a crucially important notion called the **Chomsky hierarchy** in §1.3 (see Figure 13.16; this notion was alluded to in Chapter 1). We now discuss all the machines we have studied so far as well as their **grammars** in one unified setting.[22] Each language family first discussed in Figure 13.16 is contained in all the language families later discussed.[23]

Regular languages: Regular languages are characterized by DFA as the machine type, and purely linear grammars. The presence of nondeterminism does not increase the expressive power.

Context-free languages: Context-free languages are characterized by PDA as the machine type and context-free production rules as the grammar. Each left-hand side of a production has exactly one nonterminal. The deterministic counterparts of PDA and CFL are *not equivalent* to their nondeterministic counterparts.[24]

Context-sensitive languages: Context-sensitive languages are characterized by the machine type *Linear Bounded Automaton* (LBA). Basically, LBA are Turing machines that are not allowed to write beyond the extent of the initial input.[25] It has been an **open question** (for nearly six decades) whether the family of non-deterministic LBA (NLBA) and deterministic LBA (DLBA) are equivalent.

As for production rules, context sensitive languages are based on *context sensitive grammar* (CSG) production rules that allow more than one item on the left-hand side. An example CSG production rule is:

```
a A d -> a a c d
```

[22] As discussed in Chapter 11 (§11.7), even DFA can be captured using purely right-linear (or purely left-linear) grammars.

[23] In other words, every regular language is a context-free language, a context sensitive language, and a recursively enumerable (RE) language. Every CFL is a CSL and an RE language. Finally, all CSLs are RE languages.

[24] There are inherently ambiguous languages that cannot be captured using DPDA.

[25] They have a tape that is read-only beyond the extent of the user input.

In this example, we are essentially allowing nonterminal A to be replaced by the terminal sequence a c *provided* A is "surrounded" by a and d (or, lives in the context of a and d). Context-free is a special case of context sensitive where a nonterminal can be expanded regardless of "what surrounds it." However, the catch with context sensitive productions is that the length of the left-hand sides must stand in the relation $\leq$ to the length of the right-hand side (ignoring ε productions). Thus, the productions are never "contracting."[26]

Recursively Enumerable Languages: Last but not least, recursively enumerable languages (RE)[27] are characterized by Turing machines. The grammar associated with them is similar to CSLs, except we allow both "expanding" and "contracting" productions in that there is no required length relation between the right-hand side and the left-hand side. It is known that both the deterministic and nondeterministic varieties of Turing machines are equally expressive.

13.7.1 Recursively Enumerable and Recursive Languages

A **recursively enumerable** language[28] is the language of a (general) Turing machine. A **recursive language** is the language of a Turing machine that is guaranteed to halt.

The idea of a *language of a Turing machine* "feels different" in that TMs embody general-purpose computations.[29] While regular languages and context-free languages seemed to have a purpose behind them (scanning and parsing), we now are in this territory of machines where we can't even tell whether the machine will read a given input fully. While TMs can certainly be made to recognize the structure of their input string, in a general setting, that is not the primary emphasis behind Turing machines. Part III (Concepts) will fully explain these results, but some remarks may put the reader at ease:

- In a formal sense, Turing machines define *both* a computer and a computer program.[30] Whether this reality adds to the innate understanding of "a computer" and "a computer program" in the modern context may be debatable.[31] However, these important ideas must be defined mathematically (which is what TMs help with), and not taken for granted.[32] If tomorrow brings a different form of a "protoplasmic computer," then we may, most likely, be able to far more easily prove whether it is *Turing-equivalent*—not whether it is *MacBook Pro equivalent*.

- The real fun will begin when we study Turing machines more deeply in Part III where we consider a TM *examining another TM's description (given to it as the input string)*. Such studies—and the conclusions we draw from them—are the *only way* we can prove (or disprove) whether a computer can (or cannot) do something at all. For this purpose, the

[26] It is easily seen that both linear productions and CFG productions satisfy this restriction, as they are special cases of context sensitive languages.

[27] There are non-RE languages as well—in fact, many of them! This is how the argument goes: there are only $\aleph_0$ RE languages, while there are $\aleph_1$ languages. See Appendix C for a proof.

[28] Abbreviated "RE language," not to be confused with "*regular expressions*"

[29] It could mean *the language of my spreadsheet program* or even *the language of the Obamacare website!*

[30] The tape, head, and finite-state control *mechanism* are similar to the hardware of a realistic computer. The particular finite-state controller we design for a TM is similar to an actual computer program.

[31] Some may protest, saying that this feels like analyzing Mozart's music in terms of air-pressure waves.

[32] Some modesty is always in order, considering that the universe has been around 6,000 times longer than humans have been around, and humans have been around 20,000 times longer than the notion of computation has been around. "Taking things for granted" can lead to pitfalls. Besides, if we have to tell a space denizen what a computer is, he/she/it may more readily understand if explained in terms of a Turing machine!

notion of a recursively enumerable language is central. We cannot even begin to approach such questions without this concept.

13.8 An Alternate Notation for Instantaneous Descriptions

In this chapter, we explicitly recorded the control state, the head index, and the input string as a triple (q, h, i) in discussing instantanoues descriptions (ID). Sometimes, we also included the "remaining fuel" (in Jove simulations).

For formal discussions in Chapter 16, this notation becomes rather cumbersome (even after leaving out the "fuel" component). A more compact notation is as follows:

- Denote the starting ID of a Turing machine starting in state q with tape containing w as qw.
- For a Turing machine that has $a \in \Sigma^*$ to the left of the head, has its head looking at the first symbol of $b \in \Sigma^*$, and is in control state q, we depict the ID as aqb.
  - Example 1: The flipper TM in Figure 13.2 will have the following march of IDs. We provide the Jove-style IDs and the formal notation of IDs being introduced side by side:

    ('I', 0, '010011', 100) $I010011$

    ('I', 1, '110011', 99) $1I10011$

    ...

    ('I', 6, '101100', 94) $10110I0$

    ('F', 6, '101100...', 93) $101100F$
  - Example 2: The DTM that checks for whether 101 appears in the input string (Figure 13.7) will have this cadence of IDs (here, we separate the state names with quotes so that they stand out):

    ('I', 0, '0010101', 10) $'I'0010101$

    ('I', 1, 'A010101', 9) $A'I'010101$

    ('I', 2, 'AA10101', 8) $AA'I'10101$

    ('Got1Sk0', 3, 'AAB0101', 7) $AAB'Got1Sk0'0101$

    ('Got10Sk1', 4, 'AABA101', 6) $AABA'Got10Sk1'101$

    ('Found101', 5, 'AABAB01', 5) $AABAB'Found101'01$

Exercise 13.8, DTM and NDTM design

1. Design a DTM to increment an unsigned binary number presented *little endian* (which means *least significant bit first* or LSB). Specifically, if the tape is blank, stop (the number is assumed to be unspecified). Else, the left-most bit is the LSB. Increment it, and deal with the carry appropriately. The result on the tape must be the incremented number. Simulate it in Jove. Write an explanation of why your DTM is correct. Here are some input/output behaviors:
   - The input '' leaves the tape with a sea of blanks (as it was before).

- The input '00' leaves the tape with '10' after acceptance.
- The input '01' leaves the tape with '11' after acceptance.
- The input '10' leaves the tape with '01' after acceptance.
- The input '11' leaves the tape with '001' after acceptance.

2. Design a DTM whose language is L_{1gt0}, which is the set of strings in which the number of 1's is greater than the number of 0's.

3. Design an NDTM that takes inputs of the form $w_1 \# w_2$ and checks whether w_1 is a substring of w_2. A string s is a substring of another string t exactly when one can find s beginning at some position $t[i]$ of t. For example, a, l, ap, le, and ppl are substrings of apple. We define the empty string to be a substring of any given string. *You must truly take advantage of nondeterminism to simplify your machine.* (You must not end up designing a DTM that will have to engage in more actions.)

4. Design an NDTM that takes inputs of the form $w_1 \# w_2$ and checks whether w_1 is a subsequence of w_2. A string s is a subsequence of another string t exactly when one can find s beginning at some position $u[i]$ of u, where u is obtained by deleting some of t's characters. For example, a and l are subsequences of apples. Also ap, al, and ale are subsequences of apples. More specifically, ale is found sitting in u where u=ales is obtained by deleting both p's and the s from apples. Then one finds ale sitting within ales. We define the empty string to be a subsequence of any given string. *You must truly take advantage of nondeterminism to simplify your machine.* (You must not end up designing a DTM that will have to engage in more actions.)

5. Suppose we are given a string w that consists of a finite number of binary sequences separated by the hash-mark #. By *maximal sequence* of 0's and 1's, we mean the entire sequence contained within two hash marks (#). There is also one sequence appearing at the end, after two hash marks (*i.e.*, ##). We call the last maximal sequence the *output sequence*. We call all other maximal sequences before the '##' the list of *input sequences*.[33] For instance,

- The input might be
  1010 # 0100 # 101 # 11101 # 001 ## 101001
- Here, the input sequences are 1010, 0100, 101, 11101 and 001
- The output sequence is 101001

Task for you: Design an NDTM that determines whether some concatenation of some number of input sequences (*i.e.*, taken left to right) matches the output sequence. In our example, we have 101 concatenated with 001 matching 101001 in the output. □

[33] Hereafter, we will leave out the word maximal.

Part III: Concepts

Alan Turing (left) and Alonzo Church (right)

From https://en.wikipedia.org/wiki/Alan_Turing
and https://en.wikipedia.org/wiki/Alonzo_Church respectively.

Part III: Concepts

14

Interplay between Formal Languages

> **Chapter Gist:** *We motivate the need to understand the fundamental limits of computing (§14.1). This is followed by two key notions, recursive and recursively enumerable (RE), that are introduced through examples (§14.2). More examples of recursive and RE languages are presented (§14.3). A grand finale of language containment and decidability results is presented with the help of a multi-colored Venn diagram (§14.4). Additional high-level proof sketches about RE and recursive languages finish this chapter (§14.5).*

14.1 Why Study Impossibility Results?

When somebody tells us all the features of a cool new machine, it is only natural to ask what the new machine *cannot do*. For instance, everyone knows that airplanes can fly, but they cannot fly into outer space (at least not yet). Archimedes once famously said this about the simplest of machines, namely a lever: "Give me a lever long enough and a fulcrum strong enough, and then I *can* move the earth."[1]

In the same vein, when scientists confronted the computer – a new machine in the 1930s – they of course first asked what it *can* do. You may however recall from §1.1 that Hilbert and Gödel helped settle several fundamental questions about the fundamental limits of computing—the "*can't do*" questions about computing. To study and settle this question systematically, they had to formulate the problem in terms of formal languages (Figure 13.16). In this chapter, we elaborate on these fundamental concepts that underlie computability theory.

14.1.1 Definitions: Procedure and Algorithm

In order to define what can and cannot be done using a computer, we must define two related terms: **procedure** and **algorithm**.

[1] A claim that, we know, must be shifted to the "can't" category.

> A procedure is a mechanically realizable function that is defined specific to solving a particular problem. A procedure typically takes inputs and produces outputs. It is formally defined through a Turing machine. An algorithm is a procedure that halts on all inputs. It is in fact impossible to tell whether a procedure is simply taking too long to run or is actually never going to halt.

We can now point out one thing (a very fundamental thing) that a computer cannot do:

> No computer can determine whether a procedure is an algorithm.

That is, given a Turing machine T_P describing a procedure P, there isn't a checker Turing machine that can conclusively establish whether T_P will halt on all inputs. (If at all a Turing machine can so check, then we must admit the possibility of this checker Turing machine going into an infinite loop.) This fact can also be stated another way:

> There isn't an algorithm to check whether a given procedure is an algorithm.

We will now formalize these ideas by identifying the formal language families associated with Turing machines.

14.1.2   Formal Languages Associated with Turing Machines

Formal languages defined by Turing machines belong to the family of *recursively enumerable (RE)* languages.[2] *Recursive languages* are a proper subset of RE languages, and correspond to Turing machines that *halt on all inputs*.[3]

We have used the phrase *family of languages* a few times already in this chapter. The term **family** also means *set*. However, in many contexts, it leads to smoother sentences. Instead of saying "x is contained in set L which belongs to the set of recursive languages," we prefer saying "x is contained in set L which belongs to the family of recursive languages."[4]

In this section, we will consider various computable languages L and describe procedures/algorithms to verify whether a string x is in L. Depending on what L is, we will be able to discuss the properties of various formal languages. There is a fairly comprehensive diagram that captures these languages (Figure 14.2); a quick glance at it may help now (this diagram will be discussed in detail later).

14.1.3   Allaying Confusion: Language vs. Language Family

The containment of one specific formal language within another is a totally different idea from the containment of one language *family* within

[2] Some authors also call this family *Turing recognizable (TR)*—a term that we may occasionally use.

[3] Recursive languages are also called *decidable* languages. Decidable languages are almost always desirable too.

[4] Technically, the word "family" in mathematics is used when we don't necessarily want a "set-like behavior." Thus (for some reason), we may want to allow two distinct members of a family to be identical; we know that two distinct members of a set aren't identical.

another. For instance, the universal language $Univ = \Sigma^*$ is a regular language. Now, given a context-free language L_{cfl}, it is certainly contained within $Univ$. The same is true of any context-sensitive language, recursive language, or recursively enumerable language: all are contained within $Univ$. The following "spray paint analogy" may help.

> Imagine $Univ$ as a drawing obtained by spraying black paint uniformly on a white sheet of paper. Much like $Univ$ contains every possible finite string, a paper sprayed with black paint can contain every imaginable sentence in English. The only catch is that such sentences happen to be surrounded by other droplets of black ink, and so you cannot pick out these sentences.
>
> By *leaving out* strings from $Univ$, we can obtain interesting languages that highlight the *structures* we are after. Thus, leaving out all strings that don't look like $0^n 1^n$ from $Univ$, we obtain the interesting CFL that contains (all and only those) strings of the form $0^n 1^n$ (call it L_{0n1n}). In the same sense, if we pull out the droplets "surrounding the sentences," we will reveal those sentences that remained invisible within the uniform spray.
>
> In general, *by leaving out stuff*, we can create useful things, including works of art. A block of wood contains every impressive carving, if only we knew how to remove wood correctly.

Figure 14.2 is about the containment of *language families* and not languages. Here, $Univ$ is just **one point** inside the innermost Venn diagram set labeled *Regular*. A context-free language with strings of the form $0^n 1^n$ is a point outside of *Regular*, but inside *Context-free*—and likewise for all the other language families. This is why the language family inclusions are the way depicted: less structure-endowed languages are points within inner Venn diagram sets, and more feature-laden languages are points within encompassing sets. The universal set of this Venn diagram is the union of *all intricate structures*—including even non-RE sets.

14.2   One Example of a Recursive and an RE Language

We begin with one example of a recursive language in §14.2.1 and one example of an RE language in §14.2.5. This will help us to set the stage to show more of them, plus show intricate theorems that bridge between them.

14.2.1   A Recursive Language

Consider language $L_{EmptyDFA}$ defined as follows:

$$L_{EmptyDFA} = \{x : x \text{ describes a DFA whose language is empty}\}$$

[5] String x could well be the markdown description of a DFA.

[6] Can you think of an algorithm to check whether a given DFA's language is empty? Hint: Think of the most general way to characterize a DFA that does not have any string in its language—not even ε.

[7] Hereafter, we will use D to denote a DFA and $\langle D \rangle$ to denote its description. This applies to other entities too—for instance, a Boolean function φ and its description $\langle \varphi \rangle$.

[8] Say by hiring a soothsayer or asking an oracle...

[9] The shock may be too much.

This is a recursive language because we can define a TM T that reads from its input tape the *description* of a DFA D, encoded as a string x in the tape alphabet of the TM.[5] The TM then runs an algorithm to check whether D's language is empty.[6] If so, it accepts x; else it rejects x. We take acceptance as an indicator of "yes" or "true," and rejection as an indicator of "no" or "false." Given that such a TM algorithm can be designed, we can conclude that $L_{EmptyDFA}$ is a recursive language.[7] By saying that $L_{EmptyDFA}$ is a recursive language, we mean *checking for emptiness of DFA is algorithmic.*

> *(Definition of a Decider and Semi-Decider):* A Turing machine that correctly checks for the membership of a string x in a given language L and halts, either printing "yes, $x \in L$" or "no, $x \notin L$" is said to be a **decider** for L. The algorithm implemented by such a TM is said to be a *decision procedure* for membership of x in L.
> **Deciders have recursive languages, and represent algorithms.**

Keep in mind that we will not know a priori[8] whether $x \in L$ or not. When the purveyor of a decider sells us a decider for L, *we must take a chance and run it on x.* If the purveyor is honest, we will find that her decider always halts and correctly answers whether or not $x \in L$.

14.2.2   Prerequisites to Defining the Notion of RE

Unfortunately, we cannot show you an RE language right away, unless we adequately prepare you to receive one.[9] Thus we begin with definitions, starting with that of a **semi-decider**.

> *(Definition of a Semi-Decider):* One of the coolest aspects of computer science is that *we can prove that* for some problems, only semi-deciders (or "semi-algorithms" or "half algorithms") can exist! If you wish to take a chance with a semi-decider, you'd better do this:
> - Either patiently wait for the semi-decider to "come back;" **or**
> - Have your fingers on "Control-C" to exit out of the semi-decider and answer "*MAYBE $x \in L$.*"
>
> This is because a semi-decider for membership of x in L is guaranteed to halt and print "yes, $x \in L$" only in case this assertion is true. However the only way to *find out* is to take a chance!
> **Semi-deciders have recursively enumerable (and not recursive) languages, and represent procedures.**

We will use the word 'procedure' less often, as it is not as crisp or formal as RE (semi-decider).

14.2.3   Combining Semi-deciders for L and $\overline{L}$

Ordinarily, when the purveyor of a semi-decider sells us a semi-decider for L, *we take a chance and run it on x*, keeping one pair of fingers crossed, and another pair of fingers on Control-C.[10] However, suppose we can buy these two semi-deciders:

- A semi-decider for $x \in L$ (say, SD_1)
- A semi-decider for $x \in \overline{L}$ (say, SD_2)

Then we can combine SD_1 and SD_2 to obtain a full decider for L as follows:

- If SD_1 returns, then Control-C out of SD_2 and answer "yes."
- If SD_2 returns, then Control-C out of SD_1 and answer "no."

The reader may verify that by swapping the above answers (yes for no, and vice-versa), we obtain a decider for $\overline{L}$. These arguments can be summarized as the following theorem.

> **Theorem 14.2.3:** If a language L and its complement $\overline{L}$ are both RE, then L is recursive.

14.2.4   The "no wimp" Clause

In all these constructions, one must avoid deliberately creating a semi-decider when it is possible to create a full decider. *The designer must strive to either come up with a decider or prove (or give strong reasons for) why such a decider cannot exist.*[11] We call this the "no wimp" clause because whenever someone says "L is RE," the default meaning is that "L is known to be not recursive." This convention will be applied to the exercises we shall entertain.

14.2.5   A Recursively Enumerable Language

Language L_{G1neG2} is recursively enumerable but not recursive
Consider language L_{G1neG2}, defined as follows:

$$L_{G1neG2} = \{\langle G_1, G_2 \rangle : G_1 \text{ and } G_2 \text{ are CFGs, and } L(G_1) \neq L(G_2)\}$$

This language consists of pairs of context-free grammars such that these CFGs denote different context-free languages.[12] *Now we see a distinct change in the kinds of questions being asked.*

> Instead of asking whether the syntax of the input meets certain criteria, we now ask whether a parser (PDA) loaded with CFG G_1 will *behave* the same as a parser loaded with CFG G_2 *over all parser inputs*. A Turing machine is asked to classify these CFG pairs based on their *parsing behaviors on all inputs to a parser based on these grammars.*

[10] If a purveyor *sells us a decider* when we have proof that a decider cannot exist, then we can take the purveyor to court and sue him! However, note that under "some weather conditions," a semi-decider may print "no, $x \notin L$" and halt. Thus, taking the person to court or not simply by testing out a handful of cases is not a wise idea, for, the purveyor may have "jimmied" his semi-decider to return enough "no" answers. In other words, *prove before you sue.*

[11] This is like claiming that sorting is $O(2^n)$ in an algorithms class which makes one eligible for a failing grade. While there are exponential sorting algorithms, one always strives to do better.

[12] It is in fact interesting to note that our classification is not based on the superficial structure of the production rules that represent CFGs. In fact, the production rules that represent CFGs syntactically are *regular languages*: they consist of simple repetitions of productions, where the productions themselves are repetitions of nonterminals and terminals coming after an arrow symbol. We are given the syntactic representation of G_1 and G_2 but then asked to consider $L(G_1)$ and $L(G_2)$ in their full glory, checking whether these languages are unequal.

In a sense, this is the kind of change that awaits you at every turn in the rest of this course. We will be describing machines using strings (or pairs of strings), and will attempt to classify these strings into a language based on the behaviors produced by the strings. We begin by sketching a proof for the question at hand.

Why is L_{G1neG2} Recursively Enumerable? The first part is to show that **if $L(G_1) \neq L(G_2)$**, there is indeed a test for it. This test, when implemented faithfully by a Turing machine, imparts to this TM the language L_{G1neG2}.

The test we present is perhaps not the prettiest of tests; but still it is a mechanizable test that helps establish the existence of a **semi-algorithm**. That is, *if* the grammars have different languages, the Turing machine will accept such pairs and halt.

- Keep on enumerating strings from Σ^* according to the *numeric order* introduced in §3.6. This enumeration will make sure that *every string will be eventually enumerated*.

- Parse each enumerated string using the CFG. If the CFGs indeed have a different language, they must exhibit the difference on at least one string. Thus, when that string is enumerated and parsed by the parsers of G_1 and G_2, one parser will accept while the other will reject. *Voilà, the difference between the grammars will have been exhibited!*

- All these activities can be carried out on a single Turing machine.

> The Turing machine that implements the above tests has language L_{G1neG2}, and hence this language is RE.

Why is L_{G1neG2} not recursive? This is the more interesting part of the overall question. We can't quite answer this question in full depth yet.

Intuition: We just showed that L_{G1neG2} is RE. Suppose we have a semi-decider for L_{G1eqG2}. That is, we have a semi-decider that outputs "yes" in case $L(G_1) = L(G_2)$. Here is intuitively where the difficulty lies:

- We can tell that two grammars are different just by having *one string* enumerated according to the numeric order where *one parser accepts* while the *other parser rejects*.

- However, to tell that two grammars are equivalent requires examining *all* strings in Σ^*, which is an activity that won't terminate.

Note: This is *only an indication that L_{G1eqG2} is likely not RE.*

Actual High Level Proof: The actual proof roughly goes as follows.

- In Chapter 15, we will give a proof sketch for the fact that it is impossible to tell whether, given a CFG, it has a universal language. That is, whether the CFG describes a PDA whose language is Σ^* cannot be algorithmically checked merely by analyzing the CFG.[13] That is, a decider (call it $UnivCFG$) to check whether a given CFG's language is universal cannot exist.

---

[13] Contrast this with $L_{EmptyCFG}$ which is a language of CFGs with an empty language; we will discuss this language in §14.3.

- Well, the fact that telling whether CFGs have universal languages being undecidable itself merits a detailed proof; see [42] for details. However, *given that proof*, we can rigorously argue as follows:
  - Suppose someone claims that there is an algorithm to check whether two grammars have the same language using algorithm $EqCFG$. That is, suppose $EqCFG(G_1, G_2)$ will halt and emit "yes" exactly when $L(G_1) = L(G_2)$.
  - Then we can do the following trick: ask the user for a grammar: "May I please have your grammar – call it G_{User}; I will test for you whether it has a universal language."
    * Behind the scenes, construct the Turing machine: $EqCFG(G_{Univ}, G_{User})$ where G_{Univ} is a grammar for the universal language. Notice that the second argument is the user-given grammar.
    * If $EqCFG$ determines that the grammars have the same language, it is indirectly answering **yes** – G_{User} has a universal language—this is known to be impossible.
  - Hence, $EqCFG$ cannot exist!

See Figure 14.1 for a diagram illustrating this proof. We will draw such "boxes within boxes" diagrams to explain reduction proofs. The notion of reduction itself is explained in Chapter 15.

You Call These Proofs*?!*

One thing that disturbs many who study computability is that the proofs often sound very hand-wavy! One almost always ends up using a combination of English and pseudo-code. Early-day computability theory researchers were in fact expected to write elaborate proofs. These days, with so much programming knowledge going around, simply outlining a proof at a high level will largely please even the staunchest of skeptics.[14] Moreover, when in doubt, modern-day mechanical theorem provers can easily be invoked to establish rigorous proofs. In this book, we will not have the space (or the need) to go beyond arguments based on a combination of English and pseudo-code.

14.3   *Other Examples of Recursive and RE Languages*

We now present a collection of recursive and RE languages and then provide a recap of many results with the aid of Figure 14.2.

Language L_{sat}: As our second example of a recursive language, consider

$$L_{sat} = \{\langle \varphi \rangle : \varphi \text{ is a Boolean formula that is satisfiable}\}$$

This is a recursive language because we can implement a Turing machine that first reads $\langle \varphi \rangle$ and successfully parses it to reveal the Boolean formula φ. It then invokes an algorithm to check whether φ is satisfiable.

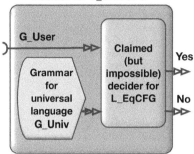

Decider for L_UnivCFG

Figure 14.1:   Suppose $EqCFG$ exists. Then we can build the decider $UnivCFG$ through the construction shown. We assemble $EqCFG$ inside a "box" where it is fed two things: (1) the user-given CFG (called G_{User}), and (2) some grammar for the universal language, called G_{Univ}. Whenever the box $EqCFG$ outputs "Yes," it means that the grammars fed to it are language-equivalent. Due to our construction, it means that G_{User} has a universal language. However, given the impossibility of realizing $UnivCFG$, we cannot realize $EqCFG$ either. This process is termed *mapping reduction*, and is a formal reduction from $UnivCFG$ to $EqCFG$, as will be explained in Chapter 15.

[14] Dear student, we are being nice to you by at least once in your life allowing you to write a proof in prose form.

This algorithm can be designed with utmost emphasis on clarity and simplicity *even if doing so renders the algorithm highly inefficient*. We are after all interested in decidability-related questions; the "Big O" complexity does not matter. As an example, the input could be the formula $p \wedge \neg q \wedge r$, and the TM can discover the satisfying instance $p = 1, q = 0, r = 1$ through an exponential enumeration of all Boolean combinations of p, q, r.[15]

Language $L_{EmptyCFG}$: As our third example of a recursive language, consider

$$L_{EmptyCFG} = \{x : x \text{ is a CFG that denotes an empty CFL}\}$$

This is a recursive language because we can implement a TM that first decodes $\langle G \rangle$ on its tape to be a syntactically legal context-free grammar G. It then analyzes this CFG to see whether its language is empty.[16] Given a CFG, go through all its productions, and in each production, mark each right-hand side terminal to be "reachable." Then if any production's right-hand side is entirely reachable, then mark the left-hand side to be also reachable. This process must terminate. Upon termination, if the **start symbol** of the CFG is found marked, the language in question is non-empty, because it can be used to expand into a terminal-only string (this process is detailed in §14.5.2). The existence of this algorithm shows that $L_{EmptyCFG}$ is indeed recursive.

It is clear that we now have begun discussing languages that look "more computationally oriented" and "less parsing oriented." That will be the flavor of the rest of this book. However, this is a completely arbitrary distinction, as formal languages and their classifications encompass all types of procedures and algorithms.

14.3.1 RE Languages that are not Recursive

We now present a few additional important languages that are only RE and not recursive. Many of the proofs (and proof outlines) will have to wait till the next chapter. We begin with a hugely important language called A_{TM}[17] that we will study in the next chapter also.

Define the language

$$A_{TM} = \{\langle M, w \rangle : M \text{ is a TM and } w \text{ its input and } M \text{ accepts } w\}$$

This is the language of pairs where M is a Turing machine description, and w an input that this TM is to be fed during its simulation of M.

Theorem 14.3.1: A_{TM} is RE.

Proof: Build a TM T that takes $\langle M, w \rangle$ and simulates M's behavior on w.[18] *In case $\langle M, w \rangle$ is in A_{TM},* the execution of T will halt, as it will discover (eventually) that M accepts w. When T detects this situation, it

[15] In a sense, undecidability is "harder" than exponentiality. Therefore, in undecidability proofs, using exponential subprocedures is totally fine (doing so does not hide any lurking undecidability).

[16] Let us use elementary productions (without the vertical bars | as separators).

[17] Its name tells you that it is your "money language."

[18] This is quite similar to writing an interpreter for Python that takes a given Python program and runs it on a given input.

itself can go to its own accept state. If M's simulation on w rejects, then T can reject. Else (if M's simulation on w loops), then T itself loops.[19] The language of Turing machine T is A_{TM} and so A_{TM} is RE. □

[19] This looping is fine because the existence of a T that loops when $w \notin L(M)$ does not change the fact that T's language is indeed $L(M)$ (T does halt and accept when $w \in L(M)$).

14.3.2 Why is it called Recursively **Enumerable**?

There is a very good reason why we call recursively enumerable languages so. After all, we introduced the synonym of Turing-recognizable early on, and that name might seem more appropriate.

> **Theorem 14.3.2(a):** For every RE language L, there is a Turing machine $Enum_L$, that eventually (in a finite amount of time) enumerates *every* string $w \in L$.

Proof: Given that L is an RE language, there is a Turing machine T whose language L is. This Turing machine can be run systematically on all the strings listed according to numeric order from Σ^*. However, one doesn't just obtain the first numeric-ordered string from Σ^*, run T till it halts in the accept state, and then proceed to run the next numeric-ordered string. Doing so would be fatal, as T may get *stuck in an infinite loop* on the first string. Instead:

- Run one step on the first string;
- Then list another numeric-ordered string;
- Run one (more) step on both strings;
- List one (more) string, and run one more step on all three strings;
- In this manner, list the ith string, and run one (more) step on all the i strings listed so far; keep doing this for increasing i;
- If and when the TM T finds one of the listed strings being accepted, it outputs that string on an "output tape."
- This way, the output tape will gradually fill up with strings from the language of T, which is L. □

This style of execution is called **dovetailing**.

> **Theorem 14.3.2(b):** If an enumerator E_L exists for a language, then L is RE.

Proof: Suppose language L has an enumerator $Enum_L$. Here is how a semi-decider T that checks for membership of a string x within L can be built. T first fires up $Enum_L$ and keeps comparing the strings emitted by it against x. If $x \in L$, it is clear that $Enum_L$ will eventually list it and the aforesaid comparison will eventually succeed. When this happens, $Enum_L$ can print "Yes, $x \in L$" and halt.[20] □

[20] Otherwise, $Enum_L$ is likely to not halt, and so won't T, which again is totally fine.

14.3.3   Alternate proof of A_{TM} being RE

Theorem 14.3.3: A_{TM} is RE.

Alternate Proof:

Approach: By building this enumerator for A_{TM}:

- Keep listing pairs $\langle A,B \rangle$ of strings from Σ^* on an "internal tape."
- Keep checking whether A is a Turing machine description (e.g., our markdown language for the TM has a parser; one can run this parser and see if it accepts A). If so, A happens to be a Turing machine description.
- Run Turing machine A on B, treating B as its input. Again, do not run to completion; instead, *engage in a dovetailed execution with all other TMs and inputs meanwhile being enumerated internally.*
- When the dovetailed simulation finds an $\langle A,B \rangle$ pair such that A accepts B, it lists the $\langle A,B \rangle$ pair on the output tape.
- This listing will produce every $\langle M,w \rangle$ such that M accepts w.
- The existence of this enumerator means that A_{TM} is RE. $\square$

14.3.4   Some More RE Languages

We now discuss a few more RE languages.

•Language $Halt_{TM}$ is RE but not recursive:

Define the language

$$Halt_{TM} = \{\langle M,w \rangle : M \text{ is a TM and } w \text{ its input and } M \text{ halts on } w\}$$

This language is closely related to A_{TM}, the only difference being that the TM that is simulating the actions of M on input w must **accept** if it is the case that M's simulation on w can *halt* (accept or reject). Chapter 15 will show that $Halt_{TM}$ is RE.

•Language $\overline{L_{UnivCFG}}$ is RE but not recursive:

Define the language

$$\overline{L_{UnivCFG}} = \{\langle G \rangle : G \text{ is a CFG and } L(G) \neq \Sigma^*\}$$

The decision problem is whether a given grammar G *doesn't have* Σ^* as its language. Exercise 14.4.1.4 asks you to describe an enumerator for $\overline{L_{UnivCFG}}$.

•Language L_{AmbCFG} is RE but not recursive:

Define the language

$$L_{AmbCFG} = \{\langle G \rangle : G \text{ is an ambiguous CFG}\}$$

The input string is a CFG and a Turing machine T is asked to check whether or not G's grammar admits *more than one parse* on any input string w. If there is such an ambiguous parse on some w, T can accept. It is clear that all such G can be systematically enumerated, *i.e.*, L_{AmbCFG} is RE.

14.4   Summary of Decidability / Semi-Decidability Results

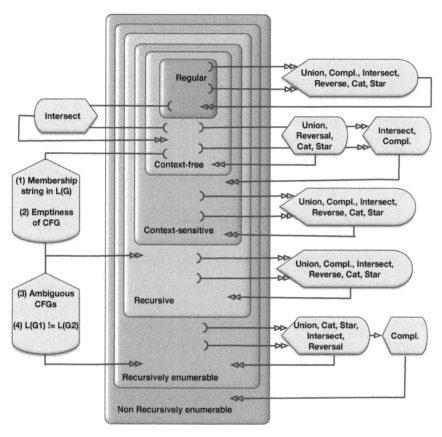

Figure 14.2: Formal languages, set operations, and recursive/RE status of four CFG questions.

Let us now summarize all the closure, decidability, and semi-decidability results captured in Figure 14.2. We will first walk top-down on the right-hand side of Figure 14.2. Then we will walk top-down on its left-hand side.

Closure: Here are the closure results, most of which have been proven in various prior chapters. (In other cases, we will point out that they are new closure results.)

- Regular languages are closed under union, complementation, intersection, reversal, concatenation, and star.
- Context-free languages are closed under reversal, concatenation, and star, **but not** intersection and complementation.
- The intersection of a regular language and a context-free language is context-free.

- Intersection and complementation of context-free languages are known to yield context-sensitive languages.
- Context sensitive languages are closed under union, complementation, intersection, reversal, concatenation, and star.[21]
- Recursive languages are closed under union, complementation, intersection, reversal, concatenation, and star.
- RE languages are closed under union, concatenation, star, intersection, and reversal.
- RE languages are *not closed* under complementation.

Decidability and Semi-Decidability: These are captured mostly on the left-hand side of the figure.

- Membership of a string in $L(G)$ (the language of a grammar) is decidable (this simply says that parsers are deciders).
- Emptiness of a CFG is decidable (we have already proved this by describing a decider).
- Ambiguity as well as grammar inequivalence are recursively enumerable but not recursive (this result was mentioned in §11.5.3).

14.4.1 Existence of Non-RE languages

Why is every language not recursively enumerable at least? Here is a summary of the proof presented in Appendix C:[22]

- There are uncountably many languages (as many languages as there are *Real Numbers*; this is denoted by the cardinal number $\aleph_1$).
- There are only countably many Turing machines (as many TMs as there are *Natural Numbers*; this is denoted by the cardinal number $\aleph_0$).
- Each RE language is the language of some TM.
- Thus, there must exist non-recursively enumerable languages.

Exercise 14.4.1, RE and Recursive

1. Describe a decider to check whether the language of a given DFA D is infinite.
2. Describe a decider to check whether a linear bounded automaton L working on an input w will go into an infinite loop.
3. Describe an enumerator for the language $Halt_{TM}$ similar to how the enumerator for A_{TM} was designed.
4. Describe an enumerator for the language $\overline{L_{UnivCFG}}$.
5. Describe a semi-decider for whether or not a grammar G_1 has a language that is *not* contained in the language of another grammar G_2. That is, we must be able to enumerate all such pairs (G_1, G_2) such that $L(G_1) \not\subseteq L(G_2)$.
6. Suppose someone claims to have a decider for $L(G_1) \subseteq L(G_2)$. Demonstrate how to build a reduction proof from $UnivCFG$. Draw a diagram similar to Figure 14.1 explaining your proof.

[21] This was never discussed thus far, so it is just an extra piece of information for you.

[22] There are mapping-reduction based proofs in [42] that actually present some of these non-RE languages—they are still countable languages! Our supplementary material at https://bit.ly/Automata_Jove under nonRELang explains one of these proofs.

7. In §15.2.2, we will argue that L_{AmbCFG} is not recursive. Assuming this result, argue that $\overline{L_{AmbCFG}}$ is not RE.

8. Elaborate on the results discussed in §14.4 in your own words (a few sentences responding to the same proofs). In particular, through discussions with others, develop an algorithm to decide whether or not a given linear bounded automaton will loop on a given input.

9. Write a proof sketch for the fact that the intersection of a regular language and a context sensitive language is context sensitive. *Hint:* Given a PDA P and a DFA D, modify the DFA product construction suitably to build a PDA modeling $L(P) \cap L(D)$. This can be done as follows. Start the machines in their respective start states. Whenever a non-ε input is to be processed, both the PDA P and the DFA D must engage in this step. Build a product state-space of the control states of the machines. The stack manipulation part carries over from P's actions. $\qquad\square$

14.5   RE and Recursive Sets: More High-Level Proof Sketches

Here are some more examples of high-level proof sketches pertaining to RE and recursive sets.

14.5.1   Language of DFA D where $L(D) = \Sigma^*$ is Recursive

Consider the language of DFA descriptions $\langle D \rangle$ such that the described DFA D has a universal language:

$$L_{UnivDFA} = \{\langle D \rangle : D \text{ is a DFA and } L(D) = \Sigma^*\}$$

We write $\langle D \rangle$ with the angle brackets highlighting that the set consists of strings representing (encoding) DFA. For instance, $\langle D \rangle$ could be Jove programs of DFA! So in a sense, $L_{UnivDFA}$ is the set of Jove descriptions that describe DFA with universal languages.

   To show that a language is recursive, we must present a Turing machine that decides and halts regardless of the input presented. Here is the algorithm that this TM can follow:

* Examine the input to see if it encodes a DFA as per the encoding conventions agreed upon. Reject if the input does not encode a DFA.
* If the input encodes a DFA, then minimize it to see if it has a single accepting state that transitions back to itself upon every symbol in Σ. If so, halt with "Accept" as the outcome. Else, halt with "Reject" as the outcome.

14.5.2 Language of CFG G where L(G) = ∅ is Recursive

Consider

$$L_{EmptyCFG} = \{\langle G \rangle : G \text{ is a CFG and } L(G) = \emptyset\}$$

- Examine the input to see if it encodes a CFG as per the encoding conventions agreed upon. Reject if the input does not encode a CFG.
- If the input encodes a CFG, then apply a bottom-up marking process that discovers whether the CFG generates some string or not. If so (the language is not empty), halt with "Reject" as the outcome. Else, halt with "Accept" as the outcome.

Determining whether $L(G) = \emptyset$**:** The bottom-up procedure to determine whether a CFG generates a string or not is now described. For simplicity, assume that the CFG production rules are of the form

$$L \rightarrow R_1 R_2 \ldots R_n$$

and not compound rules such as

$$L \rightarrow R_1 R_2 \ldots R_n \mid R_1^1 R_1^1 \ldots R_n^1$$

where multiple right-hand sides are associated with a single L.

- Initially, set the status of all nonterminals as *non-generating*. A generating nonterminal can generate a string over Σ^*. The algorithm below will discover all the nonterminals that are generating.
- For each production rule i of the form

$$L^i \rightarrow R_1^i R_2^i \ldots R_n^i$$

where all of $R_1^i, R_2^i, \ldots, R_n^i$ are terminals, mark the left-hand side nonterminals, indicating that they are *generating*. Examples of generating nonterminals discovered in this step are, for example, going to be nonterminals such as P in the rule below

$$P \rightarrow ab$$

Here, a and b are terminals. It is clear than P can generate string ab. It also includes nonterminals such as P where there is a production

$$P \rightarrow \text{''}$$

Here, the string of terminals on the rule's right-hand side is the empty string, that P can generate.

- Now locate all the rules where the right-hand sides are either generating nonterminals or terminal strings. Mark the left-hand side nonterminals as generating. This rule can induct more nonterminals into the generating category.
- Repeat scanning through the production rules until no more rule left-hand sides are classified as generating.

- If the start symbol S is generating, the language of the grammar is not empty. Else, the language of the grammar is empty.

Example: Let us further illustrate this algorithm on the grammar below, where a, b, c, d, e and f are terminals.

```
R1: S -> a B
R2: S -> a C
R3: B -> b B
R4: B -> c C
R5: C -> d D
R6: C -> ''
R7: D -> e f
```

- Rule R6 indicates that C is generating, since it can be reduced to ″
- Rule R2 indicates that S is generating, since C is generating, and a is a terminal.
- Since S is generating, the language of S is non-empty.

While not needed for our proof anymore, we can also make additional inferences; here are some of them:

- Rule R7 indicates that D is generating, since e and f are generating.
- R5 indicates that C is generating, since d is a terminal, and D is generating.
- R4 indicates that B is generating, since c is a terminal, and C is generating.
- R3 indicates that B (on the left-hand side) is generating, since b is a terminal, and B (on the right-hand side) is generating.

In the modified grammar below, we cannot conclude that S is generating, and hence the language of this grammar is empty:

```
R1: S -> a B
R2: S -> a C
R3: B -> b B
R4: B -> c C
R5: C -> d D
R6: D -> e C
```

Here, all the right-hand sides are nonterminals that have, as yet, not been shown to be generating. Hence, none of the left-hand sides – including S – can be shown to be generating.

14.5.3 *Language of LBA that halt on input w is Recursive*

Let us write a high level proof sketch for the proposition that the language $\langle L, w \rangle$ of pairs where L is a linear bounded automaton (LBA) and w is an input string is recursive. More formally, the language in question is

$$LBAHalt_{L,w} = \{\langle L,w \rangle \ : \ L \text{ is an LBA and } L \text{ halts on } w\}$$

Recall that an LBA is a Turing machine where the tape can be modified only where the initial input was laid out. Let us denote the instantaneous description (ID) of the LBA using the tuple notation (q, hi, w) where q is the control state, hi is the head index, and w is the tape contents. Thus, $w[hi]$ is the character under the LBA's "head."

Let us assume that an LBA of $|Q|$ states is started with $hi = 0$ and the input string is w. Since the head can only go through the extent of w and make changes, we can easily calculate the number of IDs possible. Specifically,

- There are $|Q|$ values possible for the first position of the ID.
- There are $|w|$ different head indices possible.
- If there are $|\Gamma|$ tape symbols possible, and given that the input alphabet Σ is a proper subset of Γ, the total number of tapes possible is upper-bounded by $|\Gamma|^{|w|}$. This assumes that all combinations of tape symbols of length $|w|$ are possible.
- Thus, the total number of distinct IDs is upper-bounded by

$$Steps_{LBA} = |Q| \cdot |w| \cdot |\Gamma|^{|w|}$$

- Thus, after starting an LBA, if it has not halted within $Steps_{LBA}$ steps, then we can assert that it will *never halt*.

Thus, the language $LBAHalt_{L,w}$ is recursive. Basically, develop a TM that simulates LBA L on w. If this simulation hasn't halted for $Steps_{LBA}$ steps, then the TM halts with "Reject" as the outcome. Else, it halts with "Accept" as the outcome.

14.5.4 Language of Turing machines whose first output is '3'

Consider the language of pairs $\langle T,w \rangle$ where $\langle T \rangle$ is a TM description and w is the input on which T is run. Let us consider the language

$$T_{w,3} = \{\langle T,w \rangle \ : \ T \text{ is a TM and } w \text{ its input, and } T \text{ prints 3 as its first output.}\}$$

We claim that $T_{w,3}$ is RE.

A semi-algorithm (or enumeration procedure) that enumerates the members of $T_{w,3}$ is the following:

- One can code-up a single Turing machine T such that T keeps enumerating TM descriptions one after the other on one of its tapes. Let these TMs be $T_1, T_2, ..., T_i, ...$
- Let the current list of TMs generated be $T_1, T_2, ..., T_i$. Run all these TMs on input w for one step. Then generate one more TM and add it to the current list of TMs. The list now becomes $T_1, T_2, ..., T_i, T_{i+1}$. Run this extended list of TMs on input w for one additional step.

- Keep this process going. Thus, the earlier-listed TMs will have run more steps than the later-listed TMs. However, eventually, every TM listed will receive any number of steps of execution, running on w.
- If/when a TM T_j prints '3', enumerate the pair $\langle T_j, w \rangle$ on an output tape. This way, we will have an enumerator for $T_{w,3}$. The existence of this enumerator tells us that $T_{w,3}$ is RE.

15

Post Correspondence, and Other Undecidability Proofs

Chapter Gist: *We begin with Post Correspondence, a "one-stop-shopping" for decidability arguments (§15.1). We sketch a proof of the undecidability of PCP (§15.2). Next, we show the undecidability of the acceptance problem (§15.3) and the halting problem via reduction (§15.4). The general idea of mapping reductions is our grand finale (§15.5).*

Dominoes

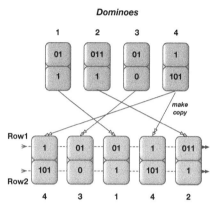

Figure 15.1: A PCP instance and *one* of its solutions. PCP helps show the undecidability of important problems such as grammar ambiguity, validity of first order logic sentences, and context-sensitive data dependence analysis.

Figure 15.2: Drosophila melanogaster (Image courtesy of pixabay.com)

15.1 Post Correspondence: "Drosophila" for Decidability

The *Post Correspondence Problem* (PCP, Figure 15.1) is like the common fruit fly (*Drosophila melanogaster*, Figure 15.2) for setting up decidability arguments. Fruit flies are very widely used for research in biology and medicine, as they share with humans about 75% of the genes that cause diseases, so scientists can learn about human genetics by studying fruit fly genetics. Similarly, the Post Correspondence Problem (PCP) is the "Drosophila of computability." Since its presentation by Emil L. Post in 1946, the undecidability of many important problems in computer science has been shown by formally connecting these problems to PCP through *mapping reductions* (covered in §15.4 and §15.5).

PCP itself is a fun puzzle to "administer to your friends" regardless of their age. As one example, consider the four dominoes on the top row, and assume access to a copying machine[1] (Figure 15.1). Each domino carries a non-empty string of 0's and 1's on the top and bottom half.[2] Your friend must find a solution (one being the six dominoes shown as the bottom row) where each solution domino is either the original domino or a copy thereof[3] such that when you scan the top row of the solution dominoes, you get the same string (1 01 01 1 011) as when you scan the bottom row of the solution dominoes (101 0 1 101 1). *You are allowed to ignore any*

[1] That can make identical copies of any domino.

[2] The nonemptiness condition is not strictly required, but does not change the problem, and is required by many PCP solvers. For dealing with ε, see Exercise 2.

[3] Domino 4 is copied once.

[4] We use a sequence for the input because then we know the index (rank) of each domino, and can write out our solution as a sequence of these indices.

[5] We can also write down the instance as [1,2,3,4] where each index serves as a shorthand for the actual domino at that position.

number (but not all) of the dominoes.

A **PCP instance** is the **sequence of dominoes** given as input.[4] Each "domino" can be viewed as a pair of bit-strings. The instance given in Figure 15.1 can be viewed as the sequence[5] [(01, 1), (011, 1), (01, 0), (1, 101)]. A **solution** is also a **sequence** of dominoes or their copies. A solution for the instance in Figure 15.1 is [(1, 101), (01, 0), (01, 1), (1, 101), (011, 1)]. A solution can also be written out as a sequence (of length 5) of indices of tiles: [4, 3, 1, 4, 2].

```
In [1]:   from PcpJupyter import *
```

1 PCP Jupyter Interface

Bringing Ling Zhao's PCP Solver to jupyter notebooks! Arguments will be exactly as they were for the command line version for the basic pcp_solve: run - number of runs ni - no iterative search di - depth increment depth - search depth

```
In [2]:   pcp_solve([('1000', '0'), ('01', '0'), ('1', '101'), ('00', '001')])
```

```
Solution 3
[4, 3, 3, 2, 3, 2, 2, 2, 4, 4, 2, 4, 3, 4, 2, 3, 2, 4, 3, 2, 2, 4, 2, 3, 2, 4, 4, 3, 2,
 2, 4, 2, 4, 3, 3, 2, 4, 4, 1, 3, 3, 2, 3, 2, 4, 2, 4, 3, 2, 3, 2, 2, 2, 3, 2, 2, 4, 2, 3,
 2, 4, 3, 2, 4, 4, 2, 4, 2, 2, 2, 4, 2, 3, 2, 4, 2, 4, 3, 4, 2, 4, 4, 2, 2, 2, 4, 2, 4, 2,
 3, 2, 4, 1, 3, 4, 1, 4, 3, 4, 2, 2, 2, 4, 2, 2, 2, 4, 1, 3, 3, 2, 4, 1, 4, 1, 4, 3, 2, 2,
 3, 2, 4, 2, 4, 2, 4, 2, 3, 2, 4, 3, 2, 4, 2, 4, 2, 4, 2, 2, 2, 4, 2, 3, 2, 4, 2, 4, 2, 4,
 2, 4, 4, 2, 2, 2, 4, 2, 4, 2, 4, 2, 4, 2, 4, 1, 4, 3, 4, 2, 4, 2, 4, 2, 4, 2, 4, 2, 3, 2,
 4, 1, 3, 4, 2, 4, 2, 4, 2, 2, 2, 4, 1, 1, 4, 1, 4, 1, 4, 1, 1, 1]
00    1    1    01   1    01   01   01   00   00   01   00   1    00   01
001   101  101  0    101  0    0    0    001  001  0    001  101  001  0

1     01   00   1    01   01   00   01   1    01   00   00   1    01   01
101   0    001  101  0    0    001  0    101  0    001  001  101  0    0

00    01   00   1    1    01   00   00   1000 1    1    01   1    01   00
001   0    001  101  101  0    001  001  0    101  101  0    101  0    001

01    00   1    01   01   1    01   01   01   00   01   00   01   1    01
0     001  101  0    0    101  0    0    0    001  0    001  0    101  0
```

Figure 15.3: The PCP instance whose optimal (minimal length) solution is 206. With "just four tiles," your friend may think that she can do it in a jiffy. See how many copies must be made and the specific clever arrangement that must be discovered! To make matters worse, there are no algorithms to solve the PCP. *None!* This means your friend may infinitely loop (taking some time off to unfriend you).

[6] The optimal solution need not be unique: for the PCP instance [(0,0),(1,1)], the optimal solution can be either [(0,0)] or [(1,1)].

[7] In a sense, the explosive growth of optimal solution lengths is an indication of PCP's undecidable nature.

The length of a solution can be artificially bloated; thus, 4, 3, 1, 4, 2, 4, 3, 1, 4, 2 is also a solution, as the top and bottom rows will still read the same—but this solution is of length 10. An **optimal** solution is one where the solution sequence is the shortest.[6]

A key observation is that in general, the optimal solution of PCP instances can be *very* long. For instance, for the instance of length four (4) in Figure 15.3, the ***optimal* solution is of length 206.** This solution was produced by the excellent PCP-solver written by Ling Zhao [47], that we have integrated into Jove (see §B.1.8 for usage details).

Why is the PCP of interest? More than this "explode in your face" behavior (which makes PCP a rather cruel puzzle[7]), **PCP instances have**

the ability to simulate A_{TM}. In other words, given an M and w, one can manufacture a PCP instance $PCP_{M,w}$ such that solving this PCP puzzle is tantamount to answering whether M accepts w! Due to this fact, many theoretical and practical problems have been shown to be undecidable by reduction *from* the PCP. PCP proves to be a much more convenient starting point for these mapping reduction arguments than A_{TM}.

More specifically, **the set of all PCP instances that have a solution is RE but not recursive.** If the optimal solutions were to be bounded in length by a function $f(n)$ of the instance length n, one could simply search for all PCP instances of length $f(n)$ or less, and reach a decision. Unfortunately, such a bound $f(n)$ does not exist.

15.2   Proof Sketch of the Undecidability of PCP

Given any alphabet Σ such that $|\Sigma| > 1$, consider the *tile alphabet* $\mathcal{T} \subseteq \Sigma^+ \times \Sigma^+$. Now consider the language

$PCP = \{S \ : \ S$ is a finite sequence of elements over $\mathcal{T}$ that has a solution$\}$.

> **Theorem 15.2:** The language PCP is not recursive (not decidable).

Proof Sketch: We will employ mapping reduction from A_{TM} but with a slight twist. Given an M and w, we will manufacture a PCP instance $PCP_{M,w}$ such that solving this PCP puzzle is tantamount to not just answering **whether** M accepts w (decision of A_{TM}) but actually showing **how** M accepts w. This is called the *computational history method*. The details of the construction are found in standard textbooks: our construction exactly follows that given in [42]; we take a simple Turing machine and its accepting computational history, which is a sequence of instantaneous descriptions (ID, §13.3). In particular, we take the history I01 -> 1J1 -> F10 given in Figure 15.4 and walk you through the construction step by step. In this figure, the first two tiles named T1 and T2 are (*#, *#*i*0*1*#*) and (*i*0, 1*j*) respectively. To explain our notations better, let us take tile T2 as an example. In Figure 15.4, we present tile T2 as
[*I*0]
[1*J*] with square brackets added and placed one above the other to appear like a domino. As per the notation of "pair of bit strings" introduced on Page 228, this tile would be (*I*0, 1*J*). In essence, the tile is (I0, 1J) representing the change of the ID I01 to 1J1 when viewed through a 2x2 "peephole" (elaborated in §15.2.1). The purpose of the decorators * and # is to make sure that the tiles fit exactly when describing an accepting computational history.[8]

[8] The details of these decorators is given in [42] and we don't elaborate how the decorators help the tiles fit.

Figure 15.4: Conversion of a simple accepting computational history of a TM to a PCP solution.

```
    T1        T2    T3  T4    T5      T6   T7    T8 T9   T10  T11 T12      T13
----------- ----- --- --- ------- --- ----- --- --- ----- --- ------- ----
[*#         ][*I*0][*1][*#][*1*J*1][*#][*F*1][*0][*#][*F*0][*#][*F*#*#][*<>]
[*#*I*0*1*#*][1*J*][1*][#*][F*1*0*][#*][F*  ][0*][#*][F*  ][#*][#*    ][<> ]
```

15.2.1 Tile Construction Basics

Given a Turing machine M, we systematically go through the transition function δ of M as well as the elements of its tape alphabet, Γ, and generate a finite set of tiles, $Tiles_M$. These tiles help describe how a TM actually moves from state to state, seeing the tape only under the head, and affecting the tape through a small peephole around its current head. All possible TM tape evolutions are captured in the tiles that model activities within peepholes. Basically, the peephole is of size 2 for a right move of the TM, and of size 3 for a left move of the TM. The intuition here is that by seeing a small region around a TM's head, one can fully determine the new state around the head of the TM after a transition. In our example, tile (*I*0,1*J*) represents a move in which the TM is in state I, sees a 0, and then moves right after changing the 0 to a 1 while also attaining state J. In a sense, we are precomputing all one-step right moves and one-step left moves as see through peepholes and encoding them into the tiles. For our example accepting computational history, we manufacture all the tiles mentioned in Figure 15.4.

All accepting computational histories start with a specific instantaneous description in which the entire initial input is present, and the TM head is "looking at" the first symbol of the given input string. In our example, this ID is I01. This tile must be the first tile of any solution, and to accomplish that, we employ the starting tile (*#,*#*I*0*1*#*). The added "decorations" surrounding I01 result in the bottom of the tile actually being [*#*I*0*1*#*], and the purpose of these decorations is to ensure that this beginning tile also can be the *first tile of any PCP solution*.

In this very simple example we are dealing with, Figure 15.4 presents the solution obtained by just lining up the tiles. In general, a hypothetical decider of PCP will have to try much harder and try any possible accepting computational history at all.[9]

In general, given a pair $\langle M,w \rangle$ that belongs to A_{TM}, we follow the

[9] Given Theorem 15.2, we know that a decider for PCP cannot exist, meaning that any program we write must admit the possibiity of going into an infinite loop.

construction outlined so far and obtain a (possibly gigantic) set of PCP tiles, say BigPuzzle. Now, if M were to accept w, the acceptance ought to be describable as a solution to BigPuzzle.[10] For the TM to accept the given input, the last few tiles allowed in the computing history must contain accepting configurations of the TM. In our example, the last few tiles contain the final state name F, and include a few "decorated tiles" that contain F10. However, we know by now that BigPuzzle can't be solved.[11]

[10] ...for which you may need a BigXMasVacation.

[11] ...and no manufacturer's refunds are possible.

The upshot of the construction just described is that if there is an algorithm for solving $PCP_{M,w}$, there would then be a decider for the language A_{TM}. Given that there *cannot* be a decider for A_{TM} (proved in §15.3), *there is no algorithm to solve the PCP puzzle* $PCP_{M,w}$. This establishes Theorem 15.2.

15.2.2 *Proving Grammar Ambiguity by Reduction from PCP*

PCP is hugely important as a stepping stone to proving many important practical problems to be undecidable (non-recursive). For instance, one can perform a mapping reduction from PCP instance P to a context-free grammar G_P in such a way that the ability to algorithmically decide the ambiguity of CFGs would make PCP decidable. Here are some more details of this construction (Exercise 5 will ask you to further elaborate this construction, with many hints provided):

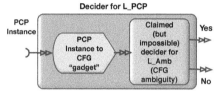

Figure 15.5: Mapping reduction from PCP to Grammar Ambiguity

- Given any PCP instance P, we obtain G_P through mapping reduction (see Figure 15.5). In this figure, the term *gadget* is used in a technical sense. It is the function that transforms an instance of PCP to an instance of a grammar whose ambiguity is checked by the "inner box."

- Hence, the algorithm to check for ambiguity must also apply to G_P. However, the construction of G_P would have ensured that the check of G_P's ambiguity would allow us to extract a solution to P.

- However, since P is any arbitrary PCP instance, and solving PCP is shown undecidable, we cannot have a decider for CFG ambiguity either.

In the same manner, PCP serves as a stepping stone for many more problems. Here are three:
- the validity of predicate logic formulae is undecidable.
- checking whether variables in a program alias is undecidable [41].
- many language containment questions for concurrent languages with synchronizations [38] are undecidable.

In all these cases, we are erecting mapping reductions *from* PCP to one of these problems.

15.2.3 *PCP in Jove*

We have included Jove notebooks to invoke Zhao's PCP solver from the comforts of Python. §B.1.8 provides details. This Jove-based interface allows you to see the solution both as an arrangement of tiles and as a listing of tiles presented as a sequence.

> **Exercise 15.2.3, PCP Solver in Jove**

1. Using the PCP solver in Jove, determine the optimal solution length for the PCP instance

$$[(100,1),(0,100),(1,0)]$$

2. Suppose someone wants to allow tiles of the form $\langle w, \varepsilon \rangle$ where $w \in \Sigma^+$ and $\Sigma = \{0,1\}$. What are four ways in which you can modify such a tile to meet the nonemptiness condition? Show one of these ways in detail by detailing the new tile you would create in lieu of $\langle w, \varepsilon \rangle$. Now explain the other three ways in words.

3. First, modify the PCP instance

$$[(01,1),(01,\varepsilon),(01,0),(1,101)]$$

to have all non-empty strings at the top and bottom of every tile (see Exercise 2). Then determine a solution of length 2 for the modified PCP instance. After that, determine a solution of length 3 for the modified PCP instance.

4. Answer the following questions with respect to the unary PCP instance:

   (a) Using Jove, solve the following unary PCP instance (*i.e.*, where the alphabet is a singleton, namely {0}):

   $$[(000,00),(0,0000),(0000,000),(0,0000),(000000,0)]$$

   (b) Argue that the unary PCP problem — PCP over a singleton alphabet ($|\Sigma| = 1$) — is decidable. Do it in two stages:

   i. Suppose for all tiles T_i in the unary PCP instance we have $len(T_i[0]) < len(T_i[1])$ or $len(T_i[0]) > len(T_i[1])$. Then what can you say about the unary PCP instance's solutions?

   ii. Suppose the above condition does not hold; that is, there are two distinct tiles T_j and T_k in the instance with $len(T_j[0]) < len(T_j[1])$ and $len(T_k[0]) > len(T_k[1])$. Then what can you say about the unary PCP instance's solutions?

5. Here is how we can build a mapping reduction from PCP to CFG grammar ambiguity; please fill in missing steps (if any) and argue that the mapping reduction actually works (achieves its purpose). Let

$$A = w_1, w_2, \ldots, w_n$$

and

$$B = x_1, x_2, \ldots, x_n$$

be two lists of words over a finite alphabet Σ. Let $a_1, a_2, \ldots, a_n$ be symbols that do not appear in any of the w_i or x_i. Let G be a CFG

$$(\{S, S_A, S_B\}, \Sigma \cup \{a_1, \ldots, a_n\}, P, S),$$

where P contains the productions

$S \to S_A,$

$S \to S_B,$

For $1 \le i \le n$, $S_A \to w_i S_A a_i,$

For $1 \le i \le n$, $S_A \to w_i a_i,$

For $1 \le i \le n$, $S_B \to x_i S_B a_i,$ and

For $1 \le i \le n$, $S_B \to x_i a_i.$

Now, argue that G is ambiguous if and only if the PCP instance (A, B) has a solution (thus, we may view the process of going from (A, B) to G as a mapping reduction). $\qquad \square$

15.3   Undecidability of the Acceptance Problem

> **Theorem 15.3:** A_{TM} is undecidable.

Proof: We prove this set to be *undecidable* through contradiction.

- **Suppose** there exists a decider A for A_{TM}. A expects to be given a Turing machine M and a string w. Notice that "giving a Turing machine to A" means "giving it a character string representing a Turing machine program." Hence, in reality, we will be feeding A the pair $\langle M, w \rangle$.

- Build a program called D as follows:[12]

  1. D takes a single argument M.
  2. As its first step, D invokes A on $\langle M, M \rangle$. (Basically, we feed $\langle M \rangle$ as both arguments of A.)
  3. If $A(\langle M, M \rangle)$ rejects, $D(\langle M \rangle)$ accepts.
  4. If $A(\langle M, M \rangle)$ accepts, $D(\langle M \rangle)$ rejects.

- Now we can ask what $D(\langle D \rangle)$ will result in. (Feel free to leave out $\langle \ldots \rangle$ when you write out this proof. These angle brackets are used mainly to highlight textual descriptions of machines.)

  - As per Step 1 above, the $D(\langle D \rangle)$ "call" turns into an $A(\langle D, D \rangle)$ call.

  - Suppose $A(\langle D, D \rangle)$ rejects. In that case, as per Step 3 above, $\boxed{D(\langle D \rangle) \text{ accepts.}}$

[12] The mnemonic significance of D comes from *diagonalization*—the proof-style being used here.

– But, according to the advertised behavior of A — which is that it is a decider for A_{TM} — the fact that $A(\langle D, D \rangle)$ rejects means that D is a Turing machine that **does not** accept $\langle D \rangle$, or that $\boxed{D(\langle D \rangle) \text{ rejects or loops!}}$ **This is the first part of a two-part contradiction that we will obtain.**

– Suppose $A(\langle D, D \rangle)$ accepts. In that case, as per Step 4 above, $\boxed{D(\langle D \rangle) \text{ rejects.}}$

– But, according to the advertised behavior of A — which is that it is a decider for A_{TM} — the fact that $A(\langle D, D \rangle)$ accepts means that D *is* a Turing machine that *accepts* $\langle D \rangle$, or that $\boxed{D(\langle D \rangle) \text{ accepts!}}$ **This is the second part of a two-part contradiction that we have obtained.**

[13] This proof is very likely to cause huge waves of confusions in students, especially when it typically gets discussed close to the end of the semester. I've added a fair amount of supplementary tutorial material at `https://bit.ly/Automata_Jove` under Halting to help students.

Therefore, we obtain a contradiction *under all cases*.[13] This is tantamount to having proven **False**. It is the claim that a decider A for A_{TM} exists that allowed us to prove **False**. Thus, by the principle of *proof by contradiction*, such a decider A cannot exist! □

15.4 Halting ($Halt_{TM}$) is Undecidable

Figure 15.6: A_{TM} to $Halt_{TM}$ Reduction.

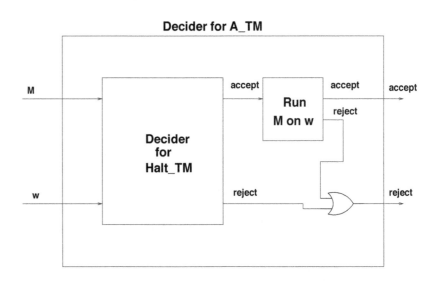

Theorem 15.4: $Halt_{TM}$ is undecidable.

Proof: (Basic idea): We perform a mapping reduction from A_{TM} (Figure 15.6). We assume that all the inner components of this figure (namely the OR-gate logic and the ability to run M on w) can be realized by programming TMs. Crucially, **we assume that** $D_{Halt_{TM}}$ **exists.** We can then build a decider for A_{TM} (the whole outer box of Figure 15.6). However, in §15.3, we proved that $D_{A_{TM}}$ cannot exist. Hence, $D_{Halt_{TM}}$ cannot exist.

(More details): Let us study the construction in Figure 15.6 in some more detail. Let

$$Halt_{TM} = \{\langle M, w \rangle : M \text{ is a Turing machine that halts on string } w\}.$$

Figure 15.6 shows $Halt_{TM}$ to be undecidable as follows:
- Suppose there is a decider for $Halt_{TM}$ called $D_{Halt_{TM}}$.
- Let's now build a decider for A_{TM} (call it $D_{A_{TM}}$). $D_{A_{TM}}$'s design will be as follows:
  - $D_{A_{TM}}$ will first feed M and w to $D_{Halt_{TM}}$, the claimed decider for $Halt_{TM}$.
  - If $D_{Halt_{TM}}$ goes to $accept_{D_{Halt_{TM}}}$, $D_{A_{TM}}$ knows that it can safely run M on w, which it does.
  - If M goes to $accept_M$, $D_{A_{TM}}$ will go to $accept_{D_{A_{TM}}}$.
  - If M goes to $reject_M$, or if $D_{Halt_{TM}}$ goes to $reject_{D_{Halt_{TM}}}$, $D_{A_{TM}}$ will go to $reject_{D_{A_{TM}}}$.

Notice that we have labeled the accept and reject states of the two machines $D_{Halt_{TM}}$ and $D_{A_{TM}}$. After one becomes familiar with these kinds of proofs, higher-level proof sketches are preferred. Here is such a higher-level proof sketch:
- This decider accepts input $\langle M, w \rangle$ and runs $Halt\_decider$ (if it exists) on it.
- If this run accepts, then we can safely (without the fear of looping) run M on w, and return the accept/reject result that this run returns; else return "reject."
- To prevent A_{TM} from existing (*i.e.*, to avoid a contradiction), we must prevent $Halt_{TM}$ from existing. □

Two observations that the reader can make after seeing many such proofs (to follow) are the following:
- One *cannot* write statements of the form "if $f(x)$ loops, then ..." in any algorithm, because termination is not detectable. Of course, one *can* write "if $f(x)$ halts, then ... ." This asymmetry is quite fundamental, and underlies all the results pertaining to halting / acceptance.
- One cannot examine the code ("program") of a Turing machine and decide what its language is. More precisely, one cannot build a classifier program Q that, given access only to Turing machine programs P_m (which encode Turing machines m), classify the m's into two bins (say "good" and "bad") according to the language of m. Any such classifier will have to classify all Turing machines as "good" or all as "bad," or itself be incapable of handling all Turing machine codes (*i.e.*, not be *total*). This result is known as Rice's Theorem.[14]

---

[14] For a lucid account of this theorem and its proof, kindly see supplementary material at https://bit.ly/Automata_Jove under RicesTheorem or my 2006 book [21].

[15] Look for supplementary material at https://bit.ly/Automata_Jove under MappingRed to illuminate mapping reductions.

[16] Important link with Chapter 16: In studying NP-Completeness, we will employ polynomial-time mapping reductions, which are denoted by $\leq_P$. A polynomial-time mapping reduction $\leq_P$ is a mapping reduction where the reduction function f has polynomial-time asymptotic upper-bound time complexity. Using the familiar notation $\mathcal{O}(...)$ for asymptotic upper-bounds, polynomial-time means $\mathcal{O}(n^k)$ for an input of length n, and $k > 1$.

Figure 15.7: Mapping reduction $A \leq_M B$. Notice that f maps points inside A to points inside B. This mapping need not be 1-1 nor onto. The only condition is that f must map points outside of A to points outside of B. Such a mapping arrow of f is also shown (acting on a shaded circle outside of A and producing a shaded circle outside of B). A mapping reduction of this kind allows us to use a membership decider in B to serve as a membership decider in A as follows: (1) accept an input $x \in A$; (2) check if $f(x) \in B$; (3) $x \in A$ iff $f(x) \in B$.

15.5   Mapping Reductions

Our previous reduction proofs were proofs by contradiction: we assumed the existence of a decider for a new problem, and using that decider, created a decider for an old (already shown impossible-to-decide) problem. The mapping reductions idea is a way to formally approach the same proofs, but has the additional property that it explicitly maps all members of the "old and impossible" problem (modeled as a language) to a carefully chosen subset of the new problem (also modeled as a language).[15]

> **Definition 15.5:** A *computable* function $f : \Sigma^* \to \Sigma^*$ is a mapping reduction from $A \subseteq \Sigma^*$ into $B \subseteq \Sigma^*$ if for all $x \in \Sigma^*$, $x \in A \Leftrightarrow f(x) \in B$.

By a *computable function*, we mean a function that can be implemented via a Turing machine.[16]

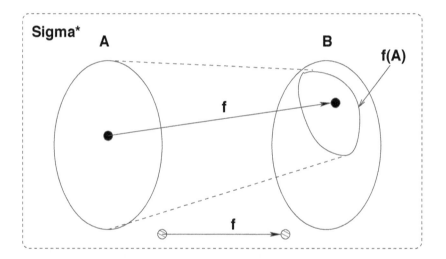

See Figure 15.7 which illustrates the general situation that A maps into a subset denoted by $f(A)$ of B, and members of A map into $f(A)$ while non-members of A map outside of B (that means they map outside of even $B - f(A)$). Also note that A and B need not be disjoint sets, although they often are. A mapping reduction can be (and usually is) a non-injection and non-surjection; *i.e.*, it can be many-to-one and not necessarily onto. It is denoted by $\leq_m$. By asserting $A \leq_m B$, the existence of an f as described above is also being asserted.

Typically mapping reductions are used as follows:

- Let A be a language known to be undecidable ("old" or "existing" language).
- Let B be the language that must be shown to be undecidable ("new"

language).

- Find a mapping reduction f from A into B.
- Now, if B has a decider D_B, then we can decide membership in A as follows:
    - On input x, in order to check if $x \in A$, find out if $D_B(f(x))$ accepts or not. If it accepts, then $x \in A$, and if it rejects, then $x \notin A$.

```
How a decider for A_TM is obtained:

Step 1: Here is the initial tape.
-----------------------------------------------------------------
| M | w |
-----------------------------------------------------------------

Step 2. Build M' and put it on the tape
-----------------------------------------------------------------
| M | w | ..build M' that incorporates M here.. |
-----------------------------------------------------------------

Step 3. Put w on the tape.
-----------------------------------------------------------------
| M | w | ..build M' that incorporates M here.. | ..put w here.. |
-----------------------------------------------------------------

Step 4. Run Halt_TM_decider on M' and w  and return its decision
-----------------------------------------------------------------
| M | w | ..build M' that incorporates M here.. | ..put w here.. |
-----------------------------------------------------------------
```

$$D_{Halt_{TM}}(M',w) = \begin{cases} accepts & \Rightarrow M' \, halts \, on \, w \Rightarrow & M \, accepts \, w \\ \\ rejects & \Rightarrow M' \, doesn't \, halt \, on \, w \Rightarrow & M \, doesn't \\ & & accept \, w \end{cases}$$

Figure 15.8: How the mapping reduction from A_{TM} to $Halt_{TM}$ works. If $\langle M,w \rangle$ is in A_{TM}, then we see that M' halts on w, or that $\langle M',w \rangle$ is in $Halt_{TM}$. On the other hand, if $\langle M,w \rangle$ is not in A_{TM}, then we see that M' does not halt on w, or that $\langle M',w \rangle$ is not in $Halt_{TM}$.

Mapping Reduction from A_{TM} to $Halt_{TM}$ We first illustrate mapping reductions by taking $A = A_{TM}$ and $B = Halt_{TM}$ with respect to Figure 15.7. Function f takes a member of A_{TM}, namely a pair $\langle M,w \rangle$, as input, and prints out $\langle M',w \rangle$ on the tape as its output. Function f, in effect, generates the *text* of the program M' from the text of the program M. Here is the makeup of M':

- $M'(x) =$
    - Run M on x.
    - If the result is "accept," then "accept".
    - If the result is "reject," then loop.

Notice that the text of M' has "spliced" within itself a copy of the text of program M that was input. Mapping reductions such as f illustrated here need not "run" the program they manufacture; they simply accept a program such as M, and a possible second input, such as w, and manufacture another program M' (and also copy over w) and then consider their task done! The reason such a process turns out to be useful is for the following reasons:

> Suppose someone were to provide a decider for $Halt_{TM}$. The mapping reduction f then makes it possible to obtain a decider for A_{TM}. When given $\langle M,w \rangle$, this decider will obtain $\langle M',w \rangle = f(\langle M,w \rangle)$, and then feed it to the decider for $Halt_{TM}$.

We have to carefully argue that f is a mapping reduction. We will be quite loose about the argument types of f (namely that it maps Σ^* to Σ^*); we will assume that any $\langle M,w \rangle$ pair can be thought to be a string, and hence a member of a suitable Σ^*. The proof itself is depicted in Figure 15.8.

Figure 15.9: Mapping reduction from A_{TM} to $\overline{E_{TM}}$. If $\langle M,w \rangle$ is not in A_{TM}, then $L(M')$ is empty. On the other hand, if $\langle M,w \rangle$ is in A_{TM}, then $L(M')$ is nonempty. This achieves the mapping reduction.

```
M'(x) {
    if x <> w then loop ; // could also goto reject_M' here
    Run M on w ;
    If M accepts w, goto accept_M' ;
    If M rejects w, goto reject_M' ; }
```

How a decider for E_TM is obtained:

Step 1: Build above M' and put it on the tape
```
-------------------------------------------------------------------
| M | w | ..build M' that incorporates M and w here.. |
-------------------------------------------------------------------
```

Step 2: Run E_TM_decider on M' and return its decision
```
-------------------------------------------------------------------
| M | w | ..build M' that incorporates M and w here.. |
-------------------------------------------------------------------
```

$$Decider_{E_{TM}}(M') = \begin{cases} accepts \Rightarrow L(M') \, is \, empty \Rightarrow M \, does \, not \, accept \, w \\ rejects \Rightarrow L(M') \, is \, not \, empty \Rightarrow M \, accepts \, w \end{cases}$$

Mapping reduction From A_{TM} to $\overline{E_{TM}}$ We show that

$$E_{TM} = \{\langle M \rangle \, : \, M \text{ is a TM and } L(M) = \emptyset.\}$$

is undecidable through a mapping reduction that maps $\langle M,w \rangle$ into $\langle M' \rangle$, as explained in Figure 15.9. Basically, having a decider that can decide whether the language of the machine M' is not empty gives us the ability to decide A_{TM}.

Mapping reduction from A_{TM} to $Regular_{TM}$ Define

$$Regular_{TM} = \{\langle M \rangle \, : \, M \text{ is a TM whose language is regular.}\}$$

Figure 15.10: Mapping reduction from A_{TM} to $Regular_{TM}$. If $\langle M, w \rangle$ is in A_{TM}, then $L(M')$ is regular. On the other hand, if $\langle M, w \rangle$ is not in A_{TM}, then $L(M')$ is not regular. This achieves the mapping reduction.

```
M'(x) {
    if x is of the form 0^n 1^n then goto accept_M' ;
    Run M on w ;
    If M accepts w, goto accept_M' ;
    If M rejects w, goto reject_M' ; }
```

$$
Decider_{Regular_{TM}}(M') = \begin{cases}
accepts \Rightarrow L(M') \, is \, regular \\
\quad \Rightarrow Language \, is \, \Sigma^* \\
\quad \Rightarrow M \, accepts \, w \\
\\
rejects \Rightarrow L(M') \, is \, not \, regular \\
\quad \Rightarrow Language \, is \, 0^n 1^n \\
\quad \Rightarrow M \, does \, not \, accept \, w
\end{cases}
$$

We can prove $Regular_{TM}$ to be undecidable by building the Turing machine M' via mapping reduction, as shown in Figure 15.10. Basically, having a decider that can decide whether the language of the machine M' is regular gives us the ability to decide A_{TM}.

15.5.1   Undecidable problems are "A_{TM} in disguise"

This chapter covered the Post Correspondence Problem and its significance. We then formally defined the notion of mapping reduction. The techniques discussed here lie at the core of the notion of "problem solving" in that they help identify which problems possess algorithms and which do not.

A closing thought to summarize the proofs in this chapter is the slogan that undecidable problems are A_{TM} in disguise. We leave you with this thought, hoping that it will provide you with useful intuitions.

Exercise 15.5.1, Mapping-Reduction Proofs

1. Draw the "boxes within boxes" style diagram (similar to Figure 15.5) corresponding to the reduction argument presented in Figure 15.8.

2. Draw the "boxes within boxes" diagram corresponding to the reduction argument presented in Figure 15.9.

3. In §15.5, we described a mapping reduction from A_{TM} to $Halt_{TM}$ by producing a machine M' such that checking for the halting of M' on input w is tantamount to checking whether the original machine M accepts w. In the same vein, describe a mapping reduction from $Halt_{TM}$ to A_{TM} by producing a machine M'' such that checking whether M'' accepts w is tantamount to answering whether M halts on w. **Hint:** Make a copy M_c of M, and modify the accept or reject label (figure out which) of M_c suitably. Now, amalgamate both machines M and M_c into one single machine (call it $M_{combined}$) such that essentially both machines get run.

Now, checking that $M_{combined}$ accepts w must be tantamount to answering whether M halts on w.

4. Draw the "boxes within boxes" diagram corresponding to the reduction argument presented in Figure 15.10.

5. Define

$$CFL_{TM} = \{\langle M \rangle : M \text{ is a TM whose language is context-free.}\}$$

Specify a mapping reduction from A_{TM} to CFL_{TM}. Please take ideas from the construction in Figure 15.10.   □

16

NP-Completeness

> **Chapter Gist:** *There are many important practical problems for which no polynomial time (P) algorithms are known. These problems can, so far, only be solved in nondeterministic polynomial time (NP), and currently this amounts to being intractable (exponential or worse). We define NP to be the class of polynomial time verifiable problems, and NP-Complete to be the hardest of all NP problems (§16.1). We present the role of NDTMs in formulating the theory of NP-Completeness in precise terms (§16.2). We take up the study of the Boolean satisfiability problem (SAT) given it has the distinction of being the first NP-Complete (NPC) problem identified (§16.3). We explain why SAT matters in practice, and also introduce a SAT-solver that can run within your own web browser (§16.8). We begin with the simpler 2-SAT polynomial time algorithm (§16.3.1). We describe a canonical problem called 3-SAT, and describe its role in showing new problems to be NP-Complete (§16.4). The idea of mapping reductions is central to this study, and we show that 3-SAT itself can be shown to be NP-Complete (§16.5). We show that the problem of finding k-cliques in a graph is NP-Hard by presenting a mapping reduction from 3-SAT to it (§16.6). We finish with some caveats and also a discussion of CoNP and allied complexity classes (§16.7).*

16.1 What Does NP-Complete Mean?

In the 1960s, computer scientists started noticing that many problems defy polynomial time algorithms; all they could come up with were exponential (or worse) algorithms. Examples of these problems included everyday scheduling problems such as the *Traveling Salesperson Problem* (**TSP**), one version of which is the following. Suppose you are asked to start from Salt Lake City, UT, travel by road and visit all the 48 US state capitals of the contiguous USA exactly once and return to Salt Lake

[1] We assume that there are fixed **time costs** to go between any two capital cities. For instance, the time it takes to go between Salt Lake City and Boise is assumed to be fixed and known. It has been estimated that it will take about 10 days to do all 48 states without traffic and without any stoppage. Our capital-to-capital costs can be assumed to be such that the 18-day figure is met by some tours and not met by others.

[2] Studies have shown that bees solve the traveling salesperson problem while covering a collection of flower patches optimally.

[3] We use the word *check* in the sense of checking whether the claim is true for a specific instance. For example, we can easily *check* that $(3, 4, 5)$ form a Pythagorean triple by checking the identity $3^2 + 4^2 = 5^2$ to be true. We will use the term *verify* to connote something deeper, such as Fermat's last theorem: there are no Pythagorean triples of the form $x^n + y^n = z^n$ for n above 2.

[4] In fact, if the certificate itself is exponentially long, even reading in the certificate will take an exponential amount of time. In this case, the cost of checking cannot be polynomial.

[5] ...somewhat irately...

[6] ...and perhaps the best known computation that can actually be carried out in P-time for many a problem. One has to keep in mind that for many problems, even the "easy checkability" in P-time has not been proven. This would make one feel grateful that at least easy checkability in P-time is an option.

[7] Many scientists think that this is a highly unlikely outcome.

[8] This is not a proof, but it already shows that sometimes all certificates we manage to come up with end up being long, causing the checking cost to go up.

[9] Or "corral"

City in **18 days or less**.[1] What is the *optimal* route you would take?[2] Is there a better algorithm than computing the cost of all 48! such tours and picking the one that finishes in 18 days or less? All algorithms so far have been **intractable** (*exponential or worse*; an example is factorial).

However, scientists also noticed that given a **claimed solution** in the form of a sequential listing of state capitals (a "tour"), they can indeed **easily check**[3] (in polynomial time) whether the cost is below 18 days. This evidence presented for checking (called a **certificate**) is compact (*i.e.,* polynomial in length), and one can simply go as per this sequence, add up the costs and check if it is below 18 or not.[4] So while the cost of solving might be intractable, the cost of checking claimed solutions is polynomial. Scientists started calling this problem class of easy-to-check and difficult-to-solve problems "NP."

At this juncture, one might protest saying that the cost of checking claimed solutions being polynomial time is not a worthwhile "consolation prize." The real goal, they might say, is to *solve* problems in polynomial time and not merely *check a given solution* in polynomial time. They might[5] ask, "who will come up with a solution to check in the first place?!"

Unfortunately, in the world of algorithm design, one sometimes has to be humble and be content with what's available and feasible. In §16.1.1, we will show that knowing that a problem is checkable in polynomial time is a valuable piece of knowledge.[6] Easy checkability has, in fact, a crucial role to play in being able to settle the P versus NP question one day, showing that:

- *Either* it is impossible to have a polynomial time algorithm for a problem in NP that is currently intractable;

- *Or* all such problems have polynomial time algorithms (hence one can stop bothering about NP)![7]

There are actually many problems for which even the cost of checking the solution appears to be intractable. Let us take the following variant of the traveling salesperson problem: '*show that* **there isn't a tour** *where the cost is* ≤ *18 days.*' Here, to answer "yes," it appears that one must list *every* tour and show that each tour takes more than 18 days. Even the certificate involved in this check case is exponentially long (*i.e.,* the sequence of all tours). This is not easily checkable (takes exponential time).[8] Thus, having easy checkability is really getting us somewhere.

16.1.1 Grouping Problems: Solving One Implies Solving All

What researchers in complexity have done is to group[9] problems into a class of problems called NP, and then identify the *hardest* problems in this class, which is the NP-Complete class. All problems in NP-Complete are equally hard; the problems outside of NP-Complete (but inside NP) are less hard. These aspects are illustrated in the Venn diagram of language

Figure 16.1: The language families P, NP, and NPC. All these set inclusions are likely to be proper.

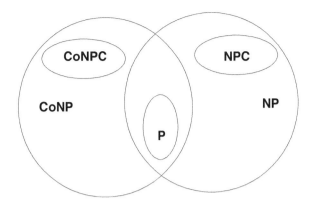

family inclusions in Figure 16.1 (for now, please ignore the families whose names start with "Co"; they are discussed in §16.7). The key property that one ensures (before calling a problem NP-Complete) is that *if even one problem in NP-Complete has a polynomial time algorithm*, then *all of NP will have a polynomial time algorithm*.[10] This behavior will be ensured in the process of showing a problem to be NP-Complete (§16.5.3). All we will be left with (*in this context*) will be polynomial time algorithms.[11] In summary, with respect to the "consolation prize" discussion earlier:

- We group polynomial time checkable (synonymous with verifiable) problems into NP. Computer science research has been inducting problems into the NP class beginning in the 1960s (detailed in §16.2).[12]

- Showing that an NPC problem can be solved in polynomial time will collapse the entire class NP, and essentially turn it into the class P.

> This chapter studies *algorithms* (not *procedures*) using TMs (*i.e.*, these TMs will halt on all inputs). Also, given the introductory nature of this chapter, we only study *time complexity*. *Space complexity* is also studied using TMs.

16.1.2 Some Historical Notes

Let us understand the ideas thus far in the context of prime numbers.

> Given a natural number of d digits, what is the algorithmic complexity of checking whether it is a prime? Is this 21-digit number prime: 147,573,952,589,676,412,927 ? (It equals $2^{67} - 1$, the 67th Mersenne number (M_{67}); the nth Mersenne number, M_n, is $2^n - 1$.)

In 1903, Frank Cole gave a lecture in which he performed the multiplication of 193,707,721 and 761,838,257,287 by hand on a chalkboard, obtaining M_{67}.[13] Thus Cole was able to prove (over the course of one lecture) that this number wasn't a prime. Yet, to obtain the factors, he

[10] ..and the "Co" families will also disappear.

[11] Clearly there will continue to be other exponential algorithms; it is only the very important NP class that will become equal to P.

[12] If one **proves** that one of these NPC problems *does not have a P-time algorithm*—the scenario that most scientists believe is likely—they would have solved one of the most important of open problems in CS. They would also win the $1 million prize money that the Clay Mathematics Institute has set apart for this challenge.

There is also another term lurking around in this area called **NP-hard**. It means *at least as hard as NP*—meaning it could be even harder than NP. In fact, some NP-hard problems are so hard that they are actually undecidable! We cover this topic in §16.7.

[13] He silently put down the chalk and walked away to a thunderous applause. https://en.wikipedia.org/wiki/Frank_Nelson_Cole

spent *"three years of Sundays."* This anecdotal evidence itself shows that proving that a number is composite (non-prime) can take very long, but checking that it is composite does not. This problem happens to be in NP.

The largest known prime number has 23,249,425 digits;[14] how hard would it be to *prove* that it is prime? Until the year 1977, there wasn't a definite algorithmic classification that applied to the entire set of primes. In 1977, Pratt [39] proved that primality checking is an NP-problem, meaning it still defied a polynomial solution, but for primes, the certificate—a proof that a number is prime—is indeed succinct, and the proof can be checked in P-time.

[15] One may prefer the term "primality testing" also, in this context. In our sense, it means the same as checking, as the answer is produced for the given instance.

[16] https://en.wikipedia.org/wiki/AKS_primality_test mentions the details including the notation $\tilde{O}(log^{12}(n))$ where n is the number itself; thus $log(n)$ obtains the number of digits. It has been improved to $\tilde{O}(log^6(n))$. The notation $\tilde{O}$ stands for "soft $\mathcal{O}$" and is explained in the references; it ignores logarithmic terms.

In 2002, Agrawal, Kayal and Saxena obtained a P-time algorithm [3] (now called the "AKS algorithm") for primality checking[15] with complexity roughly $O(d^{12})$ where d is the number of digits in the number.[16] Please note:

- Complexity classifications evolve with the state of human knowledge.
- While we can check whether a number is prime or composite in P-time (thanks to the AKS algorithm), it does not mean that *finding the factors* can be done in P-time.

Factorization is the key hard algorithm upon which cryptographic systems are built. We can easily *check* that a given number is factored correctly (as evidenced by a certificate), but the hardness of *generating* factors is still unknown—and appears to be hard. One wishes for such algorithms to remain hard—or else today's crytographic systems could be rendered useless.[17] See Cook's discussions [18] on the importance of P versus NP.

[17] Quantum computers have the ability to factor numbers in polynomial time. See a video tutorial on how this is done by Vazhirani https://youtu.be/YhjKWAMFBUU.

16.2   NPC Notions Defined Based on NDTMs

Let us consider the computation trees supplied with Figures 16.2 and 16.3 (repeated from Chapter 13) for the DTM and NDTM designed to recognize a '101' in the input string. The DTM accepts 10101 by recognizing the first 101 and rejects both 01 by going to state StuckNo0Aft1, and rejects ε ("") by going to state StuckNo1beg.

The NDTM has six executions when fed 10101: it accepts both the 101's and rejects upon badly chosen nondeterministic selections.

16.2.1   P-time

[18] If the complexity measure $\mathcal{O}(g(n))$ is ascribed to a function $f(n)$, it means that there exists some $k \in Nat$ such that $f(n) < C \cdot g(n)$ for all $n \geq k$, and $C \in Nat$ being a constant. This is the same "big Oh" one studies in a basic course on algorithms.

We define *P*-time with respect to the runtime of DTMs. The Turing machine model is **robust** in that it faithfully models *not just the computing power* but also the *computational efficiency* of realistic computers within a polynomial factor. Anything realizable on a TM with polynomial ($\mathcal{O}(n^k)$) complexity[18] can be realized on a realistic computer with complexity $\mathcal{O}(n^{k'})$ for perhaps $k' < k$.

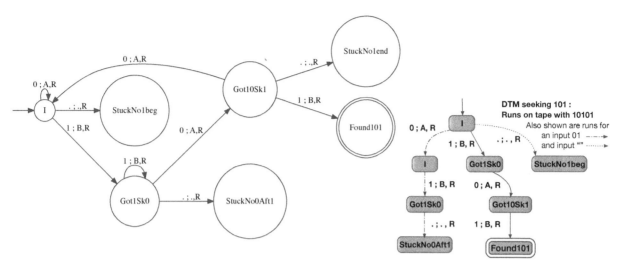

Figure 16.2: Transition diagram and computation tree for a DTM that looks for 101 within given w.

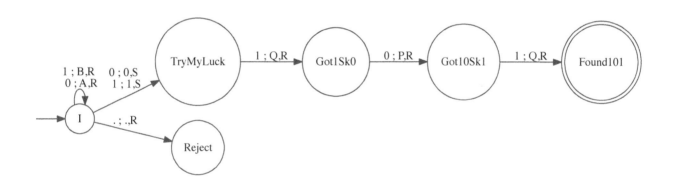

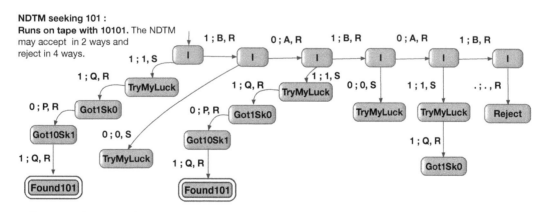

Figure 16.3: Transition diagram and computation tree for an NDTM that looks for 101 within given w.

> **Definition 16.2.1:**
> (a) For input w of length n, the **execution time of a DTM** is the number of steps taken along its **(only)** computational path.
> (b) If a DTM D_L has $\mathcal{O}(n^k)$ execution time, it is said to be a **polynomial time decider**.
> (c) **P** is the family of languages where for each language $L \in \text{P}$, there is a polynomial time decider D_L.

16.2.2   NP-time

NP-time is the upper bound of the computational cost across all the nondeterministic computational paths.

> **Definition 16.2.2:**
> (a) For input w of length n, the **execution time of an NDTM** is the **maximum** number of steps taken along **any computational path**.
> (b) If an NDTM N_L has $\mathcal{O}(n^k)$ execution time, it is said to be a **nondeterministic polynomial time decider**.
> (c) **NP** is the family of languages where for each language $L \in \text{NP}$, there is a nondeterministic polynomial time decider N_L.

Across all the executions produced by the NDTM of Figure 16.3 on input 10101, the maximum number of steps[19] taken is 6.

16.2.3   NP Verifier

To ease the construction of proofs in this area, one likes to have *two alternative definitions of NP*. Definition 16.2.2 presented the so-called *decider view of NP*. The other view is the *verifier view of NP*. This construction also equates the notion of "a certificate" to internal NDTM decisions.

> **Definition 16.2.3: A nondeterministic polynomial time verifier** is an NDTM that takes an input w along with a certificate c (together packaged as $\langle w, c \rangle$) and makes a decision in NP time. The purpose of such a verifier is to check w's acceptance status by exploiting the given certificate c.

We will provide an example of such a verifier in §16.2.4, and a proof sketch that given an NP decider we can define an NP verifier (and vice-versa) in §16.2.5.

[19] For Jove TMs, it is the maximum amount of fuel consumed along **any path.** Note that it is very easy to be somewhat mischievous and design an NDTM that branches infinitely, but with each computational path taking a finite number of steps (just put a . ; . , R transition from Reject going back to Reject). Thanks to this mischief, we can create a situation in which there isn't such a maximum. But now, recall the "no wimp clause" we stated in §14.2.4. *We must avoid creating a semidecider when a decider is possible.*

Just to drive this point home, imagine that someone added a linear transition sequence of 10^6 steps after the Reject state, with each transition labeled by . ; . , R. This would bloat the execution length to 10^6! All these "definition destroying constructions" are immaterial! The kinds of reasoning we will be engaged in will be of the form "**IF** problem P with complexity x can be solved, **THEN** via mapping reductions, we will show that a related problem Q can be solved with complexity y." In other words, we will be pursuing *relative complexity measurements* achieved through mapping reductions.

16.2.4 Examples of P-time and NP-time Deciders

In Figure 13.10, we presented a P-time decider for strings of the form
$w\#w$ where $w \in \{0,1\}^*$. This is also an NP-time decider because DTMs
are a special case of NDTMs. In Figure 13.11, we presented a decider for
strings of the form ww where $w \in \{0,1\}^*$. The fact that all the computational paths of this NDTM take only a polynomial amount of steps allows
us to state that this is indeed an NP-time decider.

To envisage an NP decider that is not in P, consider the Traveling
Salesperson problem. Here is the high-level code for an NDTM. (Note
that we don't know whether there is a P-time decider for the TSP. Also,
from now on, we will not bother to present actual DTMs or NDTMs but
only present their high-level pseudo-code):

> **NP decider for TSP:**
> - Accept the US map and the costs between the states on the in-
>   put tape. Write SLC at the end of the given input.
> - Starting from the start state (when we are in SLC), choose a
>   state capital that hasn't been considered thus far (say state S_1).
>   Write S_1 at the end of the input tape.
> - Pick another state S_2 that hasn't been considered. Again, write
> S_2 at the end of the input tape.
> - Do this for all 47 other states.
> - Call a verifier Turing machine that checks that the state se-
>   quence SLC $S_1 S_2 \ldots S_{47}$ SLC written at the end of the tape
>   meets the given cost criterion. **Accept** if so; **Reject** otherwise.
> - This ND algorithm has cost NP-time because along *any one
>   computational path*, what happens is this:
>   - We choose a linear list of states
>   - We make a subroutine call to another TM—called a **verifier
>   TM**—that simply adds up the cost of the edges present in
>   SLC $S_1 S_2 \ldots S_{47}$ SLC
>   - This computational path will only involve a polynomial num-
>   ber of steps. □

16.2.5 Decider versus Verifier Views

The aforesaid algorithm to write an NP decider for TSP used the following
trick: it converted the given problem into two subproblems:
- Write a "certificate" at the end of the user-given input.
- The TM is made to decide based on the certificate by calling an "inner"
  TM. We call this latter TM a *verifier*.

In general, an NP decider can be converted to an NP verifier, and vice-
versa. We now explain the process abstractly. Let L be a language and

N_L stand for the NDTM decider for L and V_L for a P-time verifier for L.

Given N_L, Obtain V_L: The verifier V_L that "we must output" is built as follows. Given an NDTM N_L that is an NP-time decider, ask for an additional input c that is also of polynomial length with respect to the input w that N_L already expects. Use c to pick the nondeterministic selections that N_L makes at each juncture when N_L is confronted with nondeterminism. Now, given that no path of N_L is longer than a polynomial quantity, we can completely resolve the nondeterminism via c and obtain V_L, a verifier that now takes $\langle w,c \rangle$. When N_L decides, V_L also reaches the same decision.[20]

Given V_L, Obtain N_L: The decider N_L that "we must output" is built as follows. Prepend V_L's code with a phase that can be called the *internal certificate auto-generation phase* in which the prepended code simply writes out a c on the TM tape the end of w and then feeds $\langle w,c \rangle$ to V_L. The decision of V_L is emitted as the decision of N_L. [21]

16.3   Introducing SAT Problems

The term Boolean satisfiability or SAT refers to the satisfiability of a general Boolean expression. It suffices for us to study two special cases of SAT:

- 2-SAT: the satisfiability of conjunctive normal form (CNF) Boolean formulae with exactly two literals (a Boolean variable or its negation) per clause.

- 3-SAT: the satisfiability of conjunctive normal form (CNF) Boolean formulae with exactly three literals per clause.

The reason we study 2-SAT is that it has a beautiful P-time algorithm due to Aspvall, Plass and Tarjan [5]. The reasons to study 3-SAT are several, some of which are: (1) merely going from "2 to 3" shoots up the complexity from P-time to NP-Complete. (2) it is the canonical NP-Complete problem, by understanding which deeply, one tends to understand the theory of NP-Completeness rather well. We now define the basic notions surrounding 2-SAT and 3-SAT.

> **Definition 16.3:** A variable or its negation is called a *literal*. A Boolean formula in *conjunctive normal form* (CNF) is a conjunction of clauses. It is a 2-CNF formula if it is a conjunction of 2-Clauses and a 3-CNF formula if it is a conjunction of 3-Clauses. A 2-Clause is a disjunction of two literals, and a 3-Clause is a disjunction of three literals. A 2-Clause is equivalent to a conjunction of two implications.

Notations and Examples: Let ! stand for NOT, . for AND, and + for OR. Let $\rightarrow$ stand for *implication* (we use this notation to make this operator look like a graph edge, as we will be building implication graphs).

[20] Think of c as "rudder steering control" on a boat in a vast lake. c is then a sequence of steps "take the second turn; then the first turn; then the fourth turn; ...." V_L simply allows us to input the "guess" via c. **Depending on the purpose N_L is supposed to achieve (i.e. NP-decide membership in L), a suitable set of turn instructions can be generated.** Externally, it will appear as if the "newly minted" V_L is asking for help ("Phone a friend" in "Who Wants to Be a Millionaire"), but internally it is simply doing "backseat driving" of N_L.

[21] Again, depending on the language that V_L verifies membership into, aided by certificate c, one can always construct the certificate auto-generation phase to force the same decision "out of V_L." Externally it will appear as if the "newly minted" N_L is being "smart" and deciding in NP-time. But internally, it is actually coughing up a certificate and feeding that to V_L which needs this crutch.

Let a through e be variables (they are also literals). Also !a through !e are literals. We employ **True** or 1 interchangeably, but prefer the former for *strongly connected components* (SCCs, see Figure 16.4 for a definition) and the latter for literals. Similarly, we employ **False** or 0 interchangeably, but prefer the former for SCCs and the latter for literals. Here are sample 2-CNF and 3-CNF formulae:

2-CNF: $(a + b).(!a + c).(!b + e).(f + !f)$
3-CNF: $(a + b + c).(!a + b + b).(b + d + e).(a + f + !f)$

The 2-Clause $(a + b)$ is equivalent to the conjunction of two implications:

- $!a \rightarrow b$
- $!b \rightarrow a$

This is because

- $(a + b)$ is equivalent to $!a \rightarrow b$.
- $(a + b)$ is equivalent to $(b + a)$ which can be translated to $!b \rightarrow a$.

We cannot translate a 3-Clause in this manner.

Definition 16.8

$L_{2sat} = \{\langle \phi \rangle : \phi$ is a 2-CNF formula that is satisfiable.$\}$
$L_{3sat} = \{\langle \phi \rangle : \phi$ is a 3-CNF formula that is satisfiable.$\}$

Note: Some readers may wish to look at §16.8 and actually play with a SAT solver to cultivate some familiarity with Boolean satisfiability.

16.3.1 A Warmup: 2-SAT

The fact that a 2-Clause is equivalent to a conjunction of two implications can be exploited in obtaining a P-time satisfiability checking algorithm. Figure 16.4 shows an example 2-CNF formula.

Implication Graph: We can build the implication graph for the formula F above, also shown in Figure 16.4. An implication graph is one that treats each 2-CNF clause as *two* implications. More specifically, this graph is obtained by modeling each 2-CNF clause of the form $(p + q)$ as the conjunction of two implications, namely $!p \rightarrow q$ and $!q \rightarrow p$. We treat each implication as a graph edge. As soon as we add all possible implication edges, we end up creating a graph with the following structural properties:

- The implication graph ends up having groups of nodes and edges that form *maximal strongly connected components* (maximal SCCs). In a directed graph, a strongly connected component (SCC) is a collection of nodes that are reachable from each other. A maximal SCC is one that pulls in the maximum number of nodes into each SCC such that

F = (a+b).(b+!c).(!b+!d).(b+d).(d+a)

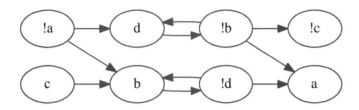

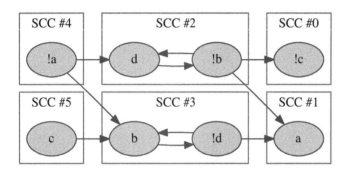

Figure 16.4: 2-CNF formula, and illustration of Aspvall et al's 2-SAT algorithm. Here, + stands for Boolean OR, . stands for Boolean AND, and ! stands for Boolean negation. The first directed graph above is what we initially obtain after we convert each disjunction such as (a + b) into a pair of implication edges. The second directed graph clusters groups of nodes and edges into *maximal strongly connected components* (maximal SCCs). In a directed graph, a strongly connected component (SCC) is a collection of nodes that are reachable from each other. A maximal SCC is one that pulls in the maximum number of nodes into each SCC such that they can all reach each other. In our example, maximal SCCs (#2 and #3) can have at most two nodes. All other maximal SCCs only have one node each.

they can all reach each other. **In Figure 16.4, we enclose all the maximal SCCs within rectangular boxes.**

- In our example, there are two maximal SCCs with two nodes each, namely SCC #2 and SCC #3. All other maximal SCCs only have one node each. Note that we cannot merge any of the maximal SCCs #0 through #5 and *still have each remain a maximal SCC*. For example, we cannot merge SCC #2 and SCC #0 because whereas !b can reach !c, !c does not reach !b.

- If any SCC includes a variable and its complement, the given formula is not satisfiable. For example, suppose an SCC contains a literal p and its negation !p. Then, such an SCC represents the conjunction of !p → p and p → !p, which is not satisfiable. This is because !p → p simplifies to p and p → !p simplifies to !p, and their conjunction cannot be satisfied. **In our example, this situation does not arise, and hence our formula is satisfiable.**

- Given that a formula is satisfiable, we can order the SCCs into a partial order, as also shown in Figure 16.4. **This is a topological sort of the SCCs.**

- **If we do all these steps correctly, we will notice that for each SCC, there is also a dual SCC.** For example, SCC #2 and SCC #3 are duals, with the negation signs in the literals flipped. Likewise, SCC #4 and SCC #1 are duals, as are SCC#5 and SCC#0.

Obtaining the Satisfying Assignment: For a collection of maximal SCCs that are situated in a partial order, we can proceed as follows in order to find a satisfying assignment. Hereafter we use "SCC" to refer to maximal SCCs.

- Consider the collection of SCCs "bottom-up" (or more precisely, as per the *reverse topological sort order*).
- If an SCC is unmarked, mark it **True**. Immediately mark its dual **False**.
- When we assign truth values to an SCC, all the literals in the SCC obtain the same assignment as the SCC. In our example,
  - we mark SCC #1 **True**, and hence mark SCC #4 **False**. This assigns $a = 1$ and $!a = 0$.
  - we mark SCC #0 **True**, and hence mark SCC #5 **False**. This assigns $!c = 1$ and $c = 0$.
- In some cases, we have a choice of making an SCC **True** or **False**. For example, SCC #3 containing $b, !d$ and SCC #2 containing $d, !b$ are incomparable in a topological sort; so they can be assigned arbitrarily.
- Pick **True** for SCC #3 containing $b, !d$. Pick **False** for SCC #2 containing $d, !b$.
- Now we've assigned all SCCs. The final assignment obtained is:
  $!c = 1, c = 0, a = 1, !a = 0, b = 1, !d = 1, d = 0, !b = 0$.

The following facts are true of this algorithm:

- An SCC assigned **False** only has **False** as predecessors.
- An SCC assigned **True** only has **True** as successors.

These facts are important in so far as they guarantee that the algorithm will never introduce a contradiction.

16.3.2   *2-SAT: Examples and Algorithm*

We now illustrate our construction on two additional examples, the first being satisfiable and the second unsatisfiable. These two examples are quite related in that they consider Boolean combinations of two Boolean variables, namely a and b.

In Figure 16.5, the graphs were obtained from the Boolean formula:

$(a + b).(a + !b).(!a + b)$

Given that the SCC #0 is encountered first in the reverse topological sort, we can assign $a = b = 1$, which assigns for the dual graph $!a = !b = 0$. In Figure 16.6, the graphs came from the Boolean formula

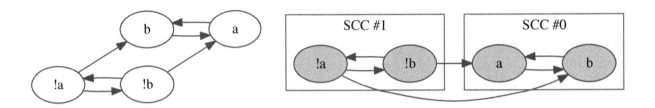

Figure 16.5: The implication graph and the SCCs for the CNF formula

$$(a + b).(a+!b).(!a + b).(!a+!b).(!a+!b)$$

Now we have only one SCC within which all the nodes fall, and thus we cannot obtain a satisfying instance for this Boolean formula.

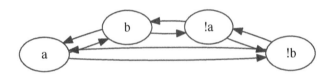

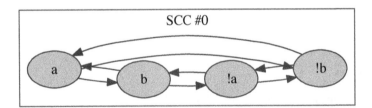

Figure 16.6: This graph and the SCCs are for the CNF formula $(a + b).(a+!b).(!a + b).(!a+!b)$

2-SAT Algorithm Recap: To recap, the algorithm to determine 2-SAT consists of the following steps:

- Obtain the implication graph.
- Divide the implication graph into maximal SCCs ("SCCs").
- If any SCC contains a literal and its negation, exit with "UNSAT".

- Else, consider the SCCs according to the reverse topological sort order.
- For an SCC without any truth assignment, assign that SCC **True**.
- Immediately assign its dual SCC **False**.
- Move up the topological sort.
- For an SCC and its dual that are not ordered, pick the assignment arbitrarily for one of the members of the dual; the other member naturally receives an inverted assignment.
- Move up the order and complete assigning all SCCs.
- We revert to assigning the next unassigned SCC to be **True**, in case there are still unmarked SCCs.
- In the end, output the assignment for all literals, by replicating the truth assignment for the SCC as really being the corresponding truth assignments for the literals in the SCC.

16.4 3-SAT and Its NP-Completeness

We could "pull off" an algorithm for 2-SAT (§16.3.2) only because of the implication graph that underlies 2-SAT. Unfortunately, there is no polynomial time algorithm for 3-SAT that we know of. The best-known result is that it is NP-Complete. We now provide two **equivalent** definitions for NP-completeness.

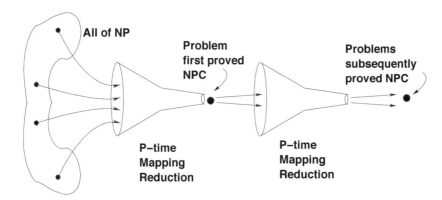

Figure 16.7: Diagram illustrating how NPC proofs are accomplished. The problem first proved NPC is 3-SAT. Definition 16.4(a) is illustrated by the "left funnel" while Definition 16.4(b) is illustrated by the "right funnel." (The funnels serve as a gentle reminder that mapping reductions need not be onto.)

Definition 16.4:(a) L is NPC if
(i) L is in NP, and
(ii) **for every language** $X \in$ NP, we have $X \leq_P L$.
(Accomplishing (ii) alone means that L is **NP-Hard**.)
Definition 16.4:(b) L is NPC if
(i) L is in NP, and
(ii) for **some other language** $L' \in$ NPC, we have $L' \leq_P L$.
(Accomplishing (ii) alone means that L is **NP-Hard**.)
The definitions are equivalent because if a language L' is NPC,

> then we have, for every $X \in$ NP, $X \leq_P L'$; then, given that $L' \leq_P L$, we have $X \leq_P L$. This is because the composition of mapping reductions is also a mapping reduction.

The P-time mapping reductions illustrated by the funnels in Figure 16.7 are the same mapping reductions defined in Definition 15.5, except the function f in that definition is a **deterministic polynomial-time** (P) reduction. In §16.6, we will concretely demonstrate how the "second funnel" works by taking a 3-SAT formula and turning it into an undirected graph.

Say one of the "problems subsequently proved NPC," modeled by language L_{new}, ends up having a P-time algorithm D_{new}.[22] We can then have a P-time algorithm for 3-SAT as follows:

- Input a 3-SAT instance ϕ modeled by language L_{3sat}.
- Apply the second mapping reduction $\leq_P$ (of the second funnel) and create an instance l_{new} of the problem modeled by L_{new}.
- Apply D_{new} on l_{new}. Because of $\leq_P$ being a mapping reduction, D_{new} accepts l_{new} **if and only if** ϕ is satisfiable.
- This gives us a P-time decider for L_{3sat}.
- Via the "first funnel" we now have a P-time decision procedure for the whole of NP. This will essentially eliminate NP and establish P=NP.

[22] Perhaps L_{new} itself is the language of graphs that have k-cliques.

16.5   3-SAT Is NP-Complete

Figure 16.8: Proof of the Cook-Levin Theorem

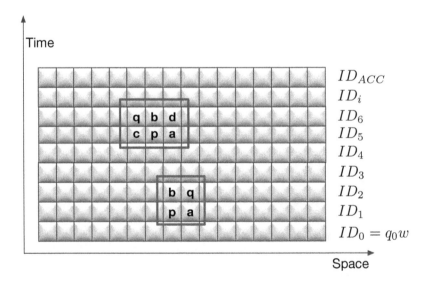

Theorem 16.5: 3-SAT is NP-Complete. (This is known as the Cook-Levin theorem, discovered independently by Cook [17] and Levin [31].)

Proof: According to both Definition 16.4(a) and (b), we have to show that 3-SAT is in NP. We will now show it using both the decider and the verifier views.

16.5.1 3-SAT is in NP

We are given a 3-SAT instance—a 3-CNF formula ϕ. The NDTM presented below can easily extract the variables in w in polynomial time.

NP-decider: Build an NDTM that chooses a random assignment (0 or 1) for each variable in ϕ. Check that this assignment satisfies ϕ. This NDTM runs in NP time.

NP-verifier: The NDTM being built also asks for an actual variable assignment c ("certificate") to follow ϕ on the tape. The NDTM then checks whether $\phi(c)$ is true, accepting exactly when so.

16.5.2 Every Language in NP Reduces to 3-SAT

To show 3-SAT is NP-Hard, we have to go by Definition 16.4(a), as there is no previous NP-Complete problem to reduce from. Every language in NP has an associated NDTM that accepts or rejects in NP-time.[23] The crux of our proof is that given one of these NDTMs, one can devise a general way to check for the acceptance of an input w by the NDTM using 3-SAT.

Let us focus our attention on the first reduction (funnel) of Figure 16.7. Given an arbitrary N_L and an arbitrary $w \in \Sigma^*$, consider the computation of N_L on w starting from the instantaneous description $ID_0 = q_0 w$. In Figure 16.8, we portray this as the bottommost layer along the space/-time diagram against ID_0 where we use the notation for instantaneous descriptions (IDs) introduced in §13.8.[24]

When this NDTM computes, it checks the cell under its tape head, changes this cell and moves right (an example is in the ID_1 to ID_2 march). It can also move left if it is not walking off the left end of the tape (an example is in the ID_5 to ID_6 march).[25] It can be seen that all changes caused by the TM's transition relation Δ affect only a window of size 3 at most. During the ID_1 to ID_2 march, an ID where the "head state" is **p** and the tape symbol under the head is **a** changes to the head state becoming **q**, the head moving right and with the **a** changed to a **b**. During the ID_5 to ID_6 march, an ID where the "head state" is **p** and the tape symbol under the head is **a** changes to the head state becoming **q**, the head moving left and with the **a** changed to a **b**. (The tape symbols **c** and **d** capture enough "stuff" around the TM head.) In Figure 16.8, we show

[23] Recall that this notion can be summarized as **reaching a decision** in NP-time.

[24] We assume that we are working with a singly infinite tape with the tape extending in "space" to the right. One can systematically translate TMs that carry out computations on doubly infinite tapes to those that use singly infinite tapes with only a polynomial increase in time.

[25] We can prevent a TM from going to the left of the leftmost cell of a singly infinite tape. Again such a transformation takes only a polynomial amount of extra cost. We omit showing such steps for simplicity to highlight our main reduction arguments.

a march of ID_1 through ID_6, show ID_i for generality, and the accepting ID is shown as ID_{ACC}.

What you really have to imagine is not just one of the NDTM choices, but **all** NDTM choices. Thus, imagine ID_1 to encompass all possible changes that ID_0 could have been subject to, for all the nondeterministic options of N_L. In general, we assume that when going from layer ID_i to ID_{i+1}, we imagine that the layer we draw for ID_{i+1} represents all possible (nondeterministic) ways in which it could have been obtained from ID_i.

Now, we know that we will "pile on" only a polynomial number of layers in this manner before a decision is reached. The reason is of course that N_L is a TM with nondeterministic runtime being NP.

Here is how SAT enters the picture:

- We can capture the evolution from ID_i to ID_{i+1} through a 3-CNF Boolean formula of polynomial length ϕ_i. The construction of this formula is described in many references [42] in this field, and we don't repeat that. Fortunately, this single formula can capture *all the nondeterministic evolutions* from layer i to layer $i+1$ in one shot.

- We can also introduce formula ϕ_0 to capture the constraints on ID_0 and formula ϕ_{ACC} to capture the constraints on the final ID containing the accepting ID.

- Thus, the entire "pile of IDs" depicted in Figure 16.8 can be captured by a formula:

$$\Phi = \phi_0 \wedge \phi_1 \wedge \ldots \phi_i \wedge \ldots \phi_{ACC}$$

This formula encodes all the nondeterministic evolutions from start to finish of the NDTM. *All the NDTM paths* are rolled into this single formula.

- Thus, given **any** NDTM N_L, we can synthesize a Boolean 3-CNF formula Φ describing **all the nondeterministic accepting computational histories** of N_L in *one fell swoop*. The formula Φ is polynomially sized and can be obtained at polynomial cost. Thus the *existence* of the $\leq_P$ mapping reduction modeled by the "first funnel" of Figure 16.7 has been demonstrated. □

16.5.3   How P=NP is Obtained if 3-SAT ∈ P?

To acid-test our construction, we now offer an algorithm to decide *any* N_L in NP in *deterministic* polynomial time, **if** we are given a SAT-solver that runs in polynomial time.[26]

- **Input:** An NDTM N_L and input string w
- **Output:** A P-time decision if $w \in L$
- **Method:** Since we don't know **how long** N_L will run on input w before a decision is reached (except the run is polynomially long for some polynomial), we devise an incremental checking method:
  - Start with $\Phi_0 = \phi_0 \wedge \phi_{ACC}$, and call the P-time SAT solver to see if

[26] This is a magical SAT-solver that does not exist; but should it exist, it will let us collapse NP down to P.

it reaches a decision, outputting this decision if so.

– If not, increase i step by step, generating

$$\Phi_i = \phi_0 \wedge \phi_1 \wedge \ldots \phi_i \wedge \phi_{ACC}$$

and checking for a decision (accept/reject).

– The decision is guaranteed in polynomial time, and this algorithm will only make a polynomial number of SAT-solver calls (before a decision is reached), with each call involving a polynomially sized formula, and each such call returning in polynomial time.[27] □

[27] That is, there isn't *a priori* knowledge on how much of the Φ_i formula to generate, before a decision is guaranteed.

16.6   Show that Clique Is NPC: Reduction from 3-SAT

The language of interest is

$$Clique = \{\langle G, k \rangle \ : \ G \text{ is an undirected graph having a } k\text{-clique}\}.$$

16.6.1   Clique is in NP

We will employ the verifier view. The NDTM being built, in addition to receiving its input which is $\langle G, k \rangle$, also asks for a certificate in the form of a list of k nodes. The NDTM then checks whether these k nodes are pairwise connected, accepting exactly when so. This checking procedure runs in polynomial time.

16.6.2   Some Language in NPC Reduces to Clique

$$\phi = (x1 + x1 + x2).(x1 + x1 + !x2).(!x1 + !x1 + x2).(!x1 + !x1 + !x2)$$

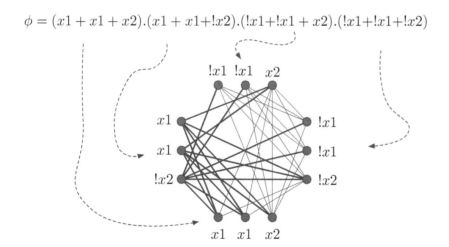

Figure 16.9: The Proof that *Clique* is NPH using an example formula $\varphi = (x1 + x1 + x2).(x1 + x1 + !x2).(!x1 + !x1 + x2).(!x1 + !x1 + !x2)$. We never connect the nodes within each clause "island" (there are four such islands, each with three nodes). Across each clause island, we draw edges in all possible ways *provided* we never connect a literal and its complement. For visual clarity, we show through dark edges all the edges emanating from the clause island for (x1+x1+!x2) going to all other clause islands. We also show the remaining edges, but using fainter lines.

To show Clique is NP-Hard, we can go by Definition 16.4(b), as we can attempt to reduce 3-SAT to Clique. All we need to do is produce a Clique $graph(\phi)$ given a 3-CNF formula ϕ with k clauses such that $graph(\phi)$

has a k-clique exactly when ϕ is satisfiable. The construction is illustrated in Figure 16.9. The basic idea is this:

- For each clause, draw an "island" of three nodes with the literals in the clause labeling the nodes.
- Never introduce any edges within an island.
- Between two islands of three nodes each, connect pairwise all pairs of literals that are compatible. Two literals are compatible if they are not of the form x and !x. Compatible literals can be satisfied simultaneously.

Observe that the example formula ϕ considered is not satisfiable as it basically is

$$\varphi = (x1 + x2).(x1 + !x2).(!x1 + x2).(!x1 + !x2)$$

where, for any chosen assignment of x1 or x2, one of the clauses will be false. Correspondingly, we cannot find a 4-clique in this graph. The existence of a 4-clique would mean that there are "compatible" literals between all the clauses. That is, we could find a setting to make all these literals true. But that would be a satisfying assignment for ϕ. □

16.7 Complexity Classes, Closing Caveats

Proving a language $L \in \mathsf{NP}$ is often the easy part of an NP-Completeness proof; yet, forgetting this part and merely showing $L \in \mathsf{NPH}$ can not only render your proof incomplete, it can also make it incorrect, as we discuss in §16.7.1. In §16.7.2, we discuss a theorem that has been used in the past by researchers to argue that a language may, after all, be in P.

16.7.1 NP-Hard Problems can be Undecidable (Pitfall in Proofs)

We will now show that the language of Diophantine equations is NP-Hard. *Diophantine*, so-called, is the language of equations that have integer roots, an example being $6x^3z^2 + 3xy^2 - x^3 - 10 = 0$. In general, it is the sum of products of powers of integer variables weighted by integer constants.

This language was shown to be **undecidable** by Yuri Matijasević in a very celebrated theorem [34].[28] This is not a contradiction because NP-Hard only means *at least as hard as* NP (it could be harder, including being undecidable). But claiming that *Diophantine* is NPC will be tantamount to the claim that something undecidable is decidable![29] Thus, in general, one must not leave an NPC proof unfinished by forgetting to show that the language in question is in NP: it may, after all, not be in NP!

We will now show[30] that *Diophantine* $\in$ NPH.

[28] One may be tempted to think that this problem is in NP: why not provide the integer roots (the values of x, y, z in the above equation) as a certificate and check that the equation is satisfied? The flaw in this argument lies in being unable to constrain the *size* of the certificate to be polynomially bounded. In particular, the values of the variables can grow without bound.

[29] Recall that all NPC problems are decidable.

[30] This proof comes from Stephen Cook's lecture notes. Please read Cook's bio at https://amturing.acm.org/award_winners/cook_n991950.cfm. He won the 1982 ACM Turing Award *"For his advancement of our understanding of the complexity of computation in a significant and profound way. His seminal paper, "The Complexity of Theorem Proving Procedures," presented at the 1971 ACM SIGACT Symposium on the Theory of Computing,* **laid the foundations for the theory of NP-Completeness.**

Theorem 16.7.1: The language $Diophantine$ is NPH:

$Diophantine = \{p : p$ is a polynomial with an integral root$\}$.

Proof: We will show that 3-SAT $\leq_P Diophantine$. This means that given a 3-SAT instance

$$\Phi = \phi_0 \wedge \phi_1 \wedge \ldots \phi_i \wedge \phi_{(N-1)}$$

we must produce a Diophantine equation $f(\Phi) = 0$ such that this equation is satisfied if and only if Φ is satisfiable. We now explain the design of f:

- Take any arbitrary clause $\phi_i = (l_{i0} + l_{i1} + l_{i2})$ of Φ. Function f turns ϕ_i into the arithmetic expression

$$E_i = (g(l_{i0}) \times g(l_{i1}) \times g(l_{i2}))^2$$

  where the mapping g works on literals as follows (this is called the *literal gadget*):

  - If the argument to g is a variable x, then $g(x) = (1 - X)$ where $X \in Nat$ is an integer variable that we introduce in order to model x.[31]

  - If the argument to g is a negated variable $!x$, then $g(x) = X$ where $X \in Nat$.

- Now, function f turns the whole formula Φ into expression $E = \Sigma_{i=0}^{(N-1)} E_i$.

- Example: Function f maps
  $\Phi = (x + y + y) \cdot (x + !y + !y) \cdot (!x + y + y) \cdot (!x + !y + !y)$ into this expression E consisting of sums of squares of three-way products of expressions:
  $((1 - X) \times (1 - Y) \times (1 - Y))^2 + ((1 - X) \times Y \times Y)^2 + (X \times (1 - Y) \times (1 - Y))^2 + (X \times Y \times Y)^2$

To argue that this is a **mapping reduction**, we must show that Φ is satisfied iff $f(\Phi) = 0$ has integral roots.

Proof that $f(\Phi) = 0$ is satisfiable iff Φ is satisfiable:

- (1) Φ is satisfiable: Then, there **exists a variable assignment** such that every clause ϕ of Φ has a literal that is true. Let $\phi_i = (l_{i0} + l_{i1} + l_{i2})$ be an arbitrary clause. Without loss of generality, let $l_{i1} = 1$.

  - If l_{i1} is an ordinary variable x, then $g(x) = (1 - X)$, and we can turn the situation "$x = 1$" into the corresponding integer assignment $X = 1$, making $g(x) = 0$.

  - If l_{i1} is $!x$, then $g(!x) = X$, and we can turn the situation "$x = 0$" into the corresponding integer assignment $X = 0$, making $g(!x) = 0$.

  Thus, corresponding to this assignment $l_{i1} = 1$, expression $E_i = 0$. This is the case for every E_i, and so $E = 0$, or that the equation $f(\Phi) = 0$ is satisfied.

- (2) $f(\Phi) = 0$ is satisfiable: This means that each three-way product term $E_i = f(\phi_i)$ must individually be 0 (since we are squaring the three-way products, we avoid the possibility of an E_i and E_j that are non-zero and cancelling each other). This means that each three-way product has one term being 0. There are two cases here:

[31] We use the uppercase convention for the integer variables that correspond to Booleans.

[32] We don't care how we assign other variables that might be present in the same clause.

– This term is of the form X, where $X = 0$: It is clear that X originated from a literal $!x$ via the g mapping. We can choose the value $x = 0$, thus satisfying the clause that $!x$ came from.[32]

– This term is of the form $(1 - X)$, where $X = 1$: It is clear that $(1 - X)$ originated from a literal x via the g mapping. We can choose the value $x = 1$, thus satisfying the clause that $!x$ came from.

This method allows us to create a satisfying instance for Φ from the integer roots of $f(\Phi)$.

16.7.2   The CoNP and CoNPC Complexity Classes

Definition 16.7.2: A language $L \in$ CoNP if $\overline{L}$ is in NP.

Similarly, L is said to be CoNPC exactly when $\overline{L}$ is in NPC. Figure 16.1 depicts these additional language classes and their likely containments. To illustrate these ideas, consider the following languages which are both subsets of Nat. The language

$$Primes = \{ n : (n > 1) \wedge (\forall p, q : (p \times q = n) \Rightarrow (p = 1 \vee q = 1)). \}$$

The language $Composites = \overline{Primes}$, where the complementation is with respect to positive naturals. $Composites$ is in NP because there exists a P-time verifier for this language, given a certificate which is a pair of natural numbers suggested to be factors. As pointed out in §16.1.2, Pratt proves that $Primes$ is also in NP; he shows this result by demonstrating that there are polynomially long proofs for primes (given a prime p, a polynomially long sequence of proof steps can serve to demonstrate that p is such). Furthermore, he showed that such a proof for $Primes$ can be checked in polynomial time. Now, $Composites$ is in CoNP because $Primes$ is in NP, and $Primes$ is in CoNP because $Composites$ is in NP. The question now is: could either of these languages be NPC?

Theorem 16.7.2 shows that *even if there exists one such language*, then NP and CoNP would become equal—a result thought to be highly unlikely. As pointed out in §16.1.2, the AKS algorithm is proof that $Primes$ is in P (and hence $Composites$ is also in P). Theorem 16.7.2 has helped anticipate the direction in which some of the open problems in this area could be resolved.

Theorem 16.7.2: $\exists L : (L \in$ NPC and $L \in$ CoNP) if and only if NP = CoNP.

Proof:

- ($\Rightarrow$) To show that if $L \in$ NPC and $L \in$ CoNP then NP = CoNP.
  - Assume L is NPC; therefore,
    * L is in NP (Definition 16.4(a), Part 1)

  ∗ Also, L is in CoNP, and thus $\overline{L} \in$ NP (Definition 16.7.2).

  ∗ Thus $\overline{L} \leq_P L$ (Definition 16.4(a), Part 2, the "every language" is $\overline{L}$).

 – To show NP = CoNP, we will take an arbitrary L' in NP and show that it is in CoNP. Then we will take an arbitrary L' in CoNP and show that it is in NP.

  ∗ Consider an arbitrary L' in NP. Then $L' \leq_P L$.

  ∗ From $L' \leq_P L$, it follows that $\overline{L'} \leq_P \overline{L}$. Also $\overline{L'} \leq_P \overline{L} \leq_P L$.[33]

  ∗ Now, since there is an NP decider for L, there is an NP decider for $\overline{L'}$ also, using the above mapping reduction chain. In other words, $\overline{L'} \in$ NP, or L' **is in** CoNP.   (∗)

  ∗ Now, consider an arbitrary L' in CoNP.

  ∗ This means that $\overline{L'}$ is in NP. Since L is in NPC, we have $\overline{L'} \leq_P L$. From this we have $L' \leq_P \overline{L}$.

  ∗ Using the fact that $\overline{L} \leq_P L$, we have $L' \leq_P \overline{L} \leq_P L$, or that there is an NP decider for L', or in other words $L' \in$ NP.   (∗∗)

  ∗ From (∗) and (∗∗), NP = CoNP.

• (⇐) To show that if NP = CoNP, then there exists an L that is in NPC and in CoNP. This is straightforward: consider any NPC language L; it would be in CoNP because L is in NP and NP = CoNP.   □

16.8   SAT in Practice

Thousands of verification, counting and optimization problems are currently being modeled in terms of Boolean satisfiability checking.[34] These include 3-SAT formulae generated during the formal verification of programs that are used in a number of safety-critical areas such as embedded systems and computer security. Despite SAT being NP-complete, heuristics invented over the last two decades have achieved several orders of magnitude increase in the efficacy of SAT-solving [9]. SAT solvers nowadays routinely deal with thousands of variables and clauses. Even "hard instances" of SAT that involve more than 50 variables and 200 clauses are routinely solved by SAT solvers.[35]

A SAT solver: Thanks to the work of Mate Soos, you can invoke a SAT solver called CryptoMiniSat in your web browser, with a SAT instance already loaded in the DIMACS format [29]. This format is explained in Figure 16.10 (along with a screenshot; the <-- are added notes for clarity). By clicking the "Play" button on the top right of this browser (near the legend "Ready"), you can invoke the CryptoMiniSat tool on the SAT instance contained in your browser.

Note that SAT solvers, in general, take general CNF formulas that have more than three literals per clause. These general CNF instances can be translated into *equisatisfiable* 3-SAT formulae.

---

[33] These are basic properties of mapping reductions. If $A \leq_P B$ then there is a polynomial-time function f such that $x \in A \Leftrightarrow f(x) \in B$ (definition of mapping reductions). This also means that $x \in \overline{A} \Leftrightarrow f(x) \in \overline{B}$ or that $\overline{A} \leq_P \overline{B}$. Mapping reductions also compose: if $x \in A \Leftrightarrow f(x) \in B$, and $y \in B \Leftrightarrow g(y) \in C$, then $x \in A \Leftrightarrow g(f(x)) \in C$.

[34] https://en.wikipedia.org/wiki/Boolean_satisfiability_problem

[35] These facts show us that the pursuit of NP-Completeness with respect to 3-SAT does not shut the door toward practical uses of SAT. Solvers based on Integer Linear Programming (ILP) are employed in literally tens of thousands of engineering tasks even though ILP is NP-Complete.

Definition 16.8: A Boolean expression E_1 is equisatisfiable with E_2 if E_1 is satisfiable exactly when E_2 is satisfiable. If $E_1 \equiv E_2$ (*i.e.*, $E_1 \rightarrow E_2$ and $E_2 \rightarrow E_1$), then of course these expressions are equisatisfiable. However, even if $E_1 \not\equiv E_2$, it is possible for these expressions to be equisatisfiable. This fact is illustrated in the callout below entitled **Equisatisfiability without Equivalence**.

Equisatisfiability without Equivalence: Given $(a + b + c + d)$, we can rewrite it into $(a + b + !p).(p + c + d)$ which is **equisatisfiable** to $(a + b + c + d)$. That is, $(a + b + c + d)$ is satisfiable if and only if $(a + b + !p).(p + c + d)$ is satisfiable. In this translation, we introduce a fresh variable p to "bridge" clauses.

These formulae are not logically equivalent. Suppose someone claims otherwise, and claims that this is a tautology: $(a + b + c + d) \equiv ((a + b + !p).(p + c + d))$. It is clear that we can falsify the implication $(a + b + c + d) \rightarrow ((a + b + !p).(p + c + d))$ by picking $a = b = 1$, $c = d = p = 0$. Thus the formulae are not logically equivalent. They are equisatisfiable: if a=b=0 and c=d=1, then we can choose p=0; if a=b=1 and c=d=0, we can choose p=1. One can work out other combinations suitably.

This idea of obtaining equisatisfiable 3-SAT formulae works for any number of variables. Consider $(a + b + c + d + e + f)$. We can rewrite it into the equisatisfiable formula $(a + b + !p).(p + c + !q).(q + d + !r).(r + e + f)$. This way, a k-CNF clause can be turned into $(k - 2)$ 3-CNF clauses.

Exercise 16.7.2, NP-Completeness

1. A SAT instance is given below in the DIMACS format.

(a) What is the CNF formula captured by this instance?

(b) By inspection, answer whether the instance is satisfiable, and why.

(c) If it is not satisfiable, then what is the minimal number of rows that must be deleted before the instance becomes satisfiable? If these rows are not unique, list the first two possible such omissions (of sets of rows), starting from the top of the given listing.

(d) Check your answer using CryptoMiniSat.

```
c A SAT instance in DIMACS format
c Your task is to determine whether this instance is satisfiable
c
p cnf 5 32
1 2 3 4 5 0
1 2 3 4 -5 0
```

```
1 2 3 -4 5 0
1 2 3 -4 -5 0
1 2 -3 4 5 0
1 2 -3 4 -5 0
1 2 -3 -4 5 0
1 2 -3 -4 -5 0
1 -2 3 4 5 0
1 -2 3 4 -5 0
1 -2 3 -4 5 0
1 -2 3 -4 -5 0
1 -2 -3 4 5 0
1 -2 -3 4 -5 0
1 -2 -3 -4 5 0
1 -2 -3 -4 -5 0
-1 2 3 4 5 0
-1 2 3 4 -5 0
-1 2 3 -4 5 0
-1 2 3 -4 -5 0
-1 2 -3 4 5 0
-1 2 -3 4 -5 0
-1 2 -3 -4 5 0
-1 2 -3 -4 -5 0
-1 -2 3 4 5 0
-1 -2 3 4 -5 0
-1 -2 3 -4 5 0
-1 -2 3 -4 -5 0
-1 -2 -3 4 5 0
-1 -2 -3 4 -5 0
-1 -2 -3 -4 5 0
-1 -2 -3 -4 -5 0
```

2. Consider the set of undirected graphs $\langle G \rangle$ with a set of nodes N and a set of edges $E \subseteq N \times N$ such that we can two-color the graph (meaning no two nodes connected by an edge have the same color).[36]

3. Using Aspvall's algorithm explained in §16.3.2, check whether the following 2-CNF formula is satisfiable. Detail the entire construction. Do not simplify the given formula (do your work on the given formula). $(!a + b) \cdot (!b + c) \cdot (!c + d) \cdot (!d + a)$

4. Check the satisfiability of the 2-CNF formula given in Exercise 3 using CryptoMiniSat.

5. Repeat Exercise 3 with $(!d + a)$ replaced by the conjunction of two clauses: $(!d + !a) \cdot (a + a)$

6. Check the satisfiability of the 2-CNF formula given in Exercise 5 using CryptoMiniSat.

7. Suppose we write a program that traverses a "tape" of n cells, numbered 1 through n. The program performs n traversals of the tape, with the ith traversal sequentially examining elements

[36] For instance, a triangle graph G with $N = \{a,b,c\}$ and $E = \{(a,b),(b,c),(c,a)\}$ cannot be two-colored.

i through n. What is the runtime of such a program in the Big-O notation?

8. A Hamiltonian cycle in a graph with respect to a given node n is a tour that begins at n and visits all other nodes exactly once, returning to n. In a 5-clique, how many distinct Hamiltonian cycles exist? How about in an n-clique?

9. Define the language $HalfClique$ to be the set of input encodings $\langle G \rangle$ such that G is an undirected graph having a clique with at least $n/2$ nodes, where n is the number of nodes in G. Show that $HalfClique$ is NPC. *Hint: Mapping reduction from Clique.*

10. The game of Sudoku has been shown to be in NPC. In practice, one can encode and solve Sudoku using SAT solvers. This is also a good way to understand the power of modern SAT solvers. Study the Sudoku solver (MIT license) written by Nicholas Pilkington at `https://gist.github.com/nickponline/9c91fe65fef5b58ae1b0`. Test it on the instance provided as well as a few that you create. Note: This solver will need Python2 (or you may adapt it for Python3).

11. In [15], Cantin et al. prove that the problem of verifying memory coherence is in NPC. Read and summarize this proof in about a page, focusing on the construction of the mapping reduction.

12. Show that a 3-CNF formula

$$\Phi = \phi_0 \wedge \phi_1 \wedge \ldots \phi_i \wedge \phi_{(N-1)}$$

is unsatisfiable *if and only if* **for any variable assignment**, there is one clause ϕ_j with all the literals true and another clause ϕ_k with all the literals false.

Illustration: Consider

$$\Phi = (x + y + y) \cdot (x + !y + !y) \cdot (!x + y + y) \cdot (!x + !y + !y)$$

For any assignment (say $x = 0, y = 1$), we have one clause whose literals are all true, and one clause whose literals are all false (these are respectively $(!x + y + y)$ and $(x + !y + !y)$). If we leave out any one clause, that is not the case, as Φ becomes satisfiable. Now finish the proof.

Figure 16.10: CryptoMiniSat, `https://msoos.github.io/cryptominisat_web/`, with DIMACS instance explained at the top, and a screenshot at the bottom.

```
c CryptoMiniSat demo, NOT the full solver

c This is uum8.smt2-stp212.cnf from SAT Competition 2016

c It should solve in about 30 seconds

c You can edit ...

p cnf 1006 3359        <-- problem CNF with 1006 vars, 3359 clauses

2 -3 -159 -214 -374 0 <-- Clause (v2 + !v3 + !v159 + !v214 + !v374)

-2 374 0               <-- Clause (!v2 + v374)

...3354 CLAUSES OMITTED...

-681 -942 -950 0       <-- Each clause ends with 0

1006 0                 <-- All clauses are implicitly conjoined

2 0                    <-- This is the 3359th clause (last one)

UNSATISFIABLE <<-- Make last line "-2 0" to get SAT instantly!
```

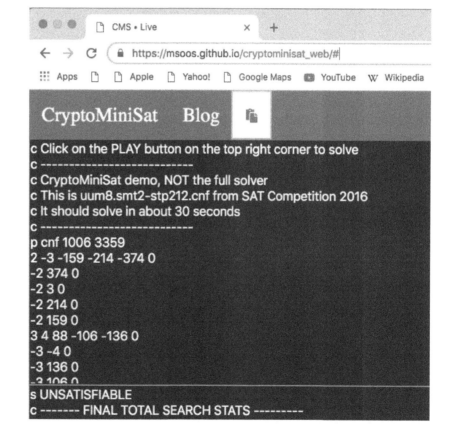

17

Binary Decision Diagrams as Minimal DFA

Chapter Gist: *We motivate the importance of efficient representation and manipulation of Boolean functions (§17.1). The Binary Decision Diagram (BDD) data structure can be viewed as an optimized representation of minimal DFA for Boolean function on-sets (§17.2). For this idea to pay off in practice, the Boolean variables must be ordered as per their semantic correlation (§17.3). We provide an intuitive overview of what BDD algorithms end up doing: as if the full exponential tree is built and common subexpressions shared (§17.4). We close off with a discussion of BDD sizes including connections with NP-Completeness (§17.5).*

17.1 Boolean Functions in Computing Theory and Practice

The representation and manipulation of Boolean functions is central to computing theory and practice.

x_1	x_2	0	$x_1 \wedge x_2$	$x_1 \wedge \neg x_2$	x_1	$\neg x_1 \wedge x_2$	x_2	$x_1 \oplus x_2$	$x_1 \vee x_2$	$\neg(x_1 \vee x_2)$	$x_1 \equiv x_2$	$\neg x_2$	$x_2 \to x_1$	$\neg x_1$	$x_1 \to x_2$	$\neg(x_1 \wedge x_2)$	1
0	0	0	0	0	0	0	0	0	0	1	1	1	1	1	1	1	1
0	1	0	0	0	0	1	1	1	1	0	0	0	0	1	1	1	1
1	0	0	0	1	1	0	0	1	1	0	0	1	1	0	0	1	1
1	1	0	1	0	1	0	1	0	1	0	1	0	1	0	1	0	1

Figure 17.1: All 2-input truth-tables

For pedagogical purposes, Boolean functions are commonly represented using truth-tables. Figure 17.1 presents a gallery of *all* possible 2-input truth-tables. We fix the listing order of inputs x_1 and x_0 to be the standard binary counting order, namely 00, 01, 10, and 11. Each 2-input Boolean function is then characterized by how we fill the output column—we call this the *personality* of the Boolean function. For example, the personality listed under $x_1 \land x_2$ is 0001 while for $x_1 \oplus x_2$, it is 0110. Given that there are 2^{2^2} (16) positions to fill (all combinations from 0000 to 1111), there are as many 2-input Boolean functions.

Unfortunately, representing N-input Boolean functions (Figure 17.2) using truth tables is impractical for larger N. For $N = 64$, we would have a truth-table of 2^{64} rows. Even to print that truth-table, we would need about 46 billion tons of paper (assuming 80 truth-table rows are printed per sheet of paper, with each sheet weighing five grams). The hardware industry needs to routinely process Boolean functions with many more inputs than 64. This is to support *formal verification* to detect bugs in the design of critical components such as microprocessors. The first efficient data structure that came to the rescue of the industry is the Binary Decision Diagram ("BDD" for short).

BDDs were introduced in [11] by Randal Bryant. Knuth, one of the giants of computer science, calls BDDs *one of the only really fundamental data structures that came out in the last twenty-five years*. They were instrumental in many of the advances in hardware verification up until (roughly) the year 2000, since when Boolean satisfiability methods have taken the front seat (with BDDs still continuing to play an important role)[1].

As we will see shortly, BDDs are polynomially sized for many of the Boolean functions that arise in practice. There are 2^{2^N} distinct Boolean functions (listing all personalities of length 2^N). This is an astronomical number: there are 256 3-input functions, 65,536 4-input functions, over 4 billion 5-input functions, and over 18 quintillion (billion billion) 6-input functions.[2] Out of these humongous numbers of Boolean functions, those that arise in practice are a miniscule fraction, and out of these, many of them tend to have *polynomially sized BDDs*. More discussions on this topic appear in §17.5.

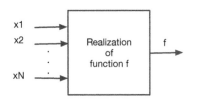

Figure 17.2: An N-input Boolean function

[1] Bryant's paper is one of the most cited of papers in Computer Science.

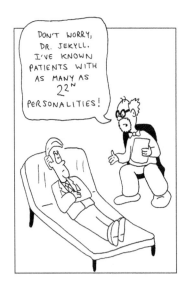

Figure 17.3: Each N-input gate is captured by its personality. Cartoon by Geoff Draper.

[2] For comparison, a human lives about 3 billion seconds.

17.2   Boolean Functions as Minimal DFA of Their On-Sets

We are studying BDDs in our book because BDDs are nothing but a small variant of minimal DFAs. In particular, BDDs are graph structures that summarize a Boolean function's on-sets (sets of inputs for which the function is true). The on-set of the And function is {11} while that for the Or function is {01, 10, 11}. The on-set of a Boolean function can be treated as a formal language. This language is {01, 10, 11} for an Or-gate and {11} for an And-gate. We will now build minimal DFA for these sets.

Minimal DFA and BDD for Xor: The minimal DFA and BDD for the Xor function are now obtained.

```
L_XOR = "(01+10)" # The regexp for the on-set of the XOR function
dotObj_dfa(min_dfa(nfa2dfa(re2nfa(L_XOR))), STATENAME_MAXSIZE=4)
```

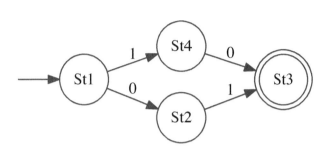

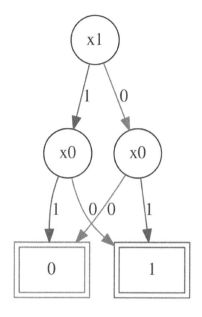

Figure 17.4: BDD for Xor

We can see that the minimal DFA[3] for Xor accepts 01 and 10. The BDD for Xor can be obtained using an online tool called PBDD[4] that can be invoked as follows (it will open the BDD tool in a new browser tab):

```
import webbrowser

# This is the URL for our PBDD tool that can be opened on a new tab
url = 'http://formal.cs.utah.edu:8080/pbl/BDD.php'
webbrowser.open(url)
```

Type in the following commands and click "build BDD" to obtain the BDD of Figure 17.4:

```
Var_Order : x1 x0
Main_Exp : x1 XOR x0
```

Basically, `Main_Exp` provides the Boolean function under study, and `Var_Order` specifies that the on-set of this function must be built for the language of two-bit words where x1 comes before x0, *i.e.*, the "x1,x0" words are in the on-set language. It can be seen that the BDD is very similar to the minimal DFA if one focuses on all paths that lead to the "1" node. The BDD also shows that the decoding goes on as per `Var_Order`: x1 is decoded first, and then x0, as shown by the edge labels. It is apparent that the "0" leaf node corresponds to the black-hole state.[5] As one

[3] In most of the DFA presented, we do not show the moves to black-hole states.

[4] Built by Dr. Tyler Sorensen when he was a BS/MS student working with this author.

[5] A state from which we can't get out, as discussed in §4.3.

example, the path 01 ends up in the 1 node (the function output is 1) while the path 00 ends up in the 0 node (the function output is 0).

Minimal DFA and BDD for Or: The minimal DFA and BDD for the Or function are now obtained.

```
L_OR  = "(01+10+11)" # Regexp for the on-set of the OR function
dotObj_dfa(min_dfa(nfa2dfa(re2nfa(L_OR))), STATENAME_MAXSIZE=4)
```

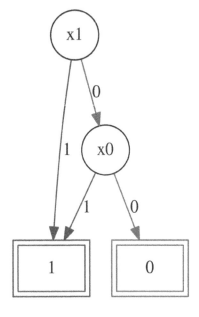

Figure 17.5: BDD for Or

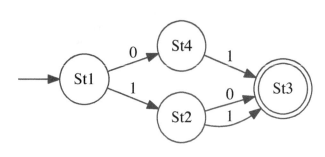

By entering these commands and clicking "build BDD," one obtains the Or BDD of Figure 17.5.

```
Var_Order : x1 x0
Main_Exp : x1 | x0
```

Again it can be seen that the BDD is very similar to the minimal DFA if one focuses on all paths that lead to the "1" node. *But there is one crucial difference*: When x1 is 1, the BDD directly jumps to the "1" node. The minimal DFA on the other hand goes to state St2, but in that state, **regardless of whether a 0 or a 1 comes**, the DFA jumps to the final state St3. It is clear that *we can simply drop all such parallel transitions* when interpreting the DFA moves as satisfying the function. In other words, the second bit is a "don't-care" and leads to state St3 no matter what. That is, x1=1 ensures that the function output is a 1, ignoring x0.

Thus far, the advantage of minimal DFA (or BDD) over truth-tables hasn't been quite apparent. We now proceed to demonstrate that for larger functions, with the right decoding order of the variables, BDDs (and minimal DFA) can indeed be far more compact. On the other hand, truth-tables are guaranteed to be exponential for any N-input Boolean function.

17.3   The Importance of Variable Ordering

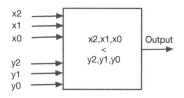

Figure 17.6: The < comparator.

Let us do some more experiments, this time taking a more practical (and non-trivial) function: that of a magnitude comparator for < (Figure 17.6). Suppose there is a 6-input Boolean function modeling a magnitude comparator that compares the binary value coming in through input ports x2,x1,x0 against the binary value coming through ports y2,y1,y0. The function is "<" where "A < B" means the usual "less than" (<) comparison. More specifically, we write "x2,x1,x0 < y2,y1,y0" and we interpret the word x2,x1,x0 using the standard positional binary notation (likewise also for y2,y1,y0). Here are some examples:

- 000 < 001: 0 is < 1 (0 is encoded in binary as 000 and 1 as 001)
- 010 < 110: 2 is < 6 (2 is encoded as 010 while 6 is encoded as 110)
- 110 < 111: 6 < 7

Let us now define a language of strings of length 6 representing the values of x2,x1,x0,y2,y1,y0 written adjacently, such that for those x,y values, the function outputs a 1. Call this language L. For instance, L contains 010101 because 010 is < 101 (i.e. 2 < 5). The reader may verify that the full L language written out as a regular expression (called R below) has 28 strings (out of the $2^6 = 64$ possible length-6 strings):

```
R =  "(000001+000011+000111+001011+001111+010011+010111+011111+\
100101+100111+101111+110111+000010+000101+000110+001010+001101+\
001110+010101+010110+011101+011110+100110+101110+000100+001100+\
010100+011100)"
```

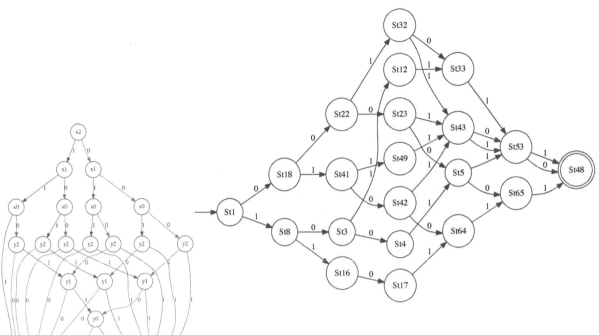

Figure 17.7: BDD for the magnitude comparator: bad input-variable order

The minimal DFA for R using the method shown earlier is above, and the BDD for it is in Figure 17.7 (for Var_Order being x2 x1 x0 y2 y1 y0). This DFA is in fact exponential in the x2,x1,x0 bits (those are the first three bits to arrive at this machine, and the machine grows exponentially with respect to these inputs). It must represent every x2,x1,x0 combination because the DFA does "not yet know" which y-bits are going to arrive. It then collapses as soon as the y bits come in.

The similarity of the DFA and BDD paths is apparent once again, with the BDD directly jumping to the "0" or "1" node when a decision is made (bypassing the don't-care decodings that minimal DFA end up passing through). However the ability to directly jump (bypassing the don't-cares) still cannot save the BDD from being exponentially big.

17.3.1 Finding a better input variable order

The main purpose of BDDs is to try and improve over truth-tables, and for that to happen the correct variable order must be presented so that "the Boolean function can decide as quickly as possible." This suggests that we pick the variable order to be x2,y2,x1,y1,x0,y0. This order makes sense because (for example) as soon as we know that x_2 is 1 and y_2 is 0, a decision can be made: < must be false. Likewise, if x_2 is 0 and y_2 is 1, again the decision is made: < must be true. Only otherwise (when $x_2 = y_2$) is it necessary to descend into the remaining bits. **This manner of keeping semantically correlated variables proximally in the BDD** ("quick decoding ability") indeed shows up as a reduced minimal DFA size (and also a reduced minimal BDD size) as we shall now demonstrate.

Let us call the regular expression obtained by interleaving the input bits "Rmix":

```
Rmix="(000001+000111+001101+011111+110001+110111+111101+000101+\
000110+010111+011011+011101+011110+110101+110110+000100+010011+\
010101+010110+011001+011010+011100+110100+010001+010010+010100+\
011000+010000)"
```

The minimal DFA for Rmix is below, and the BDD for it is in Figure 17.8 (for Var_Order being x2 y2 x1 y1 x0 y0). Again, it is easy to see that the BDD does not "trudge through" the redundant decodings. For instance, in the DFA, state St8 is reached when $x_2 = 0$, and then when $y_2 = 1$ is seen, a pathway of redundant decodings leading to the accept state is entered. Correspondingly, in the BDD, after seeing $x_2 = 0$, we reach a node which decodes y_2, and if $y_2 = 1$, the BDD jumps to the "1" leaf node.

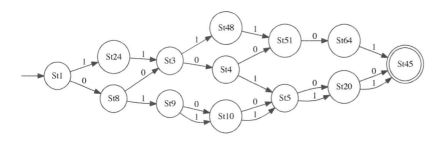

17.3.2   Functions with linearly sized BDDs

As an example of a BDD that is in fact linearly sized, see Figure 17.9 showing the BDD for a 5-input Xor function. Given that the binary Xor function is commutative and associative, we can obtain the 5-input Xor by connecting the input variables *in any order* and *without the use of parentheses*, as shown in the following PBDD commands (Var_Order does not matter, so an arbitrary one can be specified):

```
Var_Order : a b c d e
Main_Exp : a XOR b XOR c XOR d XOR e
```

Notice that Xor's parity-checking behavior is quite apparent from its BDD. *All paths from the root* described by an *even* number of 1's ends up in the 0 leaf node, while paths described by an *odd* number of 1's ends up in the 1 leaf node. (Think of a train starting from the BDD's root node, wanting to head to one of two "leaf stations 0 and 1"; the train gets

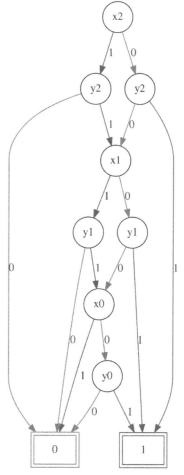

Figure 17.8: BDD for the magnitude comparator: good input-variable order

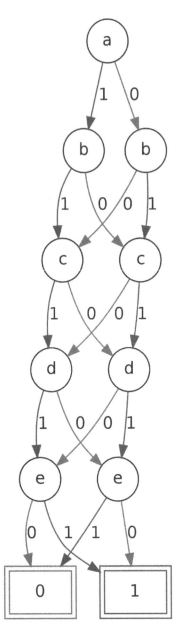

Figure 17.9: BDD for a 5-input Xor gate

shunted between two tracks based on the bits that arrive.)

How do BDDs avoid/hide the exponentiality of truth-tables? From Figure 17.8, we can see that there are cases where a BDD can skip over several levels of decoding. After seeing $y_2 = 1$, the BDD must consider x_1, while after seeing $y_2 = 0$, it can directly jump over to the 0 node. This is one way in which the exponentiality is avoided. However, the BDD for a 5-input Xor (Figure 17.9) does not skip levels. Here, we exploit the fact that in a directed graph with node sharings, even though there are a polynomial number of nodes, *there are an exponential number of paths* with each path spelling out a truth-table row. Even though the number of paths is exponential, due to the node sharings, we "forget" which path is taken to reach particular states. This helps us to represent BDDs such as an N-input Xor with linear size.

17.4   *From Minimal DFA to BDD: Intuitive Presentation*

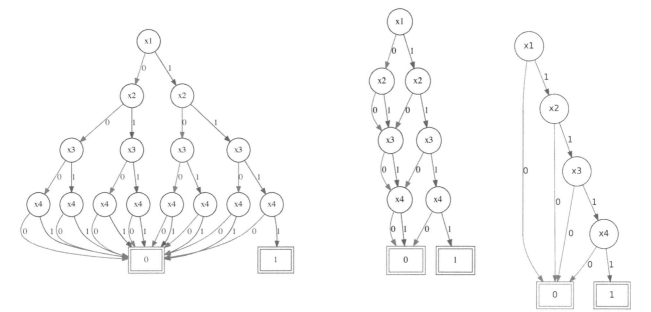

Figure 17.10: BDDs as Optimized "Decision Trees"

The uniqueness of minimal DFA for a given Boolean function (the Myhill-Nerode theorem, §6.4) does in fact apply to BDDs as well. Figure 17.10 is an attempt to portray this connection conveniently. Notice that the minimal DFA discussed so far were determined by regular expressions (two examples being "R" on Page 271 and "Rmix" on Page 273). These regular expressions are, in turn, completely determined by the sequence of inputs that make up the on-sets.

```
Var_Order : x1 x2 x3 x4
Main_Exp : x1 & x2 & x3 & x4
```

Notice the leftmost graph (almost a full "decision tree" sans the leaf-level nodes) in Figure 17.10. This is the minimal DFA for the four-input And's on-set (if one treats the 1 node as the accepting node and the 0 node as the black-hole state).

The BDD construction algorithm **does not build this graph**, as doing so would make the graph exponential. It instead builds the graph incrementally, bottom up, sharing common subexpressions, and eliminating redundant decodings (see Bryant's paper for details). For the sake of clarity, we explain these steps now as if we are building the whole graph:

• Notice that all the x_4 nodes *except for* the rightmost one have their left and right child point to the red double-rectangle 0 node.[6] So we first collapse all these x_4 nodes into a single x_4 node whose left and right children go to this 0 node. We leave the last x_4 node alone as a separate entity.

- We repeat the sharing of nodes, now pushing together the three left-most x_3 nodes into a single node. Again we have to leave the last x_3 node as a separate entity.

- In the same vein, we cannot combine both x_2 nodes, so we leave them as separate entities. This obtains the diagram in the middle of Figure 17.10.

- We now observe that this diagram has *redundant decodings*. More specifically,
  - The leftmost x_4 node has both children being 0. Thus, whether x_4 is 0 or 1, the outcome will be 0. This is because node x_4 lies under the case where x_1 is 0.
  - Thus,
    * We can simply *eliminate* the node x_4, making the left and right children of the leftmost x_3 node point directly to 0.
    * We can also make the left (0) child of the *right-hand side x_3 node* point directly to 0.
  - Now we create a situation where the leftmost x_3 node's left and right children point to 0. This causes the leftmost x_2 node to point directly to 0 (via its 0 and 1 children) and also the left child of the right-hand side x_2 node can also point to 0.
  - Proceeding in this manner, we obtain the rightmost diagram of Figure 17.10.

The above steps do not destroy the canonicity of BDDs. Thus, much like minimal DFA, BDDs are also unique (for a given function and variable order).

Myhill-Nerode Theorem in BDD Construction: For a given Boolean function, the generated BDD graphs are isomorphic.[7] This permits fast equality checking between two different Boolean functions. This is based on the Myhill-Nerode theorem already discussed in §6.4.

Bryant proposed a hash-table based representation for BDDs in such a way that isomorphic BDDs map into the same hash-table slot. This way, one can perform Boolean function comparisons in constant time. Also Bryant introduced the *Apply* operation that takes two BDDs and combines them using a Boolean operator. This operator is polynomial with respect to the sizes of the constituent BDDs. Thus if we can build polynomially sized BDDs, Boolean reasoning using BDDs can be done in polynomial time.

17.5 On BDD Sizes

For many commonly occurring Boolean functions, the BDDs involved are polynomially sized, and for these problems, Boolean reasoning becomes polynomial-time. Heuristics help choose variable orders that often ensure polynomially sized BDDs. In Figure 17.8, we chose the heuristic of clustering "closely related variables" in the variable order. Variants of this heuristic are employed in practice.

BDDs exhibit another curious fact: their size tends to blow up during BDD manipulations. Measures such as *dynamic reordering* of the variables are often able to minimize many of these bloated BDDs. Such *sifting* algorithms have been well studied in the literature.

From Chapter 16 we know that Boolean satisfiability is NP-Complete. Thus, there shouldn't be a way to get away with satisfiability checking with a lower cost even by using BDDs. This is indeed clinched by the result that discovering a good variable ordering for BDDs is NP-Complete [10].

Exercise 17.5, BDDs

1. Using PBDD produce a BDD for a 4-input Nor function over variables x_1 through x_4. How does it compare with the BDD for the 4-input And function?

2. Using PBDD, produce a BDD for a five-input Xnor circuit (similar to the 5-input Xor on the right-hand side of Figure 17.9). What are the salient differences between these BDDs?

3. Draw a decision tree for a four-input Or gate with inputs x_1, x_2, x_3 and x_4. Then apply the steps suggested in Figure 17.10. Do you obtain a linear-sized BDD at the end of the process? Check your answer by typing in the expression for a four-input Or and using PBDD.

4. Someone encodes the following PBDD file to model the situation:

```
# Implement A < B
# i.e. a2,a1,a0 < b2,b1,b0 where a2/b2 are the MSBs


Var_Order : a2, b2, a1, b1, a0, b0


Main_Exp  : ~a2 & b2 | ~a1 & b1 | ~a0 & b0
```

(a) Is the above encoding correct? Argue by generating and studying the BDD.

(b) If there is a flaw in the encoding of Main_Exp, fix the flaw, regenerate the BDD, and argue that it now stands corrected.

18

Computability Using Lambdas

> **Chapter Gist:** *We begin with a historical perspective on the Lambda calculus. (§18.1). Despite its origin as a theoretical vehicle, Lambdas are now found in everyday programming languages including C++ (§18.2). We present the three Lambda reduction rules (§18.3). Then we show how to encode numbers (§18.4) and Booleans (§18.5) in the Lambda calculus. Lambdas are ways to define anonymous functions; doing this for recursive functions is introduced (§18.6). A crucially important combinator called Y—a fixpoint combinator—is introduced (§18.7). We illustrate the use of Y in defining everyday recursive functions (§18.8), and close with a general discussion of combinators (§18.9) including just two combinators, namely S and K, that prove to be universal.*

18.1 The History of Lambda Calculus

The early part of the 20th century witnessed intense activity amongst mathematicians and logicians engaged in a quest for universal computing mechanisms. Such a universal mechanism, namely *Lambda Calculus*, was proposed by Alonzo Church who presented it to the American Mathematical Society in 1935 through his paper *An Unsolvable Problem of Elementary Number Theory*. This paper showed that the lambda notation is universal [22]. Given that Turing machines are also universal (Chapter 13), it is natural to wonder which result came first, and what the connections (if any) between these formulations are. Even though Turing's paper was written later (only around 1936), we now know that these ideas originated contemporaneously. The historical account of Turing's visit to Princeton [19] as well as the following quote from M.H.A. Newman (Turing's British mentor), support the independence of these discoveries: *I should mention that Turing's work is entirely independent; he has been working without any supervision or criticism from anyone.* Turing's PhD dissertation was, in fact, finished at Princeton, with Church serving as his thesis advisor—another glimpse of these giants acknowledging each other's work.

In the modern setting, the lambda calculus often plays much more of a *practical* role than Turing machines. Lambdas are central to functional programming languages, and also have made a much-needed entry into widely used languages such as Java and C++. Here is a C++ example employing a lambda:

```
// Find the accountant whose salary meets a test
std::find_if(
  emps.begin(), emps.end(),
  [=](const accountant& a)
    { return a.salary() >= min_wage && a.salary() < limit; }
);
```

This lambda function is designed to search the `emps` array for some accountant a whose salary meets the predicate test.

All this, however, does not mean that Turing machines are "inferior" in any way. Turing machines are, in fact, much easier to understand, program, and puzzle over.[1] They are also mathematically rigorous, and play a foundational role in our study of computability as well as space and time complexity.[2]

In this chapter, we will provide a taste of the lambda notation in its role as a universal computing formalism. We will employ a small subset of Python to help the reader easily follow along.[3]

18.2   Lambdas from a Programmer's Perspective

The lambda notation is a way to describe functions without redundantly attaching names to them. Consider someone naming the successor function `Fred` and describing it thus in C:

```
int Fred(x) { return x+1; }
```

The main criticism we would get is that we did not choose a meaningful name. If we named the function `succ` and defined it as follows, then this complaint might vanish:

```
int succ(x) { return x+1; }
```

However, there is an even more parsimonious way to describe this function in mathematics: just drop the name `succ`: it does not determine the behavior of the function, and mathematicians do not like to have *any* excess baggage in their notation:

```
function (x) { x+1; }
```

The lambda notation is a realization of such an idea: drop irrelevant names and merely describe what the function takes in as well as returns, using only a minimal number of characters and punctuation symbols. In lambda calculus, the notation adopted is, therefore:

$\lambda x.\, x + 1$

Unfortunately, this approach runs into trouble when we are presented with a recursively defined function such as this:

```
// Sum an arithmetic series. This function could carry a more
// serious name such as sumSeries or some such. It may not matter
// if we are going to get rid of the name of the function, as
// will be demonstrated shortly.
int Fred(x)
{ if (x <= 0) return 0;
  else return x + Fred(x-1); }
```

We might call this function anything at all—but the point is that without the "crutch" (a function name, say Fred in this case, or even foo if you prefer that), it appears impossible to refer back to the function. The recursive program shorn of the function name will look rather malformed:

```
function (x)
{ if (x <= 0) return 0;
  else x + ???(x-1); }
```

where the ??? indicates our inability to refer back to the function that is now devoid of its name. One can devise something like "self" for referring back to the *same* function. But then, the trouble only compounds: how does one handle mutually recursive functions where, for instance, function f might be defined in terms of function g, function g in terms of h, and function h in terms of f? If we strip all these names f, g, h from the function headers and bodies, it would become rather confusing as to which function is calling which other. While we can have many kludgy ways of handling this situation,[4] *there is actually a rather ingenious way to handle such situations without extending the lambda calculus a bit!* In other words, we can define *any* recursive program with no function names attached to function definitions, and still be able to employ arbitrary recursive structures.

This chapter is all about this idea of an **anonymous function**. We proceed layer by layer by first defining the syntax and semantics of the lambda calculus. We follow this with a discussion of how lambda reduction works, and then present how lambdas are a sufficient syntax to de-

[4] Say, putting f, g, h on say lines 27, 38, and 43, and referring to those line numbers...

scribe "anything at all." Our illustrations will not amount to a mathematical proof; however, they will be pretty convincing (we hope) as to the universality of the lambda calculus.

18.3 Syntax and Semantics of the Lambda Calculus

Computing using lambdas consists of *reducing* (or transforming) given lambda expressions into other lambda expressions.[5] To define lambda reduction properly, we now introduce the syntax of lambda expressions, as well as its three basic reduction rules.

[5] Such transforms of lambda expressions do not always result in simpler expressions. The expression can in fact grow in size. We will nevertheless use the terms reduction and transform interchangeably.

E	$\rightarrow$	x	A variable
	|	$\lambda x.E$	An abstraction
	|	$(E_1 E_2)$	An application
	|	(E)	Parenthesization

Figure 18.1: Lambda Expression Syntax

Name	Rule-Application	Comments
Alpha	$\lambda x.E[x] \rightarrow \lambda y.E[y]$	Renaming (assuming no *name clash*)
Beta	$(\lambda x.E)E_1 \rightarrow E[E_1/x]$	Call (assuming no *capture*)
Eta	$\lambda x.(Ex) \rightarrow E$	Extensionality (assuming x is not free in E)

Figure 18.2: Lambda Reductions

Syntax: The syntax of lambda expressions is given in Figure 18.1. This figure shows that a variable occurring by itself is a lambda expression. A lambda abstraction models a function, and is a lambda expression (the x in $\lambda x.E$ is the *binding occurrence* of x, with respect to uses of x within E). An application (where we apply a function to its arguments) is a juxtaposition of two lambda expressions, and is also a lambda expression. The final rule allows parentheses to be added in order to enhance readability.

Figure 18.2 presents the three basic reduction rules of the lambda calculus that go by the names of *alpha*, *beta* and *eta*, as now elaborated.

Alpha Reduction: This rule captures the fact that the formal parameters of a function only have meaning within the function body. Thus, a formal parameter x as in this illustration can be freely replaced with y. The only caveat is that we must not *clash* with another y that is already present within the body of the lambda expression, namely E. In most typical uses of alpha reduction, we will choose y to be a "fresh" variable name, thus avoiding name clashes.

Examples and non-examples:

- OK: $\lambda x.x \xrightarrow{\alpha} \lambda y.y$

- OK: $\lambda x.fx \xrightarrow{\alpha} \lambda y.fy$

- Not OK: $\lambda x.\lambda y.(+ \; x \; y) \xrightarrow{\alpha} \lambda y.\lambda y.(+ \; y \; y)$ (the y we rename x to "clashes with" (becomes captured by) the y in the inner abstraction).

Beta Reduction: The beta reduction rule is the workhorse of lambda calculus, and models function calls. Lambda expressions are (typically) applied to other lambda expressions, and transformed through the *beta rule* as follows:

$$(\lambda x.(x + 1)) \; 3 \xrightarrow{\beta} (x + 1)[3/x] = (3 + 1) = 4$$

where $(x + 1)[3/x]$ stands for *substitute 3 in place of x in the expression* $(x + 1)$. Such a substitution yields $(3 + 1)$, or 4 as the answer. Again, we ensure that we do not have free occurrences of x within E_1 (if there are, we can rename before applying the beta reduction step).

Examples and non-examples:

- OK: $(\lambda x.f x) \, E \xrightarrow{\beta} (f \; x)[E/x] = (f \; E)$

- Not OK: $(\lambda x.(\lambda y.(+ \; x \; y))) \, (\textbf{pred} \; y) \xrightarrow{\beta} (\lambda y.(+ \; x \; y)) \, [(\textbf{pred} \; y)/x] = (\lambda y.(+ \, (\textbf{pred} \; y) \, y))$. This is because the y coming in via $(\textbf{pred} \; y)$ is presumably bound in an outer scope, and should not be captured by the y in the inner scope.

  Solution: Before the beta reduction is performed, rename $(\textbf{pred} \; y)$ to $(\textbf{pred} \; z)$ (and also rename the binding occurrence of y in $(\textbf{pred} \; y)$ to z).

```
I    = lambda c: c                              # Identity
ZERO = lambda b: lambda c: c                    # Number 0
S    = lambda a: lambda b: lambda c: b(a(b)(c)) # Successor

Note that S(ZERO) reduces to lambda b: lambda c: b(c):


S(ZERO)

--> (lambda a: lambda b: lambda c: b(a(b)(c))) (ZERO) [ Definition of ''S''    ]
--> lambda b: lambda c: b(ZERO(b)(c))                 [ Beta reduction: ZERO/a  ]
--> lambda b: lambda c: b((lambda b: I)(b)(c))        [ Definition of ZERO      ]

--> lambda b: lambda c: b(I(c))          [ Eta reduction: (lambda b: I)b = I ]
--> lambda b: lambda c: b((lambda c:c)(c)) [ Definition of I                 ]
--> lambda b: lambda c: b(c)             [ Eta reduction: (lambda c: c)c = c ]


Similarly, S(S(ZERO)) reduces to lambda b: lambda c: b(b(c)):
```

Figure 18.3: Lambda as a universal notation: Church Numerals in Python.

Eta Reduction: The eta reduction rule takes advantage of the fact that the lambda expression $(\lambda x.(E\ x))\ V$ can be replaced by $(E\ V)$ via the beta rule, *provided x is not free in E*. A noteworthy aspect of the eta rule is that it allows this simplification for any lambda expression V.

18.4   Illustration: Church Numerals in Python

We will now show how to represent and manipulate numbers just using the lambda syntax. We assume that there is a number 0, and that every natural number n is described as the nth successor of 0.[6] Thus, 1 is the successor of 0 and 2 is the successor of the successor of 0—and so on. If we employ S to denote *successor* and ZERO to define 0, we can write any natural number as ZERO, S(ZERO), S(S(ZERO)), and in general, number n as $S^n(0)$ (n applications of the S). For a detailed illustration of these ideas, please study the definitions in Figure 18.3 and the definition of two helper functions, namely ChurchToNat and NatToChurch, in Figure 18.4. These examples explain how ChurchToNat(S(ZERO)) manages to print 1.[7]

This figure also introduces suitable definitions for addition (ADD) and multiplication (MUL). Take the addition of a and b: one can view the process as obtaining the ath successor of b.[8] These ideas are captured in the definitions in this figure. The reader may test these ideas out using ready-made Jove notes that are provided.

Exercise 18.4, Church Numerals

1. Study the definition of ADD (the addition operator) and MULT (the multiplication operator) of Figure 18.4 and explain how they seem to be designed. Take the approach in Figure 18.3 to unravel their working.

2. Demonstrate the ability to multiply 4 and 8, using NatToChurch to generate the required Church numerals, and ChurchToNat to print the answer.

18.5   Illustration: Booleans, Pairs, Other Functions

We now present Church encodings for Booleans, pairs, and other primitive operators in Figure 18.5. The best way to understand the interplay between these definitions is to do a few exercises that are included in our Jove notebooks. Clearly, considerable ingenuity has been shown by logicians of yesteryears, notably Alonzo Church himself.

Exercise 18.5, Booleans in Lambda

1. Take the Python pair (*True,False*) and convert it to a Church pair with *True* and *False* turned into their Church-coded forms. That is, translate (*True,False*) into PAIR(TRUE,FALSE) using Church's encodings. You may use the function BooleanToLambda

[6] This encoding stems from an area called *Peano arithmetic*.

[7] The identity function I takes an x and returns x. We represent number 0 as a lambda that takes one argument and returns the identity function I. The reason for this encoding will become clear when we perform a few example reductions in Figure 18.4. The definition of ADD and MULT are left for your exercises (with hints).

[8] or in fact, also as the bth successor of a.

```
ChurchToNat(S(ZERO))
--> S(ZERO)(increment)(0)                          [ Definition of churchToNat ]
--> (lambda b: lambda c: b(c)) (increment)(0)      [ Definition of S and ZERO ]
--> (lambda c: increment(c)) (0)                   [ Beta reduction: increment/b ]
--> (increment(0))                                 [ Beta reduction: c/0 ]
--> 1

S(S(ZERO))
--> ..
--> lambda b: lambda c: b(b(c))


def increment(n):
    """Successor of a number."""
    return n+1


def ChurchToNat(c):
    """Convert Church numeral to number."""
    return c(increment)(0)


def NatToChurch(n):
    """Returns the Church numeral equivalent of a number."""
    if n == 0:
        return ZERO
    else:
        return S(NatToChurch(n-1))


>>> ChurchToNat(S(S(ZERO)))
2
>>> ChurchToNat(ADD (S(S(ZERO))) (S(S(S(ZERO)))))
5


ADD  = lambda a: lambda b: a(S)(b)                 # Addition


MUL = lambda a: lambda b: lambda c: a((b)(c))      # Multiplication
```

Figure 18.4: How Church Numerals Work. To represent n, apply function b n times to argument c. To print the result using ChurchToNat, set b to increment and c to 0.

Figure 18.5: Booleans, Pairs, Other Functions Encoded in the Church lambda notation.

```
TRUE  = lambda a: lambda b: a # Boolean true
FALSE = lambda a: lambda b: b # Boolean false


NOT  = lambda a: a(FALSE)(TRUE)          # Boolean negation
AND  = lambda a: lambda b: a(b)(FALSE)   # Conjunction
Z    = lambda a: a(FALSE)(NOT)(FALSE)    # Equal to zero test
PAIR = lambda x: lambda y: lambda f: f(x)(y)   # Pair creator


FIRST  = lambda p: p(TRUE)           # Extracts first of pair
SECOND = lambda p: p(FALSE)          # Extracts second of pair


def LambdaToBoolean(b):
    """Returns the literal boolean equivalent of
       Church-coded boolean."""
    return b(True)(False)


def BooleanToLambda(b):
    """Returns the Church encoded boolean of a literal
       boolean."""
    if(b):
        return TRUE
    else:
        return FALSE
```

and the definition of PAIR to achieve these conversions. Now Apply AND, and turn the result back to a Python Boolean and print the same.

2. Repeat Exercise 1 except you are required to compute the OR of the given two Python Booleans, implementing OR in "Church-land" using DeMorgan's Law. Apply OR, and turn the result back to a Python Boolean and print the same (all using Jove).

18.6 Handling Recursion

Consider the recursively defined version of the standard factorial function which we call 'fact' below:

```
fact(n) = (1 if n==0 else n*fact(n-1))
```

We can rewrite the above definition into an equational form. The idea is to introduce the parameter n explicitly into the right-hand side expression.

```
fact = lambda n: (1 if n==0 else n*fact(n-1))
```

Thus, fact(n), with n set to say 5 can now be equivalently modeled as fact, with its right-hand side lambda's parameter n set to 5. Now we can read this definition of fact as if it were an equation. We can now seek a function that, when substituted for fact on either side, solves this equation.

We can now use the Beta rule of lambda calculus and write the above equation as follows (note that we are using the Beta rule "backwards"):

```
fact = (lambda F. lambda n:(1 if n==0 else n*F(n-1))) fact
```

We can immediately see that this equation is now of the form

```
fact = (G fact)
```

where G happens to be:
```
(lambda F. lambda n:  (1 if n==0 else n*F(n-1))).
```
Equations of this form are called *fixpoint equations* as now elaborated in §18.7.

18.7   Obtaining Fixpoints from Fixpoint Equations

Consider a fixpoint equation f = (G f). When such an equation is true, we say that function f is a fixpoint of G. In a sense, if we "input" f into function G, it outputs back function f. Many arithmetic functions (that input numbers and return numbers) enjoy this property; for instance, taking ! to mean *factorial*, we know that these equations are true:

- $1! = 1$

- $2! = 2$

- $cos(0.73908) = 0.73908$ (to the shown digits of accuracy)

In the lambda calculus setting, everything is a function—even "values" (numbers) are functions. A number 2018 is a zero-ary function: it takes nothing (zero arity) and yields itself (the number 2018).

We can now demonstrate that there is a standard way to extract the fixpoint of any function with the help of a single lambda-definable function known as the Y combinator.[9]

18.7.1   Y: A Fixpoint Finder

A fixpoint combinator is a combinator E such that for any lambda expression G, the identity $(EG) = G(EG)$ holds. There are an infinite number of fixpoint combinators. We will first introduce the most popular of these combinators, namely Y. We will then point out the reasons why Y cannot be directly realized in Python (for the sake of experimental edification), but then provide a variant that can indeed be experimented with in Python.

18.7.2   The Y Combinator

Define combinator Y to be

$$Y = (\lambda f.\,(\lambda x.\,f(x\,x))\,(\lambda x.\,f(x\,x)))$$

Let us show that Y is indeed a fixpoint combinator, by showing that for any lambda expression G, $(Y\,G)$ simplifies to $G(Y\,G)$ as follows:

Simplified form	Comments
$= (\lambda f\,(\lambda x.\,f(x\,x))\,(\lambda x.\,f(x\,x)))\,G$	
$= (\lambda x.\,G(x\,x))\,(\lambda x.\,G(x\,x))$	Beta reduction with f bound to G
$= G((\lambda x.\,G(x\,x))\,(\lambda x.\,G(x\,x)))$	Beta reduction with x bound to $(\lambda x.\,G(x\,x))$
$= G((Y\,G))$	Notice that $(Y\,G) = (\lambda x.\,G(x\,x))\,(\lambda x.\,G(x\,x))$

[9] We have been employing the term *the* fixpoint without batting an eyelid. While a general discussion is beyond the scope of these notes, we must say that in general, for a G function, there could be multiple fixpoints. Luckily for us, we are going to be dealing with only *total* functions – that is, recursive definitions where the function does not infinitely loop for any argument. For such functions, there is only one fixpoint. Furthermore, we will introduce two fixpoint finders, one being Y and the other being Ye—the latter meant for experimental studies in Python.

This derivation demonstrates that Y indeed fulfills its advertised role, namely that of obtaining the fixpoint of any function G.[10] In one sense, all we are saying is that the expression (YG) is the fixpoint of G. Any attempt to check this claim paints in front of your eyes the result

```
(Y G) = G(Y G)
```

as if to say "see I told you so!" Let us see how this "self-replicating trick" actually simulates recursion.

18.7.3 Expression Recursion using Y

Let us recall our old friend – the recursively defined function Fred:

```
int Fred(x) // Sum an arithmetic series
{ if (x <= 0) return 0;
  else return x + Fred(x-1); }
```

We now derive its "de-Freded form:"[11]

- Let us write the function as an equation:

  $$Fred = (\lambda f.\, \lambda x.\, if((x <= 0),\, 0,\, x + f(x-1)))\, Fred$$

- Notice that this is of the form

  $$Fred = G\, Fred,\, i.e.,\, Fred \textbf{ is a fixpoint of } G^{12}$$

  where G is given by

  $$G = (\lambda f.\, \lambda x.\, if((x <= 0),\, 0,\, x + f(x-1)))$$

- We can indeed "find" the fixpoint of G by "sending Y after G".

  Thus, we can write $Fred = (Y\, G)$, with G defined as above, or in other words

- $Fred = Y\, (\lambda f.\, \lambda x.\, if((x <= 0),\, 0,\, x + f(x-1)))$

- **Notice that there is no** $Fred$ **on the right-hand side!** That is, *we managed to successfully define what is a recursive function (namely* Fred*) without using the name* Fred *on the right-hand side.*[13]

- Thus, to call this function on say 2, which should result in $2 + 1 + 0 = 3$, we do this: $(Fred\, 2)$. We now illustrate how this "call" works.

[10] You may cry "Hogwash! It does not extract and lay out something beautiful and totally new for me to behold in front of my eyes." That is, it only seems to go through these equations and re-emerge as (Y G). This is fine, and in fact, these "rigmaroles" are what makes lambda-based functional programs run! The "recursion" you want is cleverly arranged for you by the Y combinator making a self-replicated copy of the function body whenever you want!

[11] By "de-Fredding," we mean "shaving off the word Fred" from everywhere in this expression.

[12] We do have unique least fixpoints, and Y is known to "find them." For details, please refer to Strachey's book *Denotational Semantics*. For this reason, hereafter we will use the term *'the fixpoint'* in our narrative.

[13] Mathematicians will now be found to be executing somersaults of joy, as you just attained notational parsimony by not using needless things like function names!

Simplified form	Comments
$= (Y\,G)2$	Put in the definition for $Fred$
$= (G(Y\,G))\,2$	Exploit the fact that $(Y\,G)$ is a fixpoint of G
$= (\lambda f.\lambda x.if((x \le 0),\,0,\,x + f(x-1)))\,(Y\,G)\,2$	Substitute in the definition of G
$= (\lambda x.\,if((x \le 0),\,0,\,x + (Y\,G)(x-1)))\,2$	Beta reduction with f bound to $(Y\,G)$
$= if((2 \le 0),\,0,\,2 + (Y\,G)(2-1))$	Beta reduction with x bound to 2
$= if(False,\,0,\,2 + (Y\,G)(1))$	Simplifications
$= 2 + (Y\,G)(1)$	More simplifications
$= 2 + 1 + (Y\,G)(0)$	Doing the same series of steps
$= 2 + 1 + 0$	Finally, the evaluation stops, as $0 \le 0$ is satisfied.

We observe that recursion is handled through "unraveling" using Y. In fact, $(Y\,G)$ becomes $G(Y\,G)$, and the outermost G "goes to work," keeping the inner $(Y\,G)$ "in reserve," in case the conditional is not satisfied.

This ability of the Y combinator to make replicas of a function's body is indeed the uncanny insight obtained through the above exercise. We don't need to "point at" function definitions using arbitrary names such as Fred. Thus in fact, we have arrived at a literal syntax for recursively defined functions such as Fred. In our derivation, we saw that Fred = $(Y\,G)$ and G is defined by the lambda expression given above. Thus, $(Y\,G)$ is the literal syntax ("De-Freded form") or irredundant name for the function we are interested in.[14]

[14] An exercise in §18.8 asks you to model mutual recursion by trying to simultaneously define a pair of functions in the same manner using Y. Thus, even for a system of mutually recursive definitions we can find fixpoints.

18.7.4   Reason for an alternate fixpoint finder Ye

The brief answer for needing an alternate fixpoint finder for experimental purposes using Python is simple:

> Python is "Eager".

To properly explain this term, we now spend time discussing eager versus lazy evaluation. When we apply a function f to an argument E, written $f(E)$, one could follow two approaches:

1. Evaluate E fully, obtaining some value (say v). Then apply f to v. This is the *eager evaluation* approach – alternately known as *call by value*.

2. Do not evaluate E one bit. Instead, assuming that the formal parameter of f is x, substitute E for x everywhere within the body of f. This is the non-eager evaluation approach, more properly termed *lazy evaluation* or *call by name*.[15]

Call by name has one uncanny property: in certain programs, it can avoid nontermination (infinite looping). To see this, consider a function g which takes three arguments, say x, y, and z. Let the definition of function $g(x, y, z)$ be such that it first evaluates x, and then if found true, it evaluates and returns y, ignoring z entirely. On the other hand, if x evaluates to false, g ignores y entirely and proceeds to evaluate z. Such a g function will exhibit a difference between call by value ("eager") and call by name ("lazy") when fed certain cleverly crafted arguments. In particular, if the call is $g(True, 0, InfLoop())$ (where $InfLoop()$ is a function call of no arguments that simply goes into an infinite loop), we can observe this behavior:

Lazy evaluation: Since the first argument of g is the constant $True$, we can return the answer 0 under lazy evaluation, ignoring the infinite loop entirely. Programming languages such as Haskell follow lazy evaluation by default.

Eager evaluation: The function call $g(True, 0, InfLoop())$ will be fixated on "grinding down" InfLoop() into a value v before it can proceed further with g's evaluation. Clearly, this results in the whole computation looping infinitely (which is unnecessary). Virtually all programming languages follow eager evaluation. This list includes C, C++, Java, etc.—and of course Python also!

The fixpoint finder for eager evaluation: Given a programming language that follows eager evaluation, all we need to do in our experiments with a fixpoint finder is to employ a slight variant of a fixpoint finder which we call Y_e, defined as follows:[16]

$$Y_e = (\lambda f. (\lambda x. (x\,x)) (\lambda y. f(\lambda v : ((y\,y)\,v))))$$

We will apply this fixpoint finder in our illustrations.

The notion of a "Pre" function: Functions similar to G appear everywhere in the study of recursion, and therefore it is handy to have a name for it. We will call it the *pre*-function, standing for "the prelude" to obtaining the fixpoint. The G emanating out of the recursively defined Fred

[15] Prof. Niklaus Wirth once said "In Europe, people call me Niklaus Wirth, but in the US they sometimes call me Nickel's Worth. That is because in Europe they call me by name while in the US, they call me by value."

[16] Exercise 1 in §18.8 will ask you to show that Y_e is a fixpoint finder, while Exercise 3 will ask you to find out why we need to change our fixpoint combinator to Y_e for use within Python, which is an eager language.

function can be called *Pre-Fred*, the G coming out of Fact can be called *Pre-Fact*, etc.

18.8 Illustrating the Use of Fixpoint Combinators

At long last, Figure 18.6 presents the encoding of fixpoint combinators in Python to define more familiar functions such as factorial and the nth Fibonacci number. To simplify our work, we are not encoding numbers using Church numerals, nor are we encoding Booleans using Church's encoding. While this is possible using ideas presented in §18.4 and §18.5, we avoid doing so for the sake of readability ("one new idea at a time").

Exercise 18.8, Y combinator

1. Show that Y_e is a fixpoint combinator. Also, see Exercise 3.

2. Use the Ye combinator and define fact and sum starting from prefact and presum. Demonstrate their operation on three examples (each).

3. Show why the use of the Y combinator used in §18.7.2 in place of the eager Y combinator (which we call Ye) will cause Python to loop. Show how Ye avoids this looping, by expanding out an instance of Ye application using eager (call by value) reduction rules. An astute reader might observe that the inner expression of the $\lambda v \ldots$ form has $\lambda v : ((yy)v)$ which, by the eta rule, reduces to yy. Thus, Y_e is "Y in disguise." Despite this, the extra abstraction in $\lambda v : ((yy)v)$ has a profound effect on protecting the "runaway evaluation" of $Y f$ under call by value.

4. Show how to encode mutual recursion using the Ye combinator. Use it to mutually recursively define two functions that are defined in terms of each other. For definiteness, you may work on the following example that defines fact through mutual recursion.[17]

```
fact1 = lambda n: (1 if n==0 else n*fact2(n-1))
fact2 = lambda n: (1 if n==0 else n*fact1(n-1))
```

   Hint: Take the two functions, pair them up, and define this pair through fixpoint finding.

5. Show that the set of all Y combinators is recursively enumerable. *Hint:* Any fixpoint combinator E is a lambda expression which, for any function variable G (variable G denoting a function) satisfies the identity $(EG) = G(EG)$. Take it from there.

18.9 Combinators

A lambda expression without free variables is called a *combinator*. Thus, $\lambda x.x$ is a combinator while $\lambda x.y$ is not. Combinators are "ready-to-apply

[17] We obtained this simple variant of fact simply to give you practice. Virtually no one would employ such mutual recursion in practice.

Figure 18.6: Use of the Ye Combinator (Call By Value version of the standard Y.)

```python
# For the ease of readability, we use don't use
# Church numerals below.


# Ye -- eager Y combinator
Ye = lambda f:(lambda x:x(x))(lambda y:f(lambda v:y(y)(v)))


# Pre-Factorial: performs the product of
# a natural number, and all natural numbers less than it.


# We call it pre-factorial because we need to apply
# Ye to it to obtain the real factorial


prefact = lambda fact: lambda n:(1 if n==0 else n*fact(n-1))


# Pre-sum: sums all the natural numbers less than the given
# number


presum = lambda f: lambda n: (0 if n==0 else n+f(n-1))


# Pre-Fib: returns the nth number of the series defined by
# the following definitions
#  * the first two numbers are 0 and 1
#  * the next number is defined as the sum of the prior two
#    numbers


prefib = lambda f: lambda n: 0 if n == 0
                    else (1 if n == 1 else f(n-1) + f(n-2))


>>> Ye(prefact)(5)
120


>>> Ye(presum)(10)
55


>>> Ye(prefib)(3)
2


>>> Ye(prefib)(6)
8
```

functions" once all the arguments are known.

Combinators enjoy an important formal status in the theory of computability. A system to model computable functions without the explicit use of lambdas (and lambda-bound variables) was proposed by Moses Schönfinkel in 1924. This work on *Combinatory Logic* predates even the lambda calculus, and was lost in the mists of time—till discovered by Haskell B. Curry in the 1930s.[18]

Combinatory logic is universal. In fact, two operators S and K of the combinatory logic are sufficient to express all computable functions.[19] These two combinators have very simple lambda definitions:

- $S = \lambda fgx.(fx)(gx)$. Notice that S is a combinator because all the variables in the lambda body (namely f, g, and x) are all present in its lambda abstraction's binding sites.
- $K = \lambda xy.x$. Notice that K is also a combinator.

This again serves to prove that the lambda calculus is universal (since we can represent all programs and data using S and K).

In some discussions, it is pointed out that the set of combinators $\{S, K, I\}$ are universal, where $I = \lambda x.x$ is the identity combinator. However, we do not need I separately; it can be realized as SKK, as the following derivation shows:

Simplified form	Comments
$= (\lambda f\, g\, x.\,(f\, x)(g\, x))\, K\, K$	Substitute definition for S
$= (\lambda x.\,(K\, x)(K\, x))$	Beta reduction, binding f and g to K
$= (\lambda x.\,((\lambda z\, w.\, z)\, x)\,(K\, x))$	Alpha-converted version for first K (second K is left alone)
$= (\lambda x.\,(\lambda w.\, x)(K\, x))$	Beta reduction by binding z to x
$= (\lambda x.\, x)$	Beta reduction, binding w to $(K\, x)$ (Note that $(K\, x)$ gets thrown away)
$= I$	The identity combinator is denoted by I

[18] It is sad that Schönfinkel himself did not receive any recognition when he was alive, and apparently died poor. After his death, his papers were burned by his neighbors to generate heat during winter.

[19] In contrast, it takes nature *four* nucleotides (namely G, A, C, and T) to encode life through DNA. (I am not trying to sound cocky as a computer scientist, nor imply that we are superior to nature.)

Appendices

A

A Recap of Discrete Math

Chapter Gist: *We review sets, including many standard infinite sets (§A.1). Basics of mathematical logic are reviewed (§A.2) followed by proof by contradiction and contrapositive (§A.3). Basic notions and notations pertaining to functions and relations (§A.4, §A.5) and trees (§A.6) then follow.*

A.1 Sets

Finite sets such as $S = \{0, 1, 2, 3\}$ are assumed to be familiar. The cardinality (or size) of this set, written $|\{0, 1, 2, 3\}|$, or $|S|$ is 4. The empty set is written $\emptyset$ or $\{\}$ and its cardinality is 0.

We use Nat to denote the set of counting numbers (nonnegative integers), *i.e.*, $Nat = \{0, 1, 2, \ldots\}$. We use Int to denote positive and negative whole numbers and zero, *i.e.*, $Int = \{0, 1, -1, 2, -2, 3, -3, \ldots\}$.[1] Notice that Nat and Int are infinite sets. For example, for every number $n \in Nat$, there is a number above n that is also in Nat. Yet, every number in Nat is finite. The *cardinality* of Nat (or Int) is itself not expressible as a member of Nat (or Int).[2] By contrast, the cardinality of S defined above is a member of Nat.

As another example of using these notations, let the set of even natural numbers, $Even = \{0, 2, 4, \ldots\}$. Observe that $|Even|$ is **not** a member of Nat. In Chapter C, we will prove that $|Even|$ is the same as $|Nat|$.

There is also the set $Real$ of real numbers. The cardinality of $Real$ is denoted by $\aleph_1$, which has strictly higher cardinality than $\aleph_0$. In other words, there are "many more real numbers than natural numbers." For a proof, see Chapter C.

A.1.1 Set Builder

We will in general employ the following notation for describing sets:

$$S = \{x \, : \, p(x)\}$$

[1] We aim for readability and choose Nat and Int over their more traditional $\mathbb{N}$ and $\mathbb{Z}$ respectively.

[2] These infinite cardinalities are denoted by $\aleph_0$.

This says that S is a set of x satisfying predicate $p(x)$. For example:

$$Even = \{n \,:\, n \in Nat \text{ and } n \text{ is even}\}.$$

This can also be defined as

$$Even = \{n \in Nat \,:\, n \bmod 2 = 0\}.$$

Another example of a set builder is

$$S = \{x \in \text{Nat} \,:\, x > 2 \wedge x < 8\}.$$

The '$\in Nat$' may be omitted if clear from the context. S represents the set $\{3, 4, 5, 6, 7\}$.

Observe that the set P defined below equals the set Nat itself;

$$P = \{x \in \text{Nat} \,:\, True\},$$

while the set Q defined below equals the empty set

$$Q = \{x \in \text{Nat} \,:\, False\}.$$

This is because all the elements of Nat pass the unconditional test of $True$ (for the case of defining P) while no element of Nat passes the unconditional test of $False$ (for the case of defining Q).

A.1.2 Powerset

The set of all subsets of a set S is its powerset. The notation $\mathscr{P}(S)$ denotes the powerset of set S.[3] The powerset of the set $\{1, 2\}$ is $\{\emptyset, \{1\}, \{2\}, \{1, 2\}\}$. The powerset of $\emptyset$ is $\{\emptyset\}$, because the only subset of the empty set is itself, and we have to exhibit the "set of all" such sets. The powerset of $\{\emptyset\}$ is $\{\emptyset, \{\emptyset\}\}$. Recall that $\emptyset$ and $\{\}$ mean the same thing.

Observe that
$$S = \{x \,:\, x \in \mathscr{P}(\{1, 2\}) \wedge |x| > 0\}$$

denotes the set of sets $\{\{1\}, \{2\}, \{1, 2\}\}$.

[3] Some books also use the notation 2^S. We shall stick with $\mathscr{P}(S)$.

Exercise

Consider this set S:

$$S = \{(x, y) \,:\, even(x) \wedge 0 \le x < 5 \wedge y \in \mathscr{P}(\{1, 2\})\}.$$

1. What is the *cardinality* of S? How do you arrive at this answer without explicitly listing all pairs that S contains?
2. S contains $(2, \emptyset)$ and $(4, \{1\})$. Write down six more elements in S.
3. What is a formula to compute the cardinality of the powerset of a set S that contains N elements? Test your formula on $S = \emptyset, \{1\}$, and $\{1, 2\}$.

A.1.3   Complement

All sets studied in a particular setting come from a certain universal set or "universe," say U. Given such a universe U, the complement of a set S is the set $U - S$, *i.e.*, the set subtraction of S from U. For instance, the complement of the set $Even$, the set of even natural numbers $\{0, 2, 4, \ldots\}$ is the set Odd, the set of odd natural numbers $\{1, 3, 5, \ldots\}$.

A.1.4   Equivalence Classes, Partitioning

Consider a nonempty set S (and for simplicity, if S is infinite, let its cardinality be $\aleph_0$).[4] A binary relation $\equiv$ defined over a nonempty set S is called an equivalence relation if it is reflexive, symmetric, and transitive:

- *Reflexive:* For all $a \in S$, we have $a \equiv a$.
- *Symmetric:* For all $a, b \in S$, if we have $a \equiv b$, then $b \equiv a$
- *Transitive:* For all $a, b, c \in S$, if we have $a \equiv b$ and $b \equiv c$, then we have $a \equiv c$

Any equivalence relation $\equiv$ over a set S can lead to the **partitioning** of S into equivalence classes $S_1, S_2, \ldots$ such that S_i and S_j are pairwise disjoint for $i \neq j$, and the union of all these sets is S. Intuitively, all elements within a set S_i are equivalent ($\equiv$), and all elements across two sets S_i and S_j ($i \neq j$) are not equivalent.

As a quick example, if $S = Nat$, then for $a, b \in Nat$, we have an equivalence relation $\equiv$ defined by $a\%3 = b\%3$. Under this equivalence relation, elements $\{0, 3, 6, 9, \ldots\}$ form one partition, $\{1, 4, 7, 10, \ldots\}$ form another partition, etc.

[4] For more general definitions, the reader may consider other books on Discrete Mathematics.

A.2   Mathematical Logic

We assume that you are familiar with the logical connectives $\wedge$ (and), $\vee$ (or), $\neg$ (not), $\Rightarrow$ (implication), and $\oplus$ (exclusive-or).

Predicates are Boolean formulae that are true or false. A Boolean variable x or a Boolean expression $x \wedge y$ are predicates. For integers a and b, an example predicate is $a > (b + 2)$. Two "extreme" predicates are *True* and *False*.

We begin with some basic results:

DeMorgan's Law: For Booleans a, b, c, we have

$$\neg(a \wedge b \wedge c) \equiv (\neg a \vee \neg b \vee \neg c)$$

Another equivalent statement of DeMorgan's Law is:

$$\neg(a \vee b \vee c) \equiv (\neg a \wedge \neg b \wedge \neg c)$$

Contrapositive form: For Booleans a and b, the contrapositive law is

$$(a \Rightarrow b) \equiv (\neg b \Rightarrow \neg a)$$

Quantifiers Forall and Exists: Forall ($\forall$) is used to denote repeated conjunction over the domain of quantification.

$$\forall x \in Nat : P(x) \equiv P(0) \wedge P(1) \wedge P(2) \wedge \ldots$$

Similar results apply for other domains than Nat. Exists ($\exists$) is used to denote repeated disjunction over the domain of quantification.

$$\exists x \in Nat : P(x) \equiv P(0) \vee P(1) \vee P(2) \vee \ldots$$

Negating Forall: Using DeMorgan's law and the definition of $\forall$ and $\exists$, we can write

$$\neg(\forall x : P(x)) \equiv \exists x : \neg P(x)$$

Negating Exists: By a similar line of reasoning,

$$\neg(\exists x : P(x)) \equiv \forall x : \neg P(x)$$

Negating Implication: Given that $a \Rightarrow b$ is equivalent to $\neg a \vee b$, we have

$$neg(a \Rightarrow b)) \equiv (a \wedge \neg b)$$

Thus, the implication $a \Rightarrow b$ is falsified by making a true and b false.

Stacks around implication: Suppose we have a stack of conjuncts in the antecedent of an implication and a stack of disjuncts in the consequent of an implication, as in this formula:

$$(a \wedge b \wedge c) \Rightarrow (d \vee e \vee f)$$

We can transform this to either of the following formulae:
- $(a \wedge b \wedge c \wedge \neg d) \Rightarrow (e \vee f)$
- $(a \wedge b) \Rightarrow (\neg c \vee \neg d \vee e \vee f)$

A.3 *Proof Methods: Using Contrapositive, by Contradiction*

First we illustrate proof using the contrapositive form. Suppose we are given $P \Rightarrow C$ where P is a premise and C a conclusion. Suppose we show $\neg C$. Then we can infer $\neg P$. Exercise 4 on Page 55 asks you to apply this method to first show $\neg Cond(L)$ and thus conclude $\neg Reg(L)$.

Proof by contradiction is directly related to proof using the contrapositive form. Suppose we start with an assertion P. We can then exploit any theorem of the form $P \to C$. Applying modus-ponens, we can derive C. Suppose C can be simplified to *False* (this is what *deriving a contradiction* means). Then we have $P \Rightarrow False$. The contrapositive form of this formula is $True \Rightarrow \neg P$, or simply $\neg P$.

A.4   Cartesian Product, Binary Relations, Functions

The Cartesian product of two sets A and B, written $A \times B$, is a set of all ordered pairs $\langle x, y \rangle$ where $x \in A$ and $y \in B$. If A or B is empty, the resulting product is also empty. A binary relation with domain A and codomain B is a subset of $A \times B$.

A function f from A to B, written $A \rightarrow B$, is a single-valued binary relation. A function is applied to domain elements, written $f(x)$, and yields a $y \in B$. For being single-valued, the following equation must hold:

$$\forall x \in A, \ \forall y_1, y_2 \in B: \ (f(x) = y_1 \wedge f(x) = y_2) \Rightarrow (y_1 = y_2)$$

A.5   Functions: Signature, Onto, Into, Total

Signatures are syntactic presentations of the domain and codomain of a function. They describe the *type* of the function. A function $f : A \rightarrow B$ has signature $A \rightarrow B$, which means that A is the domain ($dom(f)$) and B is the codomain of f ($codom(f)$).

The *range* of a function f (written $rng(f)$) is

$$\{ y \in B : (\exists x \in A : f(x) = y) \}$$

which means all the points in B that f can "hit" across all values fed to f. A function is *onto* if $rng(f) = codom(f)$; otherwise, it is an *into* function.

A function is said to be total if it is defined everywhere in domain A.[5] By default, all functions are assumed to be total, but for the ease of manual specification, we may sometimes partially specify a function, and then invoke a procedure to *totalize* it (provide a standard default mapping at all the undefined domain points). The signatures of functions studied in this book are briefly recapped here:

[5] For total functions, the range is guaranteed to be non-empty.

The transition function δ of a DFA (Page 43) has signature

$$\delta : Q \times \Sigma \rightarrow Q$$

The transition function δ of an NFA (Page 83) has signature

$$\delta : Q \times \Sigma_\varepsilon \rightarrow \mathscr{P}(Q)$$

The transition function Δ of a PDA (Page 165) has signature

$$\Delta : (Q \times (\Sigma \cup \{\varepsilon\}) \times (\Gamma \cup \{\varepsilon\})) \rightarrow \mathscr{P}(Q \times \Gamma^*)$$

The transition function Δ of a DTM (Page 187) has signature

$$\Delta : Q \times \Gamma \rightarrow Q \times \Gamma \times \{L, R, S\}$$

The transition function Δ of an NDTM (Page 187) has signature

$$\Delta : Q \times \Gamma \rightarrow \mathscr{P}(Q \times \Gamma \times \{L, R, S\})$$

A.6   Trees

We discuss trees in §11.9 and also in the context of parse-trees in Chapter 11. Trees are graph data structures with nodes and edges beginning with a single designated node called the root. The root itself could be also a leaf of a tree; such a tree has height 0 and has no edges. Otherwise, the tree has height $n \geq 1$ where the root has one or more children. The number of children the root has is its *branching factor*. The children are themselves roots of disjoint trees (called subtrees) with height $\leq (n-1)$, with at least one subtree having height $n-1$. For a tree of height n and a branching factor b, it can have at most b^n leaves.

B

Catalog of Jove Functions

> **Chapter Gist:** *§B.1 briefly explains Jove's top-level functions. For a fuller description, kindly download and play with Jove from* **https://github.com/ganeshutah/Jove**
> *§B.2 briefly covers lambdas—the only aspect of Python that we choose to highlight.* https://www.python-course.eu *is an excellent online resource for Python. Do check out its sections on recursive programming, lambdas, and other topics. Also see supplementary material at* `https://bit.ly/Automata_Jove` *under JoveCode for any updates.*

B.1   Jove's Top-Level Functions

We now present Jove's functions and their *very brief* documentation. For all these functions, one may type **help(function_name)** to obtain a detailed help message. We organize them with respect to the chapter for which they are relevant, and the Jove files that are involved (definitions and imports).

All Chapters

Here are handy printing and dot-generation-related utilities. For further details including some of the less-used utilities, please poke into Jove/notebooks/src/DotBashers.ipynb. The optional parameters are also documented in this notebook.

- **dotObj_dfa**(D, dfaName='do_', STATENAME_MAXSIZE=20, FuseEdges=False): Produces a dot object for a given DFA.
- **dotObj_dfa_w_bh**(D, dfaName='do_', STATENAME_MAXSIZE=20, FuseEdges=False): Produces a dot object for a given DFA. Shows black-hole states also.
- **dotObj_nfa**(N, nfaName='NO_', FuseEdges=False, visible_eps=False): Produces a dot object for a given NFA.

- **dotObj_gnfa**(G, FuseEdges=False, visible_eps=False, gnfa_name='GO_'): Produces a dot object for a given GNFA.
- **dotObj_pda**(P, FuseEdges=False, visible_eps=False, pdaName='PO_'): Produces a dot object for a given PDA.
- **dotObj_tm**(T, FuseEdges=False, tmName='TO_'): Produces a dot object for a given TM.

A few conversion utilities to generate the dot source, save it in a file, and generate PDF are listed below. Let "DO" be any dot object generated by one of the calls above:

- **DO.render**('/Users/ganesh/repos/atmm/book/CH17/DO.dot'): This will generate DO.dot as a dot-string, and also produce a DO.dot.pdf file.

B.1.1 Chapters 2 and 3

These are the routines used in the "Defining Languages: Patterns in Sets of Strings" and "Kleene Star: Basic Method of Defining Repetitious Patterns" chapters. They are defined in `jove/LangDef.py` (so you can do `from jove.LangDef import *`), as well as in `Jove/notebooks/module/Module2_LanguageOps.ipynb`. An illustration of **nthnumeric** is in `Jove/notebooks/driver/Drive_DFA.ipynb`

- **lcat**(L1,L2): Concatenates two languages.
- **lcomplem**(L,Sigma,n): Complements L with respect to Sigma* that is limited to strings of length n or less.
- **lexp**(L,n): Exponentiate L to power n.
- **lhomo**(L,f): Apply the homomorphism f (a lambda) to the language L.
- **lint**(L1,L2): Intersect L1 and L2.
- **lissubset**(L1,L2): Is L1 a subset of L2?
- **lissuperset**(L1,L2): Is L1 a superset of L2?
- **lminus**(L1,L2): Subtract L2 from L1.
- **lphi**(): Empty language ("zero" of a language), namely $\emptyset$.
- **lrev**(L): Reverse language L.
- **lstar**(L,n): Compute the star of L upto n.
- **lsymdiff**(L1,L2): Output (L1-L2) union (L2-L1).
- **lunion**(L1,L2): Union of L1 and L2.
- **lunit**(): The unit language $\{\varepsilon\}$.
- **srev**(s): Reverse a string s.
- **shomo**(s,f): Apply homomorphism f (a lambda) to string s.
- **powset**(S): Powerset of a given set S output as a list of lists.
- **product**(S1,S2): Cartesian product of sets S1 and S2.
- **nthnumeric**(N, Sigma='a','b'): Produce the Nth string in the numeric order enumeration of strings from an alphabet of size 2. This is handy for exhaustive testing up to a certain finite approximation of Σ^*, as illustrated in `notebooks/driver/Drive_DFA.ipynb` and also `notebooks/tutorial/Drive_DFA_Unit1.ipynb`.

B.1.2 Chapters 4 through 6

These are the routines used in Chapters "Basics of DFA" through "Operations on DFA". They are defined in `Def_DFA.ipynb` in `Jove/notebooks/src`. Also relevant is `Drive_DFA.ipynb` in `Jove/notebooks/driver`.

- **mkp_dfa**(Q, Sigma, Delta, q0, F): Check for partial consistency of the given DFA traits. Make and return a DFA with a partial Delta.
- **mk_dfa**(Q, Sigma, Delta, q0, F): Check for structural consistency of the given DFA traits. Make and return a DFA with a total Delta.
- **totalize_dfa**(D): Given a partially specified DFA, make it total.
- **addtosigma_dfa**(Din, addition): Expand Din's Sigma using 'addition'.
- **step_dfa**(D, q, c): Next state of q via c.
- **run_dfa**(D, s): Next state of D["q0"] via string s.
- **run_dfa_h**(D, s, q): Next state of q via string s.
- **accepts_dfa**(D, s): True iff state after s-run is in D's final.
- **comp_dfa**(D): Before we begin, make D total. Then flip the FINAL and NON-FINAL states.
- **union_dfa**(D1in, D2in): DFA for language union of D1in, D2in.
- **intersect_dfa**(D1in, D2in): DFA for language intersection of D1in, D2in.
- **pruneUnreach**(D): Given a consistent and total DFA D, returns a new (consistent) DFA with unreachable states in D removed.
- **iso_dfa**(D1,D2): Check whether D1 and D2 are isomorphic.
- **langeq_dfa**(D1, D2, gen_counterex=False): Check whether D1 and D2 are language-equivalent. gen_counterex triggers counter-example generation.
- **min_dfa**(D, state_name_mode='succinct'): Minimize the given DFA D, naming the states as specified.

B.1.3 Chapters 7 through 9

These are the routines used in Chapters "Nondeterministic Finite Automata" through "NFA to RE Conversion". They are defined in `Def_NFA.ipynb` that is located at the file path `Jove/notebooks/src`. Also relevant is `Drive_NFA.ipynb` in `Jove/notebooks/driver`.

For RE to NFA conversion via function re2nfa, take a look at `Def_RE2NFA.ipynb` in `Jove/notebooks/src` and `Drive_AllRegularOps.ipynb` in the directory path `Jove/notebooks/driver`.

- **mk_nfa**(Q, Sigma, Delta, Q0, F): Check for structural consistency of the given NFA traits. Make and return an NFA.
- **totalize_nfa**(N): Given a partially specified NFA, make it total by transitioning to state set({}) wherever Delta has gaps.
- **apply_h_dfa**(D, h): Given a DFA D and a homomorphism h on Sigma* (as a lambda from chars to chars) where Sigma is D's alphabet, return

an NFA with the homomorphism applied to D.

- **step_nfa**(N, q, c): The set of states reached via N's Delta from q upon c. EClosure is NOT performed.
- **run_nfa**(N, S, s, chatty=False): Return the set of states reached after processing string s from set of states S. 'chatty' controls printout verbosity.
- **ec_step_nfa**(N, S, c, chatty=False): Return Eclosure of all states reachable via character c from every state within S.
- **Eclosure**(N, S): Return Eclosure of the set of states S.
- **accepts_nfa**(N, s, chatty=False): True if N accepts s. Argument chatty controls verbosity of printout.
- **nfa2dfa**(N, STATENAME_MAXSIZE=20): Return a consistent DFA that is language-equivalent to N. You may supply an optional argument to shrink long state names (default for the 'shrink' to step in now is 20).
- **re2nfa**(s, stno = 0): Given a string s representing an RE and an optional state number stno, generate an NFA that is language equivalent to the RE.
- **Implementation of re2nfa:** The implementation of re2nfa involves other "helper" functions which can also be used at the top level. They are the following:
  - **mk_plus_nfa**(N1, N2): Given two NFAs, return their union.
  - **mk_cat_nfa**(N1, N2): Given two NFAs, return their concatenation.
  - **mk_star_nfa**(N): Given an NFA, return its star.
  - **mk_eps_nfa**(): An NFA with exactly one start+final state.
  - **mk_symbol_nfa**(a): The NFA for a single character.
- **rev_dfa**(D): Return a consistent NFA whose language is D's language reversed.
- **min_dfa_brz**(D): Minimize a DFA using Brzozowski's algorithm which is (reverse; determinize; reverse; determinize).
- **mk_gnfa**(Nin): Return the GNFA corresponding to the given NFA.
- **mk_gnfa_from_D**(D): Given a DFA D, turn that into a GNFA by first making the D into an equivalent N.
- **dfa2nfa**(D): Given a DFA D, make a language-equivalent NFA from it.
- **del_gnfa_states**(Gin, DelList=[]): Given a GNFA Gin with no unreachable states, delete all its states except Real_I and Real_F.

B.1.4 Chapter 10

These are the routines used in Chapter "Derivative-Based Regular Expression Matching." They are defined in `Drive_rederiv.ipynb` in `Jove/notebooks/driver`. Also relevant is `Def_rederiv.ipynb` in `Jove/notebooks/src`.

- **nullable**(E): Given an RE E represented as an AST, this function determines whether E is nullable.
- **dv**(c,E): Given a character c and an RE E represented as an AST, this function computes the Brzozowski derivative of E with respect to c, returning a new RE.
- **matches**(w,E): Given a word w and an RE E represented as an AST, this function determines whether w is included in the language of E (or in other words, RE E pattern-matches word w).

B.1.5   Chapter 11

These are the routines pertinent to Chapter "Context-Free Languages and Grammars" which is on parsing. There are three parsers introduced in this book:

- **re2ast**(s): This is the parser used in our illustration of RE derivatives. This parser can be found in `Def_rederiv.ipynb` in `Jove/notebooks/src`. Also take a look at `Drive_rederiv.ipynb` in `Jove/notebooks/driver`, which uses this parser. This function turns a regular expression (passed in as a string s) into an abstract syntax tree, and returns the tree (encoded in Python).
- **re2nfa**(s, stno = 0): This is the parser used in converting regular expressions to NFA. This parser can be found in `Def_RE2NFA.ipynb` in `Jove/notebooks/src`. Also take a look at `Drive_AllRegularOps.ipynb` in `Jove/notebooks/driver`, which uses this parser.
- **md2mc**(src="None", fname="None"): This is the parser used for our markdown language **Automd**.   md2mc converts a markdown source to a machine (mc). This parser can be found in `Def_md2mc.ipynb` in `Jove/notebooks/src`. Also take a look at `Drive_md2mc.ipynb` in `Jove/notebooks/driver`, which uses this parser. Here are the ways of using this parser:
  - **md2mc**(): prompt for a filename and read the markdown description from the provided file name.
  - **md2mc**(src=string): parse the given string, treating it as a markdown description.
  - **md2mc**(src="File", fname=filename): read the markdown description from the provided file.

B.1.6   Chapter 12

These are the routines used in Chapter "Pushdown Automata". They are defined in `Def_PDA.ipynb` in `Jove/notebooks/src`. Also relevant is `Drive_PDA.ipynb` and `Drive_PDA_Based_Parsing.ipynb` in `Jove/notebooks/driver`.
- **explore_pda**(inp, P, acceptance = 'ACCEPT_F', STKMAX=0,

chatty=False): A handy routine to print the result of run_pda.

- **run_pda**(str, P, acceptance = 'ACCEPT_F', STKMAX=0, chatty=False): Helper for explore_pda.

- **step_pda**(q_inp_stk, path, pda): The results of stepping a PDA.

B.1.7 Chapter 13

These are the routines used in Chapter "Turing Machines". They are defined in Def_TM.ipynb in Jove/notebooks/src. Also relevant is Drive_TM.ipynb in Jove/notebooks/driver.

- **step_tm**(T, q_hi_tape_fuel, path, haltList): Step a TM one step.

- **run_tm**(T, tape, fuel): Given a TM T and a tape, run the TM for fuel steps (e.g., thimbles or gallons of gas in your tank), collecting all halting configurations.

- **explore_tm**(T, tape, nsteps): A handy routine to print the result of run_tm.

B.1.8 Chapter 15

[1] For your convenience, I'll leave info on the github path in README.md within Jove/3rdparty.

These are the routines used in Chapter "Post Correspondence, and Other Undecidability Proofs" for illustrating the Post Correspondence Problem (PCP). Credits go to Ling Zhao who contributed the original software and the owner of this github page https://github.com/chrozz/PCPSolver.git for posting it on github. To run the solver, first compile it and create an executable called pcp.[1] This executable can then be conveniently invoked from Drive_pcp.ipynb defined in Jove/notebooks/driver.

- **pcp_solve**(pcp_pairs, run=None, ni=False, di=None, depth=None, tiles_per_row=15): The PCP solver takes these parameters:

  - pcp_pairs: List of tuple pairs representing pcp 'tiles'

  - run: Number of runs to perform.

  - ni: No iterative search.

  - di: Depth increment.

  - depth: Search depth.

  - tiles_per_row: Number of tiles to show in single row together (useful for creating more meaningful output).

- The above function calculates all the parameters needed by Zhao's solver, invokes it from Jove as a process, and returns the results.

B.1.9 Chapters 16 and 17

These are the routines used in Chapters "NP-Completeness" and "Binary Decision Diagrams as

Minimal DFA". The material is present on the web at URL `http://formal.cs.utah.edu:8080/pbl/BDD.php`.

B.1.10   Chapter 18

These are the routines used in Chapter "Computability Using Lambdas". They are defined in `Drive_LambdaCalc.ipynb` in `Jove/notebooks/driver`.

- **I = lambda c: c** The identity function.
- **ZERO = lambda b: I** Number 0.
- **S = lambda a: lambda b: lambda c: b(a(b)(c))** The successor function.
- **ADD = lambda a: lambda b: a(S)(b)** The add function.
- **MUL = lambda a: lambda b: lambda c: a((b)(c))** The multiplication function.
- **increment(n)** Increment n.
- **ChurchToNat(c)** Converts Church numerals to natural numbers in Python.
- **NatToChurch(n)** Converts natural numbers in Python to Church numerals.
- **TRUE = lambda a: lambda b: a** 'True' in lambda calculus.
- **FALSE = lambda a: lambda b: b** 'False' in lambda calculus.
- **NOT = lambda a: a(FALSE)(TRUE)** 'not' function in lambda calculus.
- **AND = lambda a: lambda b: a(b)(FALSE)** Boolean 'and' in lambda calculus.
- **OR = lambda a: lambda b: NOT(AND(NOT(a))(NOT(b)))**. Boolean 'or' in lambda calculus
- **Z = lambda a: a(FALSE)(NOT)(FALSE)** Zero-test of a given Church numeral.
- **PAIR = lambda x: lambda y: lambda f: f(x)(y)** Pair builder.
- **FIRST = lambda p: p(TRUE)** Extracts the first of a pair.
- **SECOND = lambda p: p(FALSE)** Extracts the second of a pair.
- **LambdaToBoolean(b)** Turns a lambda calculus Boolean to a Python Boolean.
- **BooleanToLambda(b)** Turns a Python Boolean to a lambda calculus Boolean.
- **Ye = lambda f: (lambda x: x(x))(lambda y: f(lambda v: y(y)(v)))** The eager fixpoint combinator.
- **prefact = lambda fact: lambda n: (1 if n==0 else n*fact(n-1))** How to obtain the functional expression underlying the factorial function.
- **fact = Ye(prefact)** How factorial itself is defined.

B.2 Jove's Use of Python, Including Lambda Basics

In Jove, we use three primary data types in Python: tuples, lists, and dicts (dictionaries). To keep the subset of Python within reach of a wide audience, we avoid the use of advanced features, including objects.

We employ the (mostly) side-effect free functional programming style that Python encourages, and also encourage the use of recursion.[2] Recursion and induction are two faces of the same coin.[3]

We first illustrate how functions can be defined through lambda expressions:

```
>>> lambda x: x+2
<function <lambda> at 0x1060f6d08>
>>> (lambda x: x+2)(3)
5
```

In the first example above, an anonymous function is evaluated via Python's interactive system. In the second example, this function is fed 3, yielding the value 5 as expected. One can also bind lambda functions to variable names:

```
>>> G = lambda x: x+2
>>> G(3)
5
# This function G could have been defined as follows
# def G(x):
#     return x+2
```

We now illustrate how lambda-defined functions can return functions themselves as values: First, we define functions that return other functions without the use of lambdas:

```
def G(x):
    return x+2
def F(y):
    return G
```

Next, using lambdas, one can define the same idea, where functions return other functions:

```
lambda y: lambda x: x+y
```

This expression is to be parsed as follows

[2] With recursion, one plays a game of pretense: one reduces a given problem to a simpler problem, assuming that the simpler problem has been solved and exists "in some library"—but happens to be the very function being coded! (I am indebted to David S. Warren, Emeritus Professor at Stony Brook University for this explanation of recursion given in 1982.)

[3] L. Peter Deutsch once said "To iterate is human; to recurse...divine!" Also, Mitchell Wand, Professor at Northeastern University has authored a book with "Recursion and Induction" in its title.

> lambda y: (lambda x: x+y)

The above lambda expression can be read as follows. When we feed an actual argument in place of y, we will obtain the function (lambda x: x+y) as the result. Here are a few interactive commands that will help us clarify these ideas.

```
>>> lambda y: lambda x: x+y
<function <lambda> at 0x10cd6cd08>

>>> (lambda y: lambda x: x+y) (33)
<function <lambda>.<locals>.<lambda> at 0x10cf2e268>

>>> ((lambda y: lambda x: x+y) (33)) (2)
35
```

Notice that the first definition lambda y: lambda x: x+y defines a function that takes one argument and returns a function. By applying this function to 33, we recover the inner-nested function. By feeding this function 2, we finally obtain the answer of 35.

A nest of 1-argument functions of the kind illustrated here are examples of *curried* functions.[4] For instance, we can render the addition function + into the curried form by writing it as shown above

> lambda y: lambda x: x+y

or perhaps as

> lambda x: lambda y: x+y

We will now introduce lambdas taking multiple arguments. Consider the following function defined using lambdas and bound to a variable called abcd.

```
>>> abcd = lambda a,b: lambda c,d: (a+b)*(c+d)

>>> abcd(2,3)
<function <lambda>.<locals>.<lambda> at 0x10158b158>

>>> abcd(2,3)(4,5)
45
```

Observe that abcd(2,3) binds 2 to a and 3 to b, returning

[4] Named for Haskell B. Curry.

```
lambda c,d: (2+3)*(c+d)
```

which is the inner lambda that is *specialized* to carry out a multiplication of (c+d) using (2+3) (or 5). We see this fact by feeding (4,5) to abcd(2,3). The result is (2+3)*(4+5) or 45.

.

C

There Are More Languages than RE Languages

Chapter Gist: *We will define a method to injectively (1-1) map any tuple of Nat to a Nat (§C.1). We then state the Cantor-Schröder-Bernstein (CSB) theorem that allows us to equate the cardinalities of two infinite sets. Using the CSB theorem, we can show that there are $\aleph_0$ Turing machines (also that many C programs, §C.2). We then present Cantor's diagonalization proof showing that there are $\aleph_1$ languages (§C.3) but only $\aleph_0$ recursively enumerable (RE) languages. This will finish the proof that there are (many more) non-RE languages than RE languages.*

C.1 Gödel Hash

Gödel hash is a mechanism to injectively map any tuple of Nat into Nat. It is based on this theorem:

Theorem C.1: Any natural number greater than 1 can be uniquely expressed as a product of primes.

Proof: By induction (left to the reader).

Here are examples, where we express each natural number as an N-tuple of exponents of primes (typed as lists below):

- $18 = [1,2]$ Obtained as $2^1 \cdot 3^2$
- $22 = [1,0,0,0,1]$ Obtained as $2^1 \cdot 3^0 \cdot 5^0 \cdot 7^0 \cdot 11^1$
- $24 = [3,1]$ Obtained as $2^3 \cdot 3^1$
- $256 = [8]$ Obtained as 2^8
- $5402250 = [1,2,3,4]$ Obtained as $2^1 \cdot 3^2 \cdot 5^3 \cdot 7^4$
- $75600 = [4,3,2,1]$ Obtained as $2^4 \cdot 3^3 \cdot 5^2 \cdot 7^1$

The flipside of this result is that given a finite list over Nat, we can injectively map it into Nat as follows (in a sense, we are computing the **Gödel**

Hash of these tuples and mapping them to *Nat*):

- [1,2] maps to 18, and no other list other than [1,2] maps to 18 (i.e., injectively maps)
- [1,0,0,0,1] maps to 22
- [3,1] maps to 24
- [8] maps to 256
- [1,2,3,4] maps to 5402250
- [4,3,2,1] maps to 75600

We will use this hashing method in §C.2.[1]

C.2 Cantor-Schröder-Bernstein (CSB) Theorem

The CSB theorem says that if there is an injective map from *A* into *B* (not necessarily onto), **and** there is an injective map from *B* into *A* (not necessarily onto), **then** there is a bijective map between *A* and *B*, **i.e.**, these sets have the same cardinality.

> - Illustration 1: Consider *Even* and *Nat* discussed in Appendix A. One way to injectively map *Even* to *Nat* is to divide each even number by 2. One way to injectively map each member of *Nat* to *Even* is to double the number. The existence of these injective maps ensures that
> $|Even| = |Nat| = \aleph_0$.
> - An injective map from *Odd* to *Even* is doubling. One injective map from *Even* to *Odd* is double and add 1. Thus these two sets also have cardinality $\aleph_0$.

> **Theorem C.2:** There are $\aleph_0$ Turing machines, and the same number of C programs. Hence there are also $\aleph_0$ RE languages (as each such language is the language of some TM).

Proof: We will work out the proof for C programs (denote the set of C programs by *CP*). The reader may find a similar trick for Turing machines.[2]

- Believe it or not, the shortest C program that is nonempty is `main(){}`. Let 0 map to this program.
- The next longer, "weird but legal" C program is `main(){;}`. Let 1 map to this program.
- Similarly 2, 3 and 4 map to `main(){;;}`, `main(){;;;}`, and `main(){;;;;}`.
- This is an injective map from *Nat* to *CP*.
- To find an injective map from *CP* to *Nat*, simply take the C program to be a tuple of ASCII characters, one per character in the program. Since ASCII characters can be encoded using integers from 0 to 255, taking a Gödel Hash, we have an injective map from *CP* to *Nat*.

- Hence, from the CSB theorem, $|CP| = |Nat| = \aleph_0$.

C.3 Cantor's Diagonalization Proof about Languages

We will now prove that the cardinality of all languages over $\Sigma = \{0, 1\}$ is $\aleph_1$.[3] Strings from this alphabet can be listed in a total order according to the numeric order (§3.6). Now, each language is a set of strings, and sets can be represented by infinitely long bit-vectors (characteristic vectors). Here are examples:

[3] Please convince yourself that the entire proof below will go through even if $\Sigma = \{0\}$.

- Language $\{\} \mapsto 000\ldots$: no string from Σ^* is included
- Language $\{\varepsilon\} \mapsto 100\ldots$: ε alone is included
- Language $\{0, 110\} \mapsto 010000000000010000\ldots$: That is, from the numeric order

  $\varepsilon, 0, 1, 00, 01, 10, 11, 000, 001, 010, 011, 100, 101, 110, 111, 0000, \ldots$

  The 1 bit picks out 0 and 110
- Language $\{(01)^i \mid i \geq 0\} \mapsto 00001000000000000000100\ldots$

  The strings picked out are $\varepsilon, 01, 0101, 010101, \ldots$

Now, assume that we have a bijective map from Nat to languages that looks like this:

$0 \to b_{00} b_{01} b_{02} b_{03} \ldots$

$1 \to b_{10} b_{11} b_{12} b_{13} \ldots$

$\ldots$

$n \to b_{n0} b_{n1} b_{n2} b_{n3} \ldots$

$\ldots$

Here, b_{00}, b_{01}, ..., b_{ij}, ... are all infinitely long bit sequences. Each row above models one language through a suitable "on/off" combination of b-bits.

Now consider the language denoted by the bit sequence

$$\neg b_{00} \; \neg b_{11} \; \neg b_{22} \; \neg b_{33} \; \ldots \; \neg b_{jj} \; \ldots$$

We call this the **diagonal language**. The diagonal language differs from each language listed above at least in one string. For instance,

- If the jth language has the jth string in numeric order, then b_{jj} would be 1, and in our diagonal language, this string won't be present (that is what $\neg b_{jj}$ indicates). Of course if $b_{jj} = 0$, the diagonal language would contain the jth string in the numeric order.
- Thus, however we put languages into 1-1 correspondence with Nat, one language won't be listed in such a listing.
- Thus there isn't a bijection that allows us to put Nat and languages into correspondence. That is, they have different cardinalities.
- The cardinality of the set of languages is higher. This higher cardinality is $\aleph_1$.

- However, each RE language corresponds to a Turing machine, and there are $\aleph_0$ TMs (and this many RE languages).
- **Thus there are (many) languages that are not RE.**

This proof approach is called diagonalization because it relies on deriving a contradiction based on a construction that first lists suitable candidates and discovers a non-represented candidate along the diagonal of this construction.

C.4 $|Real|$ *Is Higher than* $|Nat|$

The diagonalization argument in §C.3 almost works. Observe that if we put a "0." before the bit-strings of this form:

$b_{00}b_{01}b_{02}b_{03}\ldots$

$b_{10}b_{11}b_{12}b_{13}\ldots$

we are able to represent Real numbers in the interval $[0,1)$. Here are examples:

$0.5 = 0.1\overline{0}$ where $\overline{0}$ means *repeat 0s infinitely*

$0.25 = 0.01\overline{0}$

$0.333\overline{3} = 0.0101\overline{01}$

This 1-1 correspondence between bit-strings and Real numbers within $[0,1)$ almost works; there is one problem however. See these dual representations of Real numbers:

$0.5 = 0.01\overline{1}$

$0.25 = 0.001\overline{1}$

We cannot diagonalize and claim (by inverting the diagonal) that this number isn't present in the main listing. What if the diagonal is

$0.110\overline{0}$

and we flip the diagonal to obtain

$0.001\overline{1}$

which simply happens to be another representation for 0.25, which in binary is $0.010\overline{0}$?

The fix is simple: consider listing not just any list

$0.b_{00}b_{01}b_{02}b_{03}\ldots$

$0.b_{10}b_{11}b_{12}b_{13}\ldots$

$0.b_{20}b_{21}b_{22}b_{23}\ldots$

but actually such a list interspersed with numbers that prevent an infinite run of 1s along the diagonal. Here is one idea:

$0.b_{00}b_{01}b_{02}b_{03}\ldots$

$0.110\overline{0}$

$0.b_{10}b_{11}b_{12}b_{13}\ldots$

$0.1111\overline{0}$

$0.b_{20}b_{21}b_{22}b_{23}\ldots$

$0.111111\overline{0}$

This is just a permutation of the listing of numbers within $[0,1)$ except

$0.110\overline{0}$, $0.1111\overline{0}$, $0.1111111\overline{0}$, etc., are placed interspersed. Now, flipping the diagonal, we will never obtain the $\overline{1}$ along the complemented diagonal. This settles the "dual representation" difficulty alluded to above.

Selected References

[1] Aagaard, M., & Leeser, M. (1991). A formally verified system for logic synthesis. In *Proceedings 1991 IEEE International Conference on Computer Design*, (pp. 346–350).

[2] Adams, M. D., Hollenbeck, C., & Might, M. (2016). On the Complexity and Performance of Parsing With Derivatives. In *Programming Languages Design and Implementation (PLDI)*, (pp. 224–236).

[3] Agrawal, M., Kayal, N., & Saxena, N. (2004). PRIMES is in P. *Annals of Mathematics*, *160*(2), 781–793.

[4] Aspray, W. (1984). Interview of Alonzo Church by William Aspray. https://en.wikipedia.org/wiki/Alonzo_Church.

[5] Aspvall, B., Plass, M. F., & Tarjan, R. E. (1979). A Linear-Time Algorithm for Testing the Truth of Certain Quantified Boolean Formulas. *Information Processing Letters*, *8*(3), 121–123.

[6] Ball, T., Levin, V., & Rajamani, S. K. (2011). A Decade of Software Model Checking With SLAM. *Communications of ACM*, *54*(7), 68–76.

[7] Ball, T., & Rajamani, S. K. (2000). Bebop: A Symbolic Model Checker for Boolean Programs. In *7th International SPIN Workshop, Stanford, CA, USA*, (pp. 113–130).

[8] Bentley, J. (2000). *Programming Pearls*. Addison-Wesley. ISBN 0-201-65788-0, Second Edition.

[9] Biere, A., Heule, M., van Maaren, H., & Walsh, T. (2009). *Handbook of Satisfiability*. IOS Press. Volume 185, Frontiers in Artificial Intelligence and Applications.

[10] Bollig, B., & Wegener, I. (1996). Improving the Variable Ordering of OBDDs Is NP-Complete. *IEEE Transactions on Computers*, *45*(9), 993–1002.

[11] Bryant, R. E. (1986). Graph-Based Algorithms for Boolean Function Manipulation. *IEEE Transactions on Computers*, *35*(8), 677–691.

[12] Bryant, Randal E. (1992). Symbolic Boolean Manipulation with Ordered Binary Decision Diagrams. *ACM Computing Surveys, 24*(3), 293–318.

[13] Brzozowski, J. A. (1962). Canonical Regular Expressions and Minimal State Graphs for Definite Events. In *Proceedings of the Symposium on Mathematical Theory of Automata*, (pp. 529–561). Polytechnic Press of Polytechnic Institute of Brooklyn.

[14] Brzozowski, J. A. (1964). Derivatives of Regular Expressions. *Journal of the ACM, 11*(4), 481–494.

[15] Cantin, J. F., Lipasti, M. H., & Smith, J. E. (2005). The Complexity of Verifying Memory Coherence and Consistency. *IEEE Trans. Parallel Distrib. Syst., 16*(7), 663–671.

[16] Chomsky, N. (1959). On Certain Formal Properties of Grammars. *Information and Control, 2*(2), 137–167.

[17] Cook, S. (1971). The Complexity of Theorem Proving Procedures. In *Proceedings of the Third Annual ACM Symposium on Theory of Computing*, (p. 151–158).

[18] Cook, S. (2003). The Importance of the P versus NP Question. *JACM, 50*(1), 27–29.

[19] Edwards, J. R. (2012). An Early History of Computing at Princeton. https://paw.princeton.edu/article/early-history-computing-princeton.

[20] Ginsburg, Seymour (1966). *The Mathematical Theory of Context Free Languages*. McGraw-Hill. QA 267.5 S4 G5.

[21] Gopalakrishnan, G. L. (2006). *Computation Engineering: Applied Automata Theory and Logic*. Springer.

[22] Gordon, M. J. (1988). *Programming Language Theory and Its Implementation - Applicative and Imperative Paradigms*. Prentice Hall International series in Computer Science. Prentice Hall.

[23] Greibach, S. A. (1981). Formal Languages: Origins and Directions. *IEEE Annals of the History of Computing, 3*(1), 14–41.

[24] Hoare, C. A. (1981). The Emperor's Old Clothes. *Communications of the ACM, 24*(2), 75–83. See additional discussions around Hoare's mention of Mariner at https://en.wikipedia.org/wiki/Mariner_1.

[25] Hodges, A. (2014). Alan Turing: The Enigma. http://www.turing.org.uk/turing.

[26] Hopcroft, J.E., & Ullman, J.D. (1979). *Introduction to Automata Theory, Languages, and Computation*. Addison-Wesley.

[27] Knuth, D. E. (1962). Invited Papers: History of Writing Compilers. In *Proceedings of the 1962 ACM National Conference on Digest of Technical Papers*. ACM.

[28] Kozen, Dexter C. (1997). *Automata and Computability*. Springer.

[29] Kullmann, O. (2009).   SAT Competition 2009:   Benchmark Submission Guidelines. `http://www.satcompetition.org/2009/format-benchmarks2009.html`. (DIMACS format).

[30] Kuroda, S.-Y. (1964).   Classes of Languages and Linear Bounded Automata. *Information and Control*, (7), 207–223.

[31] Levin, L. (1973). Universal Search Problems (translation from Russian). *Problems of Information Transmission (in Russian)*, *9*(3), 115–116.

[32] Liang, S., Sun, W., & Might, M. (2014). Fast Flow Analysis with Gödel Hashes. In *14th IEEE International Working Conference on Source Code Analysis and Manipulation, (SCAM)*, (pp. 225–234).

[33] Luchaup, D., Carli, L. D., Jha, S., & Bach, E. (2014). Deep Packet Inspection with DFA-trees and Parametrized Language Overapproximation. In *IEEE INFOCOM 2014 - IEEE Conference on Computer Communications*, (pp. 531–539).

[34] Matiyasevich, Y. (1992). My Collaboration with Julia Robinson. *The Mathematical Intelligencer*, *14*(4), 38–45. Wonderful historical account by Matiyasevich.

[35] Might, M., Darais, D., & Spiewak, D. (2011). Parsing with Derivatives: a Functional Pearl. In *Proc. 16th ACM SIGPLAN International Conf. on Functional Programming (ICFP)*, (pp. 189–195).

[36] Mytkowicz, T., Musuvathi, M., & Schulte, W. (2014). Data-Parallel Finite-state Machines. In *Architectural Support for Programming Languages and Operating Systems, ASPLOS '14*, (pp. 529–542).

[37] Owens, S., Reppy, J., & Turon, A. (2009).   Regular Expression Derivatives Re-examined. *Journal of Functional Programming*, *19*(2), 173–190.

[38] Peled, D. A. (2001). *Software Reliability Methods*. Springer-Verlag New York.

[39] Pratt, Vaughan (1975). Every Prime Has a Succinct Certificate. *SIAM Journal on Computing*, (4), 214–220.

[40] Rabin, M. O., & Scott, D. (1959). Finite Automata and their Decision Problems. *IBM J. Res. Dev.*, (3), 114–125.

[41] Ramalingam, G. (1994). The Undecidability of Aliasing. *ACM Transactions on Programming Languages and Systems*, *16*(5), 1476–1471.

[42] Sipser, Michael (1997). *Introduction to the Theory of Computation*. PWS Publishing Company.

[43] Stanat, D. F., & Weiss, S. F. (1982). A Pumping Theorem for Regular Languages. vol. 14, (pp. 36–37). ACM SIGACT News. Issue 1, Winter 1982, doi:10.1145/1008892.1008895.

[44] Turing, D. (2015). *Prof: Alan Turing Decoded*. The History Press.

[45] Turing, Alan M. (1936). On Computable Numbers: With an Application to the Entscheidungsproblem. In *Proceedings of the London Mathematical Society*, vol. Series 2. Number 42.

[46] Wiles, A. (2011). NOVA Program Featuring Andrew Wiles and His Proof of Fermat's Last Theorem. `http://www.pbs.org/wgbh/nova/physics/andrew-wiles-fermat.html`.

[47] Zhao, L. (2002). *Solving and Creating Difficult Instances of Post's Correspondence Problem*. Master's thesis, University of Alberta.

Index

$3x + 1$ Problem, 200
A_{TM}, 216
A_{TM} is undecidable, 234
E_{TM}, 238
$Halt_{TM}$, 218, 234
$L_{EmptyCFG}$, 214
$Regular_{TM}$, 238
$\Sigma_\varepsilon = \Sigma \cup \{\varepsilon\}$, 83
$\hat{\delta}$, 45
q_0, 43
NP-Complete
    $\leq_P$ mapping-reductions, 254
    funnel diagram, 254
NP-Hard, 253
nondeterministic polynomial time decider, 246
3-SAT, 254
    NPC, 254

Acceptance problem is Undecidable, 234
Algorithm, 4, 184, 209
Alphabet, 15
Ambiguity, 144
    ambiguous grammar, 144
    disambiguation, 144
    inherently ambiguous language, 144, 147
    PDA, 174
Automd, 307
    automaton markdown, 64
    Jove parser for, 307
    PDA, 173

Babbage's Analytical Engine, 3
BDD, 269
    Dynamic reordering, 277
    Myhill-Nerode Theorem, 276
    PBDD tool, 269

Sifting, 277
    Variable ordering, 270
BDD variable ordering, 272
    Heuristic, 272
    Improving, 272
    NPC, 277
BDDs, 268
    Knuth, 268
Binary Decision Diagram, 269
Binary relation, 301
Boolean functions, 267
    2^{2^N}, 268
    On-set, 268
    Truth-tables, 267
Boolean satisfiability, 248

Cantor's diagonalization, 315
Cantor-Schröder-Bernstein theorem, 314
Cartesian product, 301
certificate, 242
CFG
    Ambiguity, 144
    ambiguity, PDA, 174
    conversion to PDA, 174
    hill-valley plot, 142
    inconsistency, 142
    reversal, 151
CFL
    closure, 151
    Pumping Lemma, 154
Chomsky hierarchy , 203
Chomsky's grammars, 12
    Type-0,1,2,3, 12
Chomsky's hierarchy, 13
Closure results (summary), 219
Collatz Problem, 200
Combinators, 292
    S,K,I, 292

Composite, 260
Context-free Grammar
    hill-valley plot, 142
Context-free grammar, 138
    (N, Σ, S, P), 139
    derivation sequence, 139
    inconsistency, 142
    nonterminal, 139
    parse trees, 139
    productions, 139
    start symbol, 139
    terminal, 139
Context-free language, 137
    closure, 151
    Dyck language, 137
    generalized bracketing, 138
    reversal, 151
Context-free languages
    Pumping Lemma, 154
Contrapositive, 299
CSB theorem, 314

Decidability
    Post correspondence, 227
Decidability results (summary), 219
Decider, 212
    NP, 246
    P, 244
Decider vs. Verifier, 247
Decision problem, 183
Defining a computer, 5
Definition
    Basic SAT definitions, 16.3, 248
    Co-NP, 16.7.2, 260
    Equisatisfiable, 16.8, 261
    Languages L_{2sat} and L_{3sat}, 16.8, 249
    Mapping Reduction, 15.5, 236
    NP decider, 16.2, 246
    NP Verifier, 16.2.3, 246
    NP-Complete, 16.4, 253
    P decider, 16.2, 244
    Star, 3.1, 27
DeMorgan's law, 299
Derivatives, 127
    nullability rules, 132
    pattern-matcher, 134
    rules, 129
DFA, 42
    $(Q, \Sigma, \delta, q_0, F)$, 43
    k-distinguishability, 72

acceptance (formal), 45
accepts, 43
best practices, 58
black-hole state, 43
Brzozowski minimization, 90
complementation, 67
DeMorgan's law, 78
designing, 45
goto program, 44
intersection, 67
isomorphism, 71
language equivalence, 71
lasso, 47
minimal DFA, 59
minimization, 72
Myhill-Nerode theorem, 59
pruneUnreach, 69
Pumping Lemma, 48
recognizes, 43
reversal, 90
run, 45
states, transitions, 42
step, 45
string classifier, 44
totalize, 43
union, 67
Diagonalization proof, 315
Diophantine, 258
Diophantine equations, 4
Disambiguation
    PDA, 179
DTM
    $\Delta : Q \times \Gamma \rightarrow Q \times \Gamma \times \{L, R, S\}$, 187
    $w\#w$, 193
    ww, 194
    P-time, 244

Easy, 242
Effective computability, 183
Exercise
    Ambiguous Parses, 11.6, 148
    BDDs, 17.5, 277
    Block-of-3 DFA, 5.2.1, 60
    Booleans in Lambda, 18.5, 284
    Brzozowski's DFA minimization, 7.6.1, 91
    CFG Completeness, 11.4, 144
    CFG Design, 11.10.1, 158
    CFL Pumping Lemma, 11.9.1, 157
    Church Numerals in Lambda, 18.4, 284

DFA basics, 4.2, 43
DFA exp blowup, 5.2.4, 62
DFA Jove design, 5.3.1, 64
DFA Jove, ∪,∩, 6.2, 68
DFA Lasso, 4.6, 48
DFA, DeMorgan's Law, 6.5.2, 79
DTM and NDTM Design, 13.8, 205
DTM Design, 13.5.3, 194
equal-change DFA, 5.1.1, 58
Flipper TM, 13.4.1, 189
Homomorphism, 3.5, 35
Language Operations, 2.1.3, 16
Language puzzles, >, 3.4.1, 32
Languages (identities) ∪, ∩, 2.2.6, 24
Languages (Python), 2.2, 20
Languages (review), 2.2.5, 23
Languages, 2.2, 19
Lexicographic order, 3.6, 37
Mapping-reduction proofs, 15.5.1, 239
NFA operations, 8.2, 99
NFA to DFA, 7.5, 88
NFA to RE, 9.2, 117
nfa2re: RE Size, 9.5, 123
NP-Completeness, 16.7.2, 262
Numeric order, 3.6, 37
Parse Trees, 11.5.1, 146
PCP solver in Jove, 15.2.3, 232
PDA Design, 12.6.1, 180
Postage stamp, 8.8.5, 114
RE and recursive, 14.4.1, 220
RE Derivatives, 10.3, 134
RE, Error Correction, 8.6.1, 107
Regular or not?, 4.7.1, 52
Regularity preserving, 4.9, 54
Slippery concepts, 2.3, 25
Star properties, 3.2, 30
Sylvester's Formula, 8.8, 111
Turing Machine Simulation, 13.2, 186
Y combinator, 18.8, 292
Zero and One for Concat, 2.2.2, 21
Zero, One, Exp, 2.1.4, 18

Family of languages, 210
Fermat's Last Theorem, 4
Final state, 43
Formal verification, 268
   BDDs, 268
Frobenius number, 111
   $Fr(p,q) = pq - p - q$, 111

Sylvester's Formula, 111
Function, 301
   onto, 301
   range, 301
   signature, 301
   total, 301

Gödel Hash, 313
   unique prime factorization, 313
Gödel's proof, 184
Generalized NFA (GNFA), 115
GNFA
   handle loops, 117
   merge paths, 117
   Real_F, 116
   Real_I, 116

Halting problem, 235
Hilbert's problems, 4
Hill-valley plot, 142
   completeness, 142

Impossibility, 210
Inconsistency, 142
Intractable, 242

Kleene star, 27

Lambda, 280
   Alpha reduction, 282
   Anonymous function, 281
   Beta reduction, 283
   Booleans, 284
   Church numerals, 284
   Eager evaluation, 290
   Eta reduction, 283
   Fixpoint, 287
   Fixpoint finder, 287, 288
   in C++, 280
   Lazy evaluation, 290
   Numbers, 284
   Pairs, 284
   Recursion, 287
   Reductions, 282
   Syntax, 282
   Y combinator, 288
   Y Lambda Expression, 287
Lambda Calculus, 279
   Combinators, 292
   history, 279
   Universality, 279

Lambda calculus
  versus TMs, 280
Language, 18
  complementation, 31
  Concatenation, 20
  context-free, 204
  context-sensitive, 204
  Empty, 22
  enumerate in lexicographic order, 35
  enumerate in numeric order, 35
  Exponentiation, 22
  family, 210
  Finite approximation, 19
  homomorphism, 33
  Intersection, 24
  recursive, 204
  recursively enumerable, 204
  regular, 204
  reversal, 32
  setminus, 32
  star, 27
  symmetric difference, 32
  Union, 24
  Unit, 22
  Zero, 21
Language family, 210
Lexicographic order, 35
Lifelong learning, 13
Linearity, 150
  mixed linearity, 150
  purely left-linear, 150
  purely right-linear, 150
Logical connectives, 299

Mapping reduction, 236
  $\leq_m$, 236
  Diophantine example, 258
Markdown
  PDA, 173
Minimization
  DFA, Bzozowski, 90

NDTM
  $\Delta : Q \times \Gamma \to \mathscr{P}(Q \times \Gamma \times \{L, R, S\})$, 187
  ww, 193
  NP-time, 246
NFA
  δ, 83
  $\delta : Q \times \Sigma_\varepsilon \to 2^Q$, 83
  $\hat{\delta}(q, x)$, 86
  alphabet, 82

determinize, 90
Eclosure, 85, 86
fork, 81
initial state **set**, 83
language, 84
language of, 86
string transfer function, 86
subset construction, 87
to DFA, 87
to RE, 115
Non-RE languages, 220
NP, 246
  certificate, 242
NP decider, 246
NP-hard
  Diophantine, 258
NP-time, 246
NPC
  $\leq_P$ mapping-reductions, 254
  NP-Hard, 253
  3-SAT, 254
  Easy, 242
Numeric order, 35, 214

On-set
  Minimal DFA, 268

P decider, 244
P-time, 244
Pascal's calculating machine, 3
Pattern, 8
  context-free, 9
  context-sensitive, 10
  recursively enumerable, 10
  regular, 9
PCP
  undecidable, 229
PDA, 161
  $(Q, \Sigma, \Gamma, \Delta, q_0, z_0, F)$, 164
  $\Delta : (Q \times (\Sigma \cup \{\varepsilon\}) \times (\Gamma \cup \{\varepsilon\})) \to \mathscr{P}(Q \times \Gamma^*)$, 165
  acceptance, 165
  bottom-of-stack marker, #, 162
  deterministic, 165
  disambiguation, 179
  edge-label type, 161
  empty-stack acceptor, 165
  final-state acceptor, 165
  formal description, 164
  from CFG, 174
  ID, 165

input, 161
  instantaneous description, 165
  markdown (Automd), 173
  nondeterministic, 165
  of ambiguous grammars, 174
  stack, 161
  stack alphabet, Γ, 162
Post correspondence, 227
Predicate, 299
Primes, 260
Problem, 4
Procedure, 4, 184, 209
Proof
  by contradiction, 50, 300
  by contrapositive, 50, 300
Pumping Lemma, 47
  contrapositive, 50
  transformations, 54
  why all splits, 51
Pushdown Automata
  see PDA, 161

Quantifiers, $\forall$, $\exists$, 299

RE, 211
  A_{TM}, 216
  $Halt_{TM}$, 218
  L_{AmbCFG}, 218
  L_{G1neG2}, 213
  $T_{w,3}$, 224
  $\overline{L_{UnivCFG}}$, 218
  derivatives, 127
  language examples, 218
  nullable, 128
Recursion
  Y combinator, 289
  Y versus Ye, 290
Recursive, 211
  $LBAHalt_{L,w}$, 223
  $L_{EmptyCFG}$, 216, 221
  $L_{EmptyDFA}$, 211
  $L_{UnivDFA}$, 221
  L_{sat}, 215
Recursively enumerable, 211
Regular expression
  derivatives, 127
Regular expressions, 95
  conversion to NFA, 96, 100, 103
  error-correcting NFA, 106
  mindfa check, 108
  Operators +, concat, star, 96

postage-stamp problem, 111
Primitive ε, ϕ, 96
RE (abbreviation), 95
security attacks, 101
syntax for NFA, 95
Regular languages
  ultimately periodic set, 112

SAT, 248
  3-CNF, 261
  3-SAT, 261
  2-CNF, 261
  2-SAT, 261
  2-SAT Algorithm, 252
  equisatisfiable, 261
Semi-decider, 212
Sets, 297
  Int, $\mathbb{Z}$, 297
  Nat, $\mathbb{N}$, 297
  builder, 297
  cardinality, 297
  complement, 299
  infinite, 297
  powerset, 298
Star
  limit of series, 28
  three definitions, 27
State
  accepting, 43
  final, 43
String, 16
  Concatenation, 16
  Exponentiation, 17
  homomorphism, 33
  Unit, 17
  Zero, 17
Symbol, 15
Syntax Checking, 7

Textual syntax
  importance in CS, 182
Theorem
  A_{TM} is RE, 14.3.1, 216
  A_{TM} is RE, 14.3.3, 218
  A_{TM} is undecidable, 15.3, 233
  $Halt_{TM}$ is undecidable, 15.4, 234
  $L^{*^*} = L^*$, 3.5.1, 35
  $\aleph_0$ RE languages, C.2, 314
  3-SAT is NP-Complete, 16.5, 254
  CFL closure, 11.7.2(a), 151
  CFL Pumping Lemma, 11.9, 156

Closure Results, 9.7, 123
Derivative-based String Matching, 10.3, 134
Diophantine NPH, 16.7.1, 258
Enumerator exists means RE, 14.3.2, 217
Fundamental Theorem of Arithmetic, C.1, 313
Kleene's Theorem, 8.3, 101
Left-linear grammars, 11.7.2(b), 152
Myhill-Nerode theorem, 6.4, 72
NPC and CoNP, 16.7.2, 260
RE means enumerator exists, 14.3.2, 217
Reg Pumping Lemma, 4.6.1 , 49
Regular means exists NFA, 7.5, 88
Right-linear grammars, 11.7.1, 150
Semi-deciders for L and $\overline{L}$, 14.2.3, 213
Ultimate periodicity, 8.8.1, 112
Undecidability of PCP, 15.2, 229
TM
$(Q, \Sigma, \Gamma, \Delta, q_0, B, F)$, 186
$3x + 1$ Problem, 200
ε input, 186
Collatz Problem, 200
doubly-infinite tape, 186
Halting problem, 184
ID, 188
Instantaneous description, 188
robustness, 244
Transition function, 45
$\hat{\delta}$, 45
for strings, 45

Traveling Salesperson Problem, 241
Trees, 302
Truth-tables
Personalities, 267
TSP, 241
Turing
Benedict Cumberbatch, 184
Church-Turing thesis, 184
Turing machine, 5
simplified forms, 11
Turing-complete, 185

Ultimately periodic set, 111
Regular languages, 112
Undecidable problem
Halting, 184
Universal, 184, 279
Lambda calculus, 279
one queue, 185
Turing machines, 184
two counters, 185
two stacks, 185

Verifier
certificate, 246
Verifier TM, 247

Word, 16

Y
Lazy, 290
Ye
Eager, 290